# Field-Effect Transistor Devices

**Scrivener Publishing**
100 Cummings Center, Suite 541J
Beverly, MA 01915-6106

*Publishers at Scrivener*
Martin Scrivener (martin@scrivenerpublishing.com)
Phillip Carmical (pcarmical@scrivenerpublishing.com)

# Field-Effect Transistor Devices

## Post-CMOS Theory and Applications

Edited by

**P. Suveetha Dhanaselvam**

**Naima Guenifi**

**Shiromani Balmukund Rahi**

and

**J. Ajayan**

Scrivener Publishing

WILEY

This edition first published 2026 by John Wiley & Sons, Inc., 111 River Street, Hoboken, NJ 07030, USA and Scrivener Publishing LLC, 100 Cummings Center, Suite 541J, Beverly, MA 01915, USA

For more information about Scrivener publications please visit www.scrivenerpublishing.com.



**Wiley Global Headquarters**
111 River Street, Hoboken, NJ 07030, USA

For details of our global editorial offices, customer services, and more information about Wiley products visit us at www.wiley.com.

The manufacturer's authorized representative according to the EU General Product Safety Regulation is Wiley-VCH GmbH, Boschstr. 12, 69469 Weinheim, Germany, e-mail: Product_Safety@wiley.com.



***Library of Congress Cataloging-in-Publication Data***

ISBN 9781394311453

Cover image: Generated with AI using Adobe Firefly
Cover design by Russell Richardson

Set in size of 11pt and Minion Pro by Manila Typesetting Company, Makati, Philippines

Printed in the USA

10 9 8 7 6 5 4 3 2 1

# Contents

# Preface

In the semiconductor industry, the key term CMOS stands for "Complementary Metal Oxide Semiconductor". This is one of the most popular technologies in the integrated circuits (ICs) design industry and it is broadly used today to form integrated circuits (ICs) in numerous applications. CMOS technology used for design and development of integrated circuits (ICs) that are widely used in digital circuits. Today's computer memories, CPUs, and cell phones make use of this technology due to several key advantages. In CMOS technology, both n-type and p-type metal-oxide-semiconductor field-effect transistors (MOSFETs) are used together in a complementary way to form a digital logic gate. The combination of both types of MOSFETs provides better power efficiency and noise immunity compared to other technologies.

As conventional CMOS technology scaling results in smaller and smaller transistor sizes, existing CMOS technology starts to encounter certain fundamental physical limitations. These limitations can potentially result in drastic reduction in performance and energy efficiencies of these transistors, which could make it untenable to continue scaling them. This in turn, would severely limit the benefits of further innovations in IC design. In this scenario, there have been several alternatives proposed at the device level. However, each of these devices comes with varying energy-performance tradeoffs and optimal design points in comparison to CMOS technology. This is the dominant semiconductor technology for microprocessors, microcontroller chips, memories like RAM, ROM, EEPROM and application-specific integrated circuits (ASICs). It is also used in various applications, such as sensors, image processing, and wireless communications. The conventional CMOS technology faces physical limits (short-channel effects, power density) as transistors shrink, requiring new physics and materials to continue performance improvements beyond traditional scaling.

The post-CMOS technologies explore alternatives and complements to traditional silicon CMOS for future computing. It focuses on new materials (2D materials, nanotubes), devices (TFETs, spintronics, memristors), and integration (photonic, neuromorphic, 3D stacking) to overcome CMOS scaling limits, enabling lower power, higher speed, and novel functions like in-memory computing or quantum systems. The key areas include Tunnel FETs, Graphene/CNTs, Spintronics, Photonics, Topological Materials, and Molecular Electronics, often used in hybrid systems with CMOS. Various beyond-CMOS device technologies including tunneling FET (TFET), hybrid phase transition FET (HyperFET), carbon nanotube FET (CNTFET), silicon nanowire FET (SiNWFET), symmetrical tunneling FET (SymFET), phase-change memory (PCM), spin-transfer torque magnetic tunnel junction (STT-MTJ), resistive random access memory (RRAM) have been considered.

The shrinking of transistors to dimensions below 100 nm enables hundreds of millions transistors to be placed on a single chip. The increased functionality and reduced cost of large variety of integrated circuits and systems has brought its own benefit to the end users and above all the semiconductor industry. A low cost of manufacturing, increased speed of data transfer, computer processing power and the ability to accomplish multiple tasks simultaneously are some of the major advantages gained as a result of transistor scaling. The packing density of IC on microelectronics roughly estimated with the help of Moore's Law (an empirical compact mathematical formulation; $N = \frac{D^2 P_E}{F^2}$. In this empirical formula $N$ is the number of transistors per chip, $F$ is the minimum feature size, $D$ is the chip area, and $P_E$ is transistor packaging efficiency measured per minimum feature area). This empirical formula plays a n crucial for observation on the rate of growth of semiconductor technology. Post CMOS FET refers to filed effect transistor that are being explored as potential future digital logic technologies beyond the current limits of CMOS technology, addressing challenges like scaling and performance. Transistor scaling is a fundamental concept in today's electronics era, driving significant exponential growth in power-efficient technology. This book explores the essential concept and characteristics of CMOS devices and emerging semiconductor technologies that play a crucial role in the industry. It presents a comprehensive yet accessible approach, aiming to deepen understanding of semiconductor scaling and characterization. It is specifically for young researchers, engineers, and students, providing a better understanding of recent developments for the semiconductor inductory.

Nowadays, integrated circuits are realized using CMOS technology. However, the demand for ever smaller, more efficient circuits is now pushing the limits of CMOS. The post-CMOS refers to the possible future digital logic technologies beyond the CMOS scaling limits. The demand for ever smaller and portable electronic devices has driven metal oxide semiconductor-based (CMOS) technology to its physical limit with the smallest possible feature sizes. This presents various size-related problems such as high power leakage, low-reliability, and thermal effects, and is a limit on further miniaturization. To enable even smaller electronics, various nano-FET devices including carbon nanotube transistors, graphene transistors, and tunnel transistors, junctionless FETs, FeFETs, NCFETs, collectively called post-CMOS devices are emerging that could replace the traditional and ubiquitous silicon transistor.

In Chapter 1, the transition from single-gate to multigate field-effect transistors (MG-FETs) represents an important evolution in semiconductor technology, aiming to improve device performance and manage the challenges posed by further miniaturization of components. The multigate field-effect transistors offer significant advantages in several important applications, where ultra-low power consumption, energy efficiency, and low voltage operation are crucial. While they may not yet fully match the drive currents and switching speeds of conventional MOSFETs in all scenarios, their unique operational characteristics make them highly valuable in specific applications, especially as the demand for power efficiency and performance at low voltages solutions grows. Recently, junctionless LMG-FETs (JLMG-FETs) are propped as an advanced class of transistor design to overcome the conventional MG-FET challenges and provide better scalability behavior and low elaboration cost. They are unlike conventional FETs, junctionless FETs do not have p-n junctions. Instead, the entire structure is uniformly doped, simplifying fabrication and enhancing control over channel body. By eliminating junctions and leveraging multigate architectures, these devices offer enhanced electrostatic control, and improved scalability. However, further improvements regarding the analog parameters should be performed in order to understand the impact of each design parameter on the device performances for both double-gate and gate-all - around (GAA) FET structures. In this context, in **chapter 1** the scaling capabilities of the JL-MG-FETs has been investigated. To do also, extensive 2D numerical simulations are conducted using Silvaco commercial tool to model the behavior of JL-DG and GAA MOSFETs including the effect of high-k material as a gate-oxide on the subthreshold properties and drain current. A detailed analysis and simulations of their scaling capabilities will be presented and all results will be discussed. By leveraging 2D numerical simulations, our aim is to gain valuable insights into the scaling capabilities and performance benefits of JL-DG and GAA MOSFETs, paving the way for next-generation low-cost nanoscale semiconductor devices.

In Chapter 2 due to high power and high speed applications, high electron mobility transistors (HEMT) have been the area of focus for many research groups in recent years. Compared to conventional field effect transistors (FET), HEMT devices have demonstrated superior performance considering the operational frequency, output power density, and power added efficiency. High electron mobility transistor (HEMT) technology, considerable performance improvements have been brought to the fields of power electronics, radio frequency applications, and high-speed digital circuits. GaN-based complementary metal oxide semiconductor (CMOS) technology combines the wide-band gap characteristics of GaN with the low static power consumption advantages of CMOS, showing significant potential in promoting the development of integrated circuits (ICs). In recent years, numerous research organizations have focused on high electron mobility transistors due to their high power and speed applications, which provide a wide range of applications in electronics, optical and thermal properties. A comparison of field effect transistors with conventional HEMTs reveals that field effect transistors enhance power added efficiency, power density, and operating frequency. Materials into randomly and vertically layered heterostructures. The recently reported vapor phase development of 2D materials offers the path for directly constructing vertical and lateral heterojunction. This study provides insights into layered 2D heterostructures, as well as a brief introduction to 2D material and heterostructure preparation methods. These remarkable 2D heterostructures offer a wide range of possible applications.

In Chapter 3, the nanosheet field-effect transistor (NSFET) represent a transformative advancement in semiconductor technology, addressing the challenges of traditional scaling limits while delivering enhanced performance for next-generation electronics. Chapter 3 provides a comprehensive overview of nanosheet FETs, focusing on their structural innovations, fabrication methodologies, and operational principles. The discussion begins with the evolution of transistor architectures, leading to the development of nanosheet designs that leverage horizontal nanostructures for superior electrostatic control. The key fabrication techniques, including lithography, epitaxy, and selective etching, are examined, emphasizing their role in achieving the precise dimensional control required for nanosheet production. This chapter also delves into the advantages of nanosheet FETs, such as reduced power consumption, higher current drive, and scalability beyond FinFET technologies. Furthermore, challenges related to process complexity, material compatibility, and device variability are critically analyzed. By exploring the latest research trends, practical applications, and potential future directions, this chapter aims to serve as an essential resource for researchers, engineers, and students working in the field of advanced semiconductor devices.

Chapter 4 is dedicated to the semiconductor devices have undergone significant advancements, propelling innovation across sectors such as computing, telecommunications, automotive, and healthcare. The key developments in material science, fabrication techniques, and device architecture have led to improvements in performance, miniaturization, and energy efficiency. Transitioning from traditional silicon to materials like gallium nitride (GaN) and silicon carbide (SiC) has enabled the creation of devices that excel in high-power and high-frequency applications, offering greater efficiency in power electronics and faster communication systems. The introduction of three dimensional (3D) semiconductor architectures, including Fin FETs and gate-all-around (GAA) transistors, has successfully tackled the challenges of scaling down devices, resulting in better performance and reduced power consumption. Furthermore, the integration of artificial intelligence (AI) and machine learning (ML) in the design and optimization processes of semiconductor devices has accelerated innovation, allowing for the development of more complex and intelligent electronic systems. Emerging fields like quantum computing and neuromorphic engineering are also driving the evolution of semiconductor technology, pushing the boundaries of conventional computing and enabling new possibilities. As the need for higher performance, lower energy consumption, and greater functionality continues to grow, ongoing research in semiconductor technologies is expected to lead the next generation of technological advancements, shaping the future of electronics. In summary, the progress in semiconductor devices is central to the rapid advancement of modern technology, resulting in more powerful, efficient, and intelligent systems that will continue to meet the increasing demands of the digital age.

In Chapter 5, nanoscale device design, short channel effects (SCEs) significantly deteriorate the performance of conventional MOSFETs, necessitating innovative structural and material engineering approaches. Multigate architectures, such as double-gate, triple-gate, and surrounding-gate, MOSFETs have been widely explored to enhance gate control and suppress SCEs. In parallel, gate and channel engineering strategies, including multimaterial gate structures and halo doping profiles, have further improved device electrostatics. This work presents a novel device architecture, underlapped Triple Material Surrounding Gate (U-TMSG) MOSFET, which integrates the benefits of underlap engineering with triple-material gate technology. An analytical model of the device is developed using conformal mapping techniques, providing closed-form expressions for the surface potential and electric field distribution. The model is validated through TCAD simulations, and results demonstrate reduced leakage current, improved ON-state current, and enhanced electrostatic integrity compared to existing structures. The study highlights the U-TMSG MOSFET as a promising design-oriented solution for future nanoscale applications.

In Chapter 6, the carbon nanowires (CNWs) are a type of nonmaterial with a one-dimensional structure, meaning they have a very small diameter typically in the nanometer range and a much longer length. These nanowires are composed entirely of carbon atoms and are related to other carbon-based nonmaterial's, such as carbon nanotubes and graphene. In comparison to regular nanowires, carbon nanowires have benefits in thermal conductivity, mechanical strength, chemical stability, electrical conductivity, and nanoscale quantum effects. They are especially well-suited for cutting-edge applications where endurance and performance are crucial, such as electronics, energy storage, biomedical engineering, and materials research. This survey work provides an extensive summary of the production, properties, and uses of carbon nanowires. It examines several fabrication processes, highlighting the benefits and drawbacks of each, such as chemical vapor deposition, laser ablation, and electrochemical approaches. The numerous uses of CNWs in industries including biomedicine, energy storage, nanoelectronics, and sensor technologies are also covered in this study. Ongoing research is focused on optimizing the properties of CNWs, improving their synthesis processes, and exploring new applications, particularly in advanced electronics and nanotechnology. As a frontier material in nanotechnology and materials science, CNWs hold the potential to drive significant innovations in high-tech industries, paving the way for future advancements.

In Chapter 7, the junctionless transistors has represented a significant breakthrough in semiconductor technology, offering a simplified yet highly effective alternative to traditional transistors. Unlike conventional metal-oxide-semiconductor field-Effect transistors (MOSFETs), which rely on p-n junctions to control the flow of current, Junctionless transistors operate using a uniformly doped channel, either n-type or p-type. This architecture eliminates the need for junction formation, thus simplifying the fabrication process and reducing manufacturing costs, especially at nanoscale dimensions. By controlling current purely through the electric field applied to the gate, Junctionless transistors provide enhanced electrostatic control and help mitigate common short channel effects (SCEs) such as Drain-Induced Barrier Lowering (DIBL), threshold voltage roll-off, and increased leakage current. This makes them highly suitable for applications where low power consumption, high efficiency, and high performance are crucial. As the semiconductor industry continues to push the limits of scaling, particularly with the shift towards sub-nanometer technologies, the need for innovative transistor architectures becomes even more pressing. The traditional scaling of MOSFETs has reached its physical limits, requiring new device designs like gate-all-around (GAA) structures. In this context, cylindrical surrounding gate (CSG) transistors are gaining attention due to their superior electrostatic control over the channel, significantly reducing SCEs. Combining Junctionless architecture with GAA CSG structures results in devices that are highly efficient and scalable, paving the way for next-generation low-power, high speed semiconductor devices. This book chapter focuses on the design, modeling, and analysis of Junctionless Cylindrical Surrounding Gate (CSG) Silicon Nanowire Transistors. Various gate materials, including single, double, and triple materials, are explored to improve the device's performance. The study begins by analyzing Junctionless Single Material CSG (JLSMCSG) silicon nanowire transistors, where an analytical model for drain current, transconductance, and transcapacitance is developed using the Pao-Sah gradual channel approximation. By solving the 2D Poisson's equation, the model evaluates the potential distribution in the channel, and the results are validated through MATLAB simulations and the Sentaurus TCAD numerical simulator. Subsequently, the work extends the analysis to Junctionless Double Material CSG (JLDMCSG) and Junctionless Triple Material CSG (JLTMCSG) transistors. Introducing multiple gate materials in these designs enhances electrostatic field control and reduces SCEs such as DIBL and tunneling effects. The study demonstrates that the use of double and triple material gates leads to a significant reduction in threshold voltage and improved transconductance, with the JLTMCSG design showing the most substantial improvements. The analytical modeling for these devices examines the channel potential and investigates the impact of design parameters such as oxide thickness, gate length, and doping concentration on device performance. A key focus of the chapter is the subthreshold behavior of these junctionless transistors, which is crucial for lowpower applications. The ON/OFF switching characteristics, an essential feature for digital circuit design, are analyzed, demonstrating that the JLTMCSG device offers enhanced switching capabilities. The comparative analysis between single, double, and triple material gate designs reveals that the introduction of additional gate materials leads to proportional reductions in SCEs and improved overall device efficiency. In conclusion, this chapter highlights the potential of Junctionless transistors, particularly in combination with CSG structures, as promising candidates for future semiconductor devices. The findings establish that these devices meet the growing demands of the industry for low-power, high-speed, and cost-effective solutions in mobile, computing, and digital applications.

In Chapter 8, current toxic and polluted scenario, biosensing in various fields, particularly in medicine, is playing a crucial role as numerous diseases affecting human body. The real-time monitoring and detection of these diseases may save many human lives as early detection of these diseases is essential for diagnosis. The human body expel out many volatile organic compounds (VOCs) through stools, urine, saliva, breath, etc. These VOCs provide important insights of an individual's metabolism and by analyzing their level and concentrations; one can identify the diseases and abnormalities associated with that specific human body. Unique VOC signatures show information about various health conditions such as diabetic, respiratory system diseases, kidney functionality and even cancer. To detect composition and concentration of these VOCs, portable and handy biosensors are required as they are convenient to use and can integrate with other systems. Although various invasive and non-invasive techniques exist, such as needle-based electrochemical sensors, biopsies, implantable biosensors, chromatography and mass spectrometry (GC-MS). But FET-based sensors are appropriate for the application of biosensing due to its high efficiency, cost effectiveness, wide detection range and compact size. These biosensors are basically an analytical tool that can transform a biochemical signal into a measurable electrical quantity such as change in current, transconductance, etc. This chapter provides a detailed and analytical explanation of the principles underlying bio-field effect transistors (bio-FETs) and their sensing mechanisms are provided along with the VOCs associated with specific disease and their origin. This chapter also covers different VOCs, such as ammonia ($NH_3$), nitrogen oxide ($NO_2$), acetone ($CH_3COCH_3$), hydrogen sulfide ($H_2S$), which are commonly associated with specific diseases and health conditions. In addition, it explores the variety of nonmaterial's employed as sensing layers in bio-FETs, such as graphene, transition metal dichalcogenides (TMDCs), polymers, and metal oxides. Each of these materials offers unique properties, such as high surface area, tunable electronic characteristics, and strong interaction with target analytes, which enhance their sensitivity and selectivity for detecting specific VOCs. Their integration into bio-FET platforms enables the development of highly sensitive, real-time detection systems for medical diagnostics and environmental monitoring.

Chapter 9, presents the design of a two-stage ultra-low-power operational amplifier to find the best design, the suggested operational amplifier is examined from a number of angles. It has a 76° phase margin, a maximum gain of roughly 75.33 dB, and a dissipation of 2.08 mW at 1.5 V supply voltage. For biological applications, where high gain at little power is needed, it is hence ideal. A circuit simulator based on SPICE has been used to construct the suggested operational amplifier. Multiplexing is done by passing one analog or digital input signal into a single line from a variety of input signals using an electrical device known as a multiplexer. The design of a low power consumption 90 nm CMOS 4:1 multiplexer using n-Pass Time Division Techniques is covered in this study. This study compares and simulates a few critical multiplexer factors, including as size, transistor count, and power dissipation. We shall identify $V_{dd}$ as the constant power supply in static CMOS logic. Since energy recovery does not occur as it does in the case of static CMOS logic, the capacitor loses all the energy that is across the resistor, retaining very little of it. The design of a CMOS-based fourth-order low pass biquad filter for biomedical applications is covered in this short. Cascading the second order P-FVF and N-FVF biquad results in the construction of the fourth-order filter structure. The recommended filter functions in a region below the threshold. 114.4 Hz is the cut-off frequency of this circuit. The proposed design is shown in the Cadence environment with 90-nm CMOS technology.

In Chapter 10, perovskite materials have emerged as a ground-breaking class of semiconductors, offering unprecedented opportunities to revolutionize modern technology. With their unique crystal structure, tunable electronic and optical properties, and cost-effective fabrication processes, perovskites are poised to address many of the limitations faced by traditional semiconductor materials. This chapter explores the science behind perovskites, highlighting their exceptional properties such as high charge carrier mobility and defect tolerance. It delves into their transformative applications, including high-efficiency photovoltaics, bright and flexible light-emitting diodes (LEDs), sensitive photodetectors, and their potential in spintronics and quantum computing. Despite their remarkable advantages, challenges such as environmental stability, toxicity, and scalability remain significant hurdles to widespread adoption. Recent advances in material engineering, lead-free alternatives, and hybrid designs are paving the way for more robust and sustainable perovskite-based technologies. This chapter provides a comprehensive overview of the current state of perovskite research, its challenges, and its potential to shape the future of semiconductor technologies.

Chapter 11 presents the design and simulation of a high-performance, lead free perovskite solar cell (PSC) using SCAPS-1D. The proposed device features $CH_3NH_3SnI_3$ as the absorber layer, chosen for its superior optoelectronic properties and environmental sustainability, along with strontium titanate (STO) as the electron transport layer (ETL) and $CuSbS_2$

(CAS) as the hole transport layer (HTL). Numerical optimization yielded outstanding photovoltaic parameters: a short-circuit current density (JSC) of 40.62 $mA/cm^2$, an open-circuit voltage (VOC) of 1.0405 V, a fill factor (FF) of 86%, and a power conversion efficiency (PCE) of 36.35%. The STO/CAS heterostructure exhibited enhanced charge transport and suppressed recombination losses, contributing to the device's high efficiency. These findings underscore the potential of lead-free perovskites in sustainable photovoltaics, providing a framework for future development in high-efficiency solar cell architectures.

In Chapter 12, the myocardial infarction (MI), commonly known as a heart attack, occurs when blood flow to the heart is blocked due to plaque buildup in the coronary arteries. If the plaque ruptures, it can lead to blood clot formation, further restricting circulation and increasing the risk of severe complications. With high mortality and disability rates worldwide, early and accurate detection is essential for timely intervention, reducing heart damage, and improving patient survival rates. Detecting the exact location of the blockage is equally important, as it enables targeted treatment and better medical outcomes. This chapter introduces a Very Large Scale Integration (VLSI) architecture designed for real-time MI detection and precise infarction localization. By analyzing electrocardiogram (ECG) signals, a crucial diagnostic tool in cardiology, the system classifies ECG data into 12 categories, each corresponding to specific infarction patterns. It delivers remarkable accuracy of 99.90%, sensitivity of 99.49%, and specificity of 99.94%, ensuring precise diagnosis while minimizing false positives and negatives. The architecture is implemented using a 180 nm Bulk CMOS technology node, balancing performance, power efficiency, and compact design. It occupies a small area of 1.69 mm2 and operates with low power consumption of 268.9 μW at a clock frequency of 250 KHz, making it highly suitable for portable and wearable medical applications. Additionally, its real-time processing capability allows for rapid diagnosis, supporting timely medical intervention and improving patient prognosis. With its combination of high accuracy, low power consumption, and compact design, this VLSI system presents a reliable and efficient solution for MI detection and localization. Its integration into medical devices has the potential to enhance early diagnosis, enable faster treatment, and significantly reduce MI-related mortalityrates, ultimately improving healthcare outcomes on a global scale.

In Chapter 13, the growing demand for portable electronic devices highlights the importance of minimizing power consumption in circuit design. While dynamic power usage declines significantly with lower supply voltages, there is a challenge posed by leakage power because of the inverted sub-threshold slope. Reducing the supply voltage results in a lowered pinch-off point, and with the settled sub-threshold slope, this leads to the exponential escalation in leakage current. This chapter presents an extended source double gate (ESDG) dielectric modulated (DM) TFET label free biosensor using TCAD simulator. The nanogap cavity is considered near the source and drain regions of fixed $HfO_2$ layer to put up large number of biomolecules. Due to the extended source in ESDG TFET, the vertical tunnelling can be achieved along with lateral tunnelling and this leads to enhanced sensitivity (Sn). The Sn is evaluated for fully filled nanogap cavity under the both neutral and charged (positive/negative) biomolecules with wide variation in dielectric constant (k). It has been reported the subthreshold swing (SS), input characteristics, energy band diagram, and $I_{ON}/I_{OFF}$ ratio by varying k from 1 to 12 for uncharged biomolecules within the nanogap. The sensitivity considering the steric hindrance for four diverse step profiles like increasing, decreasing, concave, and convex are reported by varying k. Furthermore, sensitivity of ESDG-DMTFET is compared with the sensitivity of existing FET/TFET based biosensors.

In Chapter 14, a new 2D analytical model for the V-shaped stacked graphene nanosheet field effect transistor (GNFETs) is developed in this study. Using suitable boundary conditions, the linear approximation method (LAM) is used to determine the 2D Poisson equation for the first time. To assess the performance of the device, the exact analytical equations for the variables drain current, transconductance, threshold voltage, electric field, and channel potential are derived. The device operates at cryogenic temperatures as a result of the high electron fluxes into the channel caused by the effect of the Nanosheet channel contact with the gate. It offers a significant scaling of device with a large capacity for drive current. With a perpendicular layered structure, the V-shaped nanosheet achieves low threshold voltage and high transconductance. When TCAD simulation software and analytical model results are compared, a good degree of agreement has obtained.

In Chapter 15, the design and performance analysis of 1-bit full adder circuit with MOSFET and FinFET using 28 T and 10 T Gate Diffusion Input (GDI) techniques is carried out. The three technology nodes, 180, 90, and 45 nm are taken into consideration for constructing MOSFET-based 1-bit full adder circuits. A 32 nm FinFET-based 1-bit full adder circuit with 28 T and 10 T GDI is also designed and analysed for its functionality using timing diagram. The behavior of both MOSFET- and FinFET-based full adder circuits are investigated by varying the parameters, voltage and operational frequency. The performance of the circuit is assessed using the metrics, power consumption, and propagation delay. It is observed that

10 T GDI-based FinFET full adder shows an average power consumption of 11.97 nW at 0.5 V with 5 GHz operational frequency at a propagation delay of 66.67 ps. The MOSFET-based full adder at 45 nm technology node shows a stable output with a power consumption of 176 nW at 0.6 V, 5 GHz with a propagation delay of 0.244 ns. It is also observed from the simulation results that 10 T GDI FinFET-based full adder is more superior than 28 T FinFET full adder in terms of both average power consumption and propagation delay. It is also understood that MOSFET-based full adder are not advantageous for low -power applications as they consume more power and propagation delay.

In Chapter 16, the intention of developing a biopotential amplifier that has a low amount of noise during its operation. In order to enhance the quality of the electroencephalogram (EEG) signal, LNAs are a vital component. This allows the EEG data to be used effectively in a wide range of biomedical applications *via* their efficient utilization. When compared to the electrical activity that is generated by the movement of the brain, the electroencephalogram (EEG) signal is representative of this activity. One of the most important elements of an EEG signal monitoring circuit is an operational amplifier that consists of two individual stages. Within the context of the technology that is now accessible, the most significant factors that must be satisfied for each and every circuit are the enhancement of power efficiency and the reduction of noise. The collection of biosignals is of the greatest significance because of their low frequency (range from 0.1 Hz to 10 kHz) and minute amplitude (which is typically within a few millivolts). In the process of producing medical electronics for the purpose of health monitoring, amplifiers are used as a component. Nevertheless, in order to obtain accurate bio-signal measurement, it is required to have a high CMRR and precise amplification in order to decrease the quantity of input noise and enhance the signal strength. This is necessary in order to achieve the desired accuracy. These are the performance requirements that are being addressed, and they include things like power, gain, noise margin, CMRR, SQNR, and plenty of other things. In order to do this, it is recommended that our biopotential amplifier concept be implemented using technology that is 90 nm in size. A gain of about 82 dB, a phase margin of 48.6°, and a low power consumption of 5.1 μW were achieved by this design. These characteristics are essential for electroencephalogram (EEG) devices that are powered by batteries or portable applications. A low-noise amplifier (LNA) that has been created makes it possible to capture high-quality electroencephalogram (EEG) and electrocardiogram (ECoG) data interfaces, and human–computer interaction. In the future, it is predicted that research and technological improvements will unearth further possibilities in the use of electroencephalogram (EEG) signals and the impact that these signals have on the field of neuroscience research as well as clinical applications.

In Chapter 17, the continuous scaling of transistors into the nanoscale regime has led to the exploration of novel device architectures beyond conventional MOSFETs, with tunnel field-effect transistors (TFETs) emerging as promising candidates due to their steep subthreshold swing and low-power operation. In this work, a progressive modeling approach has been adopted to enhance TFET performance. Initially, a conventional dual-gate TFET is modified by introducing heterojunction material at the source and channel interface, resulting in a heterojunction TFET (HJ-TFET). To further improve carrier transport across the tunneling region, the device is engineered with heterogate dielectrics, forming a heterogate dielectric heterojunction TFET (HGD-HJ-TFET). Building on this structure, an ultrathin graphene nanoribbon layer of thickness 0.5 nm is deployed over the intrinsic silicon channel of length 20 nm, yielding the proposed graphene-based heterojunction heterogate dielectric TFET (GR-HGD-HJ-TFET). The incorporation of graphene enhances carrier mobility and tunneling efficiency, thereby reducing leakage current and improving on-state current ($I_{ON}$). The device parameters such as gate oxide thickness ($t_{ox}$), body thickness, and heterogate dielectric lengths has been optimized through analytical modeling and simulation. The results confirmed that the proposed GR-HGD-HJ-TFET exhibits superior electrostatic control and performance compared to conventional and previously reported TFET structures, establishing its potential for future low-power nanoelectronic applications.

In chapter 18, the relentless pursuit of miniaturization and energy efficiency in semiconductor devices has pushed conventional MOSFETs to their physical and operational limits, signaling the end of Moore's Law. To address these challenges, researchers have turned to innovative Field-Effect Transistor (FET) architectures that leverage novel materials and operating principles. Among these, Ferroelectric FETs (FeFETs), tunnel FETs (TFETs), and junctionless FETs (JL-FETs) have emerged as promising candidates, each offering unique advantages such as non-volatility, steep subthreshold slopes, and simplified fabrication. However, no single FET technology can simultaneously address all the requirements of modern electronics, including low power consumption, high performance, and scalability. This has led to the development of hybrid FET architectures, which combine the strengths of multiple FET technologies to create devices that surpass the capabilities of their individual counterparts. This chapter provides a comprehensive exploration of hybrid FETs, beginning with an overview of the limitations of conventional MOSFETs and the motivations for adopting advanced FET architectures. The foundational principles of FeFETs, TFETs, and JL-FETs are discussed in detail, highlighting their unique mechanisms such as polarization-induced threshold control in FeFETs, band-toband tunneling in TFETs, and electric field modulation in JL-FETs. By bridging the gap between conventional and emerging transistor technologies, hybrid FETs represent a transformative approach to semiconductor design, offering a pathway to next-generation electronics that are faster, smaller, and more energy-efficient.

# Acknowledgment

The editors of book *Field-Effect Transistor Devices: Post-CMOS Theory and Applications* would like to acknowledge our almighty GOD, family members, contributors, readers and publishers. We would also like to acknowledge and thank all who are directly or indirectly involved in the book project. This book is based on research work carried out by a number of subject experts. Many people have contributed greatly to the book on *Field-Effect Transistor Devices: Post-CMOS Theory and Applications.* We are also grateful for a number of friends and colleagues in encouraging us to start working the work on this book. We, as the editors, would like to acknowledge all of them for their valuable help and generous ideas in improving the quality of this book. With our feelings of gratitude, we would like to introduce them in turn. First, we acknowledge the authors and reviewers of each chapters of this book. Without their outstanding expertise, constructive reviews, and devoted effort, this comprehensive book would have lacked content. Second, we acknowledge the Wiley and Scrivener staff for their constant encouragement, continuous assistance, and untiring support. Without their technical support, this book would not have been completed. Third, we acknowledge the editor's family for being a source of continuous love, for the unconditional support, and for the prayers—not only for this work, but throughout our life. Last, but far from least, we express our heartfelt thanks to the Almighty for bestowing upon us the courage to face the complexities of life and complete this work.

November 2025
**Editors**
**P. Suveetha Dhanaselvam**
**Naima Guenifi**
**Shiromani Balmukund Rahi**
**J. Ajayan**

# 1

# Junctionless Multigate Field Effect Transistor

**R. Ouchen[1], T. Berghout[2], F. Djeffal[1]* and H. Ferhati[1,3]**

[1]*LEA, Department of Electronics, University of Batna 2, Batna, Algeria*
[2]*LAME, University of Batna 2, Batna, Algeria*
[3]*ISTA, University of Larbi Ben M'hidi, Oum El Bouaghi, Algeria*

## *Abstract*

The transition from single-gate to multigate field-effect transistors (MG-FETs) represents an important evolution in semiconductor technology, aiming to improve device performance and manage the challenges posed by further miniaturization of components. Multigate field-effect transistors offer significant advantages in several important applications where ultra-low power consumption, energy efficiency, and low voltage operation are crucial. While they may not yet fully match the drive currents and switching speeds of conventional MOSFETs in all scenarios, their unique operational characteristics make them highly valuable in specific applications, especially as the demand for power efficiency and performance at low voltages solutions grows. Recently, junctionless LMG-FETs (JLMG-FETs) are propped as an advanced class of transistor design to overcome the conventional MG-FET challenges and provide better scalability behavior and low elaboration cost. They are unlike conventional FETs, junctionless FETs do not have p-n junctions. Instead, the entire structure is uniformly doped, simplifying fabrication and enhancing control over channel body. By eliminating junctions and leveraging multigate architectures, these devices offer enhanced electrostatic control, and improved scalability. However, further improvements regarding the analog parameters should be performed in order to understand the impact of each design parameter on the device performances for both double-gate and gate-all-around FET structures. In this context, the objective of this chapter is to investigate the scaling capabilities of the JL-MG-FETs. To do so, extensive 2D numerical simulations are conducted using Silvaco commercial tool to model the behavior of JL-DG and GAA MOSFETs including the effect of high-k material as a gate-oxide

**Corresponding author*: faycal.djeffal@univ-batna2.dz; ORCID: 0000-0003-0742-5864

P. Suveetha Dhanaselvam, Naima Guenifi, Shiromani Balmukund Rahi and J. Ajayan (eds.) Field-Effect Transistor Devices: Post-CMOS Theory and Applications, (1–32) © 2026 Scrivener Publishing LLC

on the subthreshold properties and drain current. A detailed analysis and simulations of their scaling capabilities will be presented and all results will be discussed. By leveraging 2D numerical simulations, our aim is to gain valuable insights into the scaling capabilities and performance benefits of JL-DG and GAA MOSFETs, paving the way for next-generation low-cost nanoscale semiconductor devices.

***Keywords*:** CMOS, junctionless, multigate, machine learning, classification, optimization

## 1.1 Introduction

The semiconductor industry has been lead to develop devices that are much more compact as well as energy-efficient. Beside the huge electronic progress of semiconductor industry over the last 50-years, the down-scaling of field-effect transistor (FET) below 10nm has been anticipated by the famous International Technology Roadmap for Semiconductors (ITRS). CMOS technology is an organization of two types of MOSFET such as P-type and N-type. CMOS technology has dominated silicon industry due to various advantages. As complementary metal–oxide–semiconductor (CMOS) technologies are scaled down into the nanometer range, a number of major non-idealities must be addressed and overcome to achieve a successful logic and analog circuits design, this technology is used in most very large scale integrated (VLSI) or ultra-large scale integrated (ULSI) circuit chips. The CMOS technology evolution has been guided by the well known International Technology Roadmap for Semiconductors (ITRS). Technology roadmaps aid predict requirements for future technologies and guide ongoing technology research and development. It is completely revised in odd-numbered years and updated in even-numbered years. The short-channel effects (SCEs) and the gate-leakage-current are major challenges for the down-scaling of metal oxide semiconductor FET (MOSFET). As a solutions, the classical MOSFETs devices have been improved by increasing the number of gates, including emerging material in the channel and using high-k gate materials. Initially with rapid down-scaling MOSFET devices which gave rise to short channel effects: Drain Induced Barrier Lowering (DIBL), threshold voltage roll off, high, leakage current and limits the subthreshold voltage to 60mV/decade. Advanced aggressive scaling of conventional metal oxide semiconductor field effect transistor requires the use of advanced processing with multiple additive technology boosters, such as ferroelectric material, negative capacitance technique and technology. Therefore, the scientific community is moving towards four kinds of transistors: carbon-nanotube devices, nanowire FET, junctionless transistors and tunnel FET, to attain very low-power dissipation and excellent circuits' performances. In order to overcome the disadvantages of conventional MOSFET device a promising device Tunnel Field Effect Transistor was reported. Because of the p-i-n configuration, many new devices can be constructed by combining TFET with MOSFET, to solve different problems. Very low OFF-current of TFET enables ultralow power operation of analog design, and steeper switching slope of TFET enables greater transconductance ($g_m$) even at a low power. In ideal condition for room temperature, its numerical value has been obtained as 60 mV/decade. Practically for classical MOSFET and its family member like double gate MOSFET, FinFET, gate all around (GAA) FET, Nanosheet FET, its value is larger than 60 mV/decade.

The scaling of single-gate MOSFETs transistor has had a profound impact on the miniaturization of semiconductor devices, driving the exponential growth in computing power and electronic functionality over the past several decades. However, as device dimensions continued to shrink, this scaling introduced a number of challenges that have influenced the trajectory of semiconductor miniaturization [1–5]. The continuous scaling of single-gate MOSFETs faced physical and material limitations, such as the increasing difficulty of reducing the channel length without compromising performance [6, 7]. These scaling limits threatened the continuation of Moore's Law [8, 9]. Therefore, the optimization of gate structures in semiconductor devices has evolved significantly, with the transition from single-gate to multigate field-effect transistors (FETs) being a key strategy in overcoming the limitations of traditional MOSFET designs. By addressing the limitations of traditional MOSFETs, multigate FETs have enabled continued scaling in advanced semiconductor technologies. This evolution has been instrumental in sustaining the progress of Moore's Law and driving the development of next-generation electronic devices. By utilizing multiple gates to control the current in the channel, multigate FETs offer superior electrostatic control, reduced short-channel effects, and improved performance, enabling further miniaturization of semiconductor technology [6].

In a double-gate FET, the transistor features two gates that control the channel from both the top and bottom (or from opposite sides). These gates can be independently biased or connected to the same voltage source, depending on the desired operation [10]. The channel is typically a thin layer of semiconductor material sandwiched between the two gates, allowing both gates to influence the flow of charge carriers within the channel. Double-gate FETs are particularly useful in low-power devices where precise control of the channel is critical. They are also used in some memory devices and analog circuits where enhanced control over the transistor characteristics is needed.

Multigate FETs, including FinFETs and GAA FETs, extend the concept of double-gate FETs by adding more gates around the channel. This architecture can include three gates (tri-gate) or completely surround the channel (GAA) [11–16]. In a FinFET, the channel is shaped like a vertical fin, and the gate wraps around three sides of the fin, offering improved control over the channel compared to multigate and/or single-gate designs. While the gate in the GAA FETs is completely surrounds the channel providing the most comprehensive control. The channel can be in the form of a nanowire or nanosheet, with the gate enveloping it entirely.

Multigate FETs are widely used in advanced processors, mobile devices, and other integrated circuits where performance, power efficiency, and scaling are critical. They are the dominant transistor architecture in modern semiconductor manufacturing for technology nodes below 20 nm [15–19].

Another advanced class of transistors designed to overcome the restrictions of traditional MOSFETs and provide better scalability, performance, and power efficiency is the junctionless LMG-FETs (JLMG-FETs) [20–22]. They are unlike conventional FETs, junctionless FETs do not have p-n junctions. Instead, the entire structure is uniformly doped, simplifying fabrication and enhancing control over short-channel effects. Junctionless multigate FETs represent a promising advancement in semiconductor technology, addressing many of the challenges faced by traditional single-gate FETs. By eliminating junctions and leveraging multigate architectures, these devices offer enhanced electrostatic control, reduced leakage currents, and improved scalability.

The aim of this chapter is to examine key advancements in transistor structure design, focusing on the scaling capabilities and performance advantages of DG and GAA FETs—areas that are pivotal in cutting-edge semiconductor research. Also, as semiconductor technology advances, these innovative structures are vital for continued miniaturization, reducing power consumption, and enhancing device speed, thus driving the development of next-generation semiconductor devices.

## 1.2 From Single-Gate to Multigate FETs

Because MOSFET transistors are fundamental components in the development of integrated circuits, they play a crucial role in achieving high levels of flexibility and performance. The versatility of MOSFET devices extends across a wide domain of microelectronic applications, from memory storage to portable calculators, making them indispensable in various fields. This allows for more efficient use of space, resulting in improved performance and the ability to integrate more features within a smaller footprint.

### 1.2.1 The Era of Single-Gate FETs

In the early 1960s, MOSFET device was developed [23–26]. It quickly became the backbone of digital electronics due to its simplicity, low power consumption, and scalability. The traditional MOSFET features a single gate that controls the current flux in the channel by modulating the conductivity of a semiconductor channel as shown in (Figure 1.1). The single-gate MOSFET enabled the development of integrated circuits (ICs), leading to the rapid growth of the semiconductor industry and the realization of Moore's Law [27–31]. Throughout the 1970s to the 1990s, the semiconductor industry focused on scaling down the dimensions of single-gate MOSFETs to increase transistor density and improve performance.

As MOSFETs were scaled below 1 micron, engineers began to encounter significant challenges, such as short-channel effects (SCEs), where the control of the gate over the channel diminishes, leading to issues like leakage currents and threshold voltage instability [32–34]. To mitigate some of these challenges, the inclusion of high-k dielectrics in the late 1990s helped to reduce gate leakage and enhance performance, but further scaling required more radical innovations [35].

Figure 1.1(a), shows the typical structure of a four-terminal bulk Si-MOSFET. Electrical carriers can flow from the source (where they are injected) to the drain depending on the voltage applied to the gate. The gate bias is that one which controls the conductivity of the path between source and drain, namely active region. The gate terminal is formed from metal or poly-Si and it is separated from the transistor channel by a thin dielectric layer (gate oxide). N- and P-type channel devices can be formed depending on the doping of the source, drain and substrate regions. Following, we give the operation regimes of an n-type Si-MOSFET. In this case, the substrate of the transistor is p-type doped, whereas source and drain are highly n+- doped. When a positive voltage is applied to the gate terminal, due to electrostatics forces holes are repelled from the channel, forming a depletion layer and as the gate voltage takes higher positive values more electrons are attracted to this region forming an inversion layer on the surface of the semiconductor. In other words, the concentration of the electrons is equal with that of holes in the substrate. Once the polarity of the inversion layer is the same with the polarity of the source and drain regions, above a specific gate voltage (threshold voltage) and for voltages

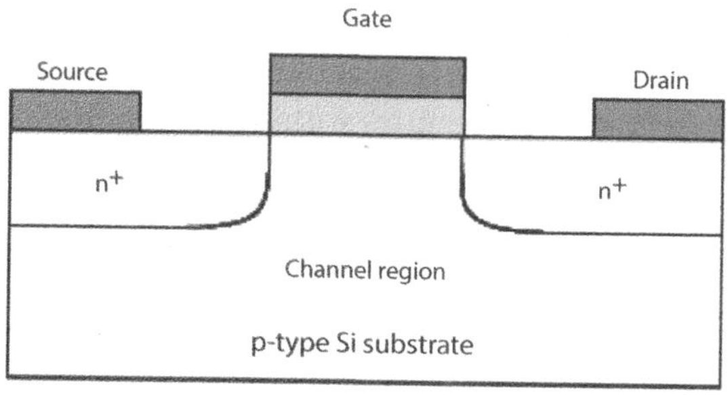


(a)

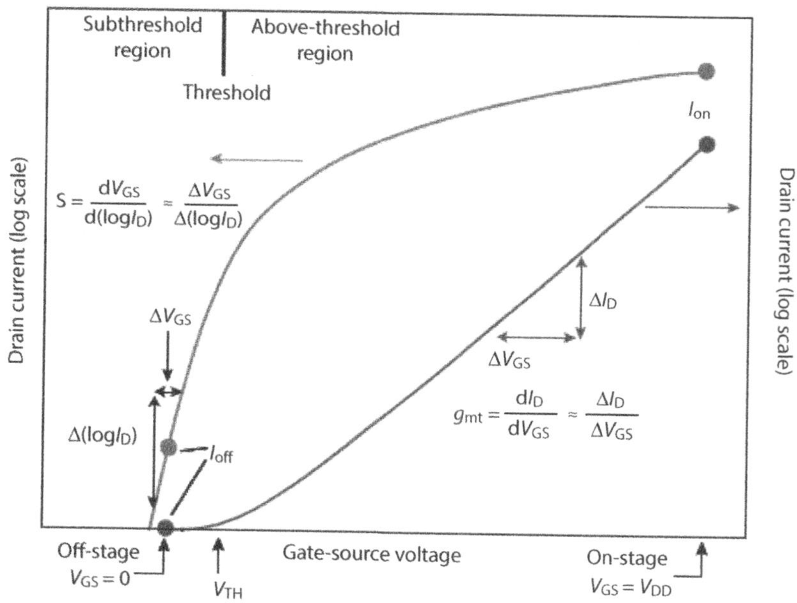


(b)

**Figure 1.1** (a) Cross-section of an n-channel Si MOSFET. (b) n-MOSFET transfer characteristics showing $I_{DS}$ (on a logarithmic scale on the left and a linear scale on the right) versus the gate source voltage ($V_{GS}$).

applied to the drain terminal higher than that applied to the source one (usually the source terminal is kept grounded) current (drain current, $I_D$) can flow between the source and drain region. In MOSFET two different mechanisms that govern the drain current, diffusion and drift. The former one dominates when $V_{GS} < V_{TH}$, where $V_{GS}$ is the gate to source voltage and $V_{TH}$ is the threshold voltage, respectively. Under this condition, known also as subthreshold regime, current flows due to the difference in electron concentration and the drain current varies with the gate to source voltage exponentially [82–90] follows as Equation 1.1:

$$I_{D\text{-}off} \approx \frac{qV_{GS}}{\eta k_B T} \tag{1.1}$$

Above the threshold voltage ($V_{GS} > V_{TH}$) the drift mechanism dominates, and the drain current is governed by the electric field. In this case, we have two different operation regimes, the linear and saturation regime, depending on the applied drain voltage. When the drain voltage is smaller than the gate overdrive ($V_{DS} < V_{GS} - V_{TH}$), the drain current varies linearly with the gate voltage follows [85–93] as Equation 1.2:

$$I_{D\text{-}Linear} = \frac{W_G}{L_G}\mu C_{ox}\left(V_{GS} - V_{TH} - \frac{V_{DS}}{2}\right)V_{DS} \tag{1.2}$$

where $W_G$ and $L_G$ are the gate width and length, μ is the mobility and $C_{OX}$ is the gate dielectric capacitance. When the drain voltage is bigger than the gate overdrive ($V_{DS} \geq V_{GS} - V_{TH}$) the drain current is independent of the drain voltage follows as Equation 1.3:

$$I_{D\text{-}saturation} = \frac{W_G}{2L_G}\mu C_{ox}(V_{GS} - V_{TH})^2 \tag{1.3}$$

### 1.2.2 Double-Gate Devices

In 1980s, the idea of using two gates to control the channel more effectively has been emerged as a theoretical solution to the scaling limitations of single-gate MOSFETs. So, Double-gate FET structure is advanced architecture in the field of transistor design (Figure 1.2), and it is first key to enable continued scaling of semiconductor devices [36, 37]. Those transistors promised better electrostatic control. A decade after, researchers developed experimental double-gate FETs, demonstrating that two gates could indeed provide better control over the channel, reducing leakage currents and improving drive strength. The double-gate transistors have emerged as promising devices for nano-scale circuits due to their better scalability compared to bulk CMOS. Among the various types of double-gate devices, quasi-planar SOI FinFETs are easier to manufacture compared to planar double-gate devices. The double-gate devices with independent gates (separate contacts to back and front gates) have recently been developed. The double-gate devices with symmetric and asymmetric gates have also been demonstrated. Such device options have direct implications at the circuit level. Independent control of front and back gate in double-gate devices can be effectively used to improve performance and reduce power in sub-50nm circuits.

In series development of conventional CMOS technology, double-gate (DG) field effect devices, particularly MOSFETs and FinFETs, utilize two gates (top and bottom/side) to sandwich a thin semiconductor channel, offering superior electrostatic control over the channel compared to single-gate transistors. The upgraded FET architecture reduces short-channel effects (SCEs), lowers power consumption, and increases drive current ($I_{ON}$), making them ideal for high-performance, nanoscale, and RF applications.

### *1.2.2.1 Planar DG-FET*

The planar DG FET design associated with two gates on opposite sides of a thin semiconductor channel, typically fabricated on a silicon substrate [38]. The planar DG device is analogous to conventional planar bulk MOSFET technology but with reduced geometric issues (Figure 1.2). Moreover, the planar DG design avoids the "crystal plane orientation uncertainty" problem that arises when device orientations differ. Another benefit of the planar DG MOSFET is that the two gates can be easily biased independently, as they

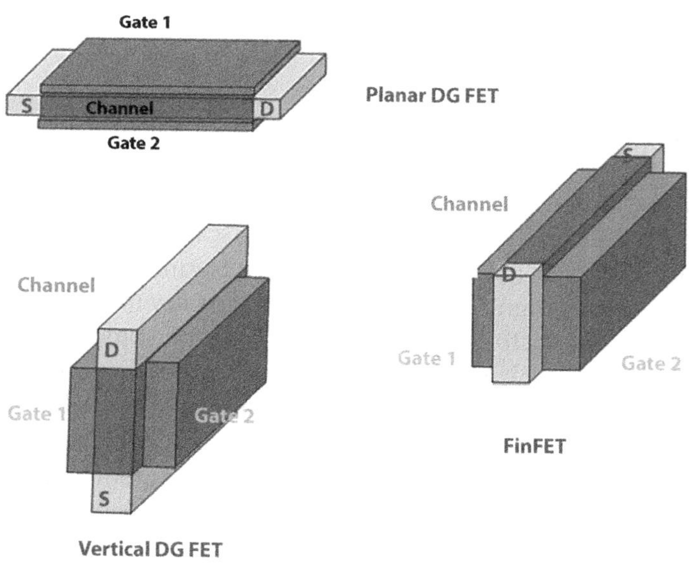


**Figure 1.2** 3D presentation of (a) planar DG FET, (b) the FinFET and (c) the vertical DG FET.

are fabricated in separate processing steps. Both gates control the channel simultaneously, providing enhanced control over the electrostatic potential in the channel, which reduces short-channel effects. The planar double-gate FETs are suitable for logic circuits where planar integration is preferred.

#### *1.2.2.2 Vertical DG FET*

In the vertical DG FET, the channel is oriented vertically, with gates on opposite sides of the channel. This structure allows for a more compact layout compared to planar designs (Figure 1.2). The vertical orientation can lead to improved electrostatic control and is often used in power transistors or devices where vertical integration is beneficial. Power electronics and 3D integrated circuits where vertical stacking is advantageous [39, 40].

### 1.2.3 Multigate Structures

To enhance addressing undesirable effects, more new FET structural design (Figure 1.3), the multigate (MG) FET, was proposed and studied [41, 42]. In these devices, SCEs are more effectively controlled by the dimensions of the channel, resulting in less noticeable swing factor degradation. Notably, the ideal SS, assuming no short-channel degradation, is known to be 60 mV per decade [43].

Figure 1.3 illustrates the evolution of MOSFET structures, including double-gate, tri-gate, pi-gate, omega-gate, and gate-all-around designs. Double-gate and tri-gate FinFETs are particularly popular because of their straightforward design. Although the gate-all-around (GAA) device was conceptualized before the FinFET, FinFETs were more practical for large-scale production.

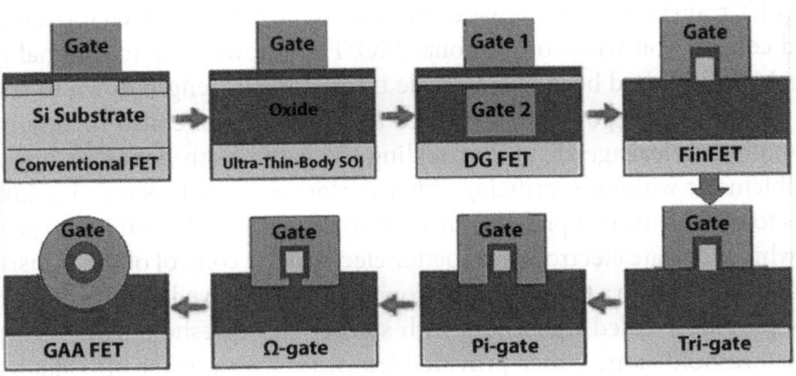


**Figure 1.3** Multigate structures evolution. Image based on Ho, B., *et al.*, [81].

### 1.2.3.1 Fin Field-Effect Transistor

In 1999, researchers at California University were introduced a new transistor structure, the Fin Field-Effect Transistor (FinFET) (Figures 1.2 and 1.3). The FinFET featured a thin, fin-like channel [44]. They offered significantly improved control over the channel compared to single- and double-gate MOSFETs. So, they reduced short-channel effects, leakage currents, and allowed for better scaling to smaller nodes. However, by the early 2000s, FinFETs began to be adopted by major semiconductor manufacturers. They became the standard transistor architecture for technology nodes at 22 nm and below. And, they are widely used in modern processors, mobile devices, and other advanced integrated circuits (ICs), particularly in technology nodes of 14, 10, 7 nm, and beyond [45, 46]. As technology nodes reached 7 and 5 nm, FinFETs began to face limitations in further scaling, prompting the semiconductor industry to explore GAA FETs for future nodes.

For decades, the relentless miniaturization of planar CMOS (Complementary Metal-Oxide-Semiconductor) transistors drove the exponential performance improvements predicted by Moore's Law. However, as feature sizes shrank below the 22nm node, planar transistors began to encounter severe physical and electrical limitations. Short-channel effects, increased leakage currents, and reduced gate control became major barriers to further scaling. Planar CMOS devices, which rely on a single gate controlling a thin channel, struggle to maintain performance and efficiency at nanoscale dimensions. The thinner gate oxide layers required for scaling led to increased gate leakage, while the reduced channel length compromised the transistor's ability to fully switch off. These challenges resulted in higher static power consumption and degraded switching behavior [82–85].

The next major transistor innovation was the introduction of FinFET (tri-gate) transistors on Intel's 22-nm technology in 2011. In Figure 1.4, FinFET transistor (a): schematic diagram (b) layout (c): I-V characteristics and comparison with conventional MOSFET shown. The traditional planar MOSFETs had been able to scale transistor gate length down to about 32 nm to deliver good performance and density while also maintaining low off-state leakage ($I_{OFF}$). But scaling the gate length below 32 nm was problematic without sacrificing either performance or leakage. A solution was to convert from a planar transistor structure to a 3D FinFET structure in which the gate electrode had better electrostatic control of the transistor channel formed in a tall narrow silicon fin. This improved electrostatic control provided scaled transistors with steeper sub-threshold slope. Steeper sub-threshold slope either provided transistors with lower off-state leakage or allowed threshold voltage to be reduced, which enabled improved

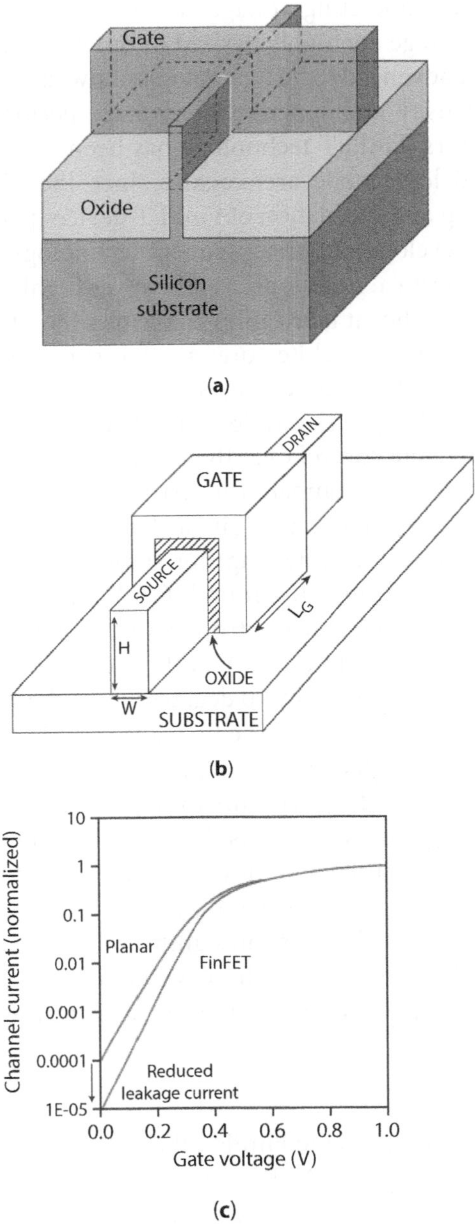


**Figure 1.4** FinFET transistor (a): schematic diagram (b) layout (c): I-V characteristics and comparison with conventional MOSFET.

performance at low operating voltage. The operating integrated circuits (ICs) at a lower voltage are highly desired in order to reduce active power consumption. All advanced logic technologies now use FinFET transistors for their good density and superior low-voltage performance compared to planar transistors. FinFET technology has been widely adopted as the leading advanced logic node architecture since the 22 nm generation. Accurate modeling and simulation of FinFET devices play a crucial role in accelerating the development of subsequent technology nodes. The 7 nm node represents a critical milestone, as it involves highly complex physical phenomena and significant fabrication challenges [86–93].

The most common 3D architectures are Tri-gate, π-gate, Ω-gate, Gate-All-Around, and FinFETs. These names come from the shape of the transistor channel, and the gate electrode is situated on more than one side of the channel. Better gate control over the channel can be achieved through π- and Ω-gate structures compared to Tri-gate ones, while an improvement to mobility and reduction of gate leakage are also possible after an appropriate treatment (H2 annealing) for smoothing the fin surface and corners. Regarding the Gate-All-Around (GAA) FETs, the device channel is completely enclosed the gate electrode with the ability to maintain the electrostatic control down to 3nm. A FinFET can be viewed as a planar device with the channel turned by 90 degrees where the width of the gate is much smaller than the fin height. The fin width is an important parameter of a FinFET due to its impact on device performance. The off-state current reduces as the fin width decreases due to improved gate control over the charge in the channel and also affects the carrier mobility and the threshold voltage. FinFET technology is massively adopted as the advanced logic node technology since 22nm generation. Modelling and simulation of FinFETs and accurate predictions can significantly speed up the next technology node development. 7nm technology node is a critical point due to its extremely complicated physics and technical difficulty in fabrication.

### *1.2.3.2 Tri-Gate FET*

Another structure similar to the FinFET, the tri-gate FET also has a three-sided gate structure (Figure 1.3). However, the term "tri-gate" often refers to the specific implementation where the gate completely encloses the fin from three directions, sometimes with more emphasis on planar integration [47]. The gate's comprehensive control over the channel reduces leakage currents and enhances drive strength, making it highly efficient device. They are used in high-speed processors and other ICs that require balanced performance and power efficiency [48].

### 1.2.3.3 *GAA FETs*

GAA FETs, where the gate fully surrounds the channel, were conceptualized as an evolution of multigate FETs (Figure 1.3). Early prototypes in the 2000s demonstrated even greater control over the channel than FinFETs. GAA FETs evolved into two main variants: nanowire and nanosheet FETs. Nanowire GAA FETs feature a cylindrical channel surrounded by the gate, while nanosheet GAA FETs have a flat, ribbon-like channel surrounded by the gate [49, 50].

GAA FETs provide the best electrostatic control among all FET structures, virtually eliminating short-channel effects and minimizing leakage currents. They are highly scalable and are seen as the future of transistor technology for nodes below 5nm. Consequently, leading semiconductor companies like Samsung and TSMC have announced plans to use GAA FETs in their upcoming 3- and 2-nm processes, marking the next major transition in transistor architecture. Therefore, future of multigate FETs may involve 3D integration, where multiple layers of transistors are stacked vertically to further increase transistor density and performance. And, the use of new materials, such as 2D materials (e.g., graphene) and high-mobility semiconductors, can enhance the performance of multigate FETs.

In GAA FETs technology, the channel has been used in the form of a nanowire, nanosheet, or nanotube, with the gate wrapping around all sides. The **nanowire GAA FET** having a cylindrical nanowire channel surrounded by the gate [51, 52]. It offers superior electrostatic control and high scalability. The second subtype is the **nanosheet GAA FET**; it uses a flat, ribbon-like channel surrounded by the gate and provides a larger effective channel width,

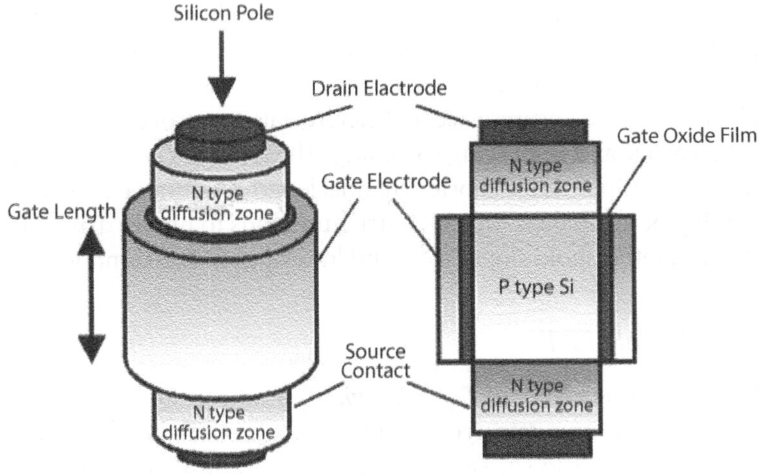


**Figure 1.5** The surrounding gate MOSFET device.

leading to higher drive currents compared to nanowires. These two types FETs are emerging as a key technology for nodes at 5nm and smaller, with potential use in high-performance computing, AI, and other cutting-edge applications.

#### *1.2.3.4 Surround-Gate Transistor*

The surround-gate transistor (SGT) is a variant of the GAA FET where the gate fully encircles the channel (Figure 1.5), similar to the GAA structure but sometimes applied in different geometries or materials [53]. It offers maximal gate control over the channel, ideal for minimizing short-channel effects and enhancing device performance. Widely used in advanced memory devices, high-speed processors, and applications requiring extreme miniaturization.

#### *1.2.3.5 Omega-Gate FET*

An "Omega-gate FET" (Field Effect Transistor) refers to a modern transistor design, often a nanowire or FinFET, with a gate wrapping around the channel (like the Greek letter Ω) for superior electrostatic control, leading to better performance and reduced short-channel effects. The Omega-Gate FET (Field-Effect Transistor) is an advanced transistor design, often for nanowires, where the gate wraps around the channel in an "Ω" (omega) shape, maximizing gate control for better performance, smaller size, and improved electrostatic properties compared to older designs like FinFETs, crucial for scaling down electronics. It provides superior electrostatic confinement, better drive currents, and sharper switching (subthreshold swing), making it ideal for next-generation integrated circuits, especially for high-performance computing and mobile devices. This FET structure has a gate that forms an omega (Ω) shape, wrapping around three sides of the channel [54]. It is a variant of the FinFET, with more extensive gate coverage (Figure 1.3). The partial wrap-around ensures excellent electrostatic control over the channel, reducing leakage and improving switching. The Omega-Gate FET allows for smaller device dimensions (shorter channel lengths, thinner channels) while maintaining performance. This FET structure leads to higher drive currents ($I_{on}$), better subthreshold slopes (SS), and lower power consumption.

#### *1.2.3.6 Pi-Gate FET*

The Pi-gate (π-gate) MOSFET is an advanced Silicon-On-Insulator (SOI) transistor with a unique gate structure that wraps around the channel, resembling the Greek letter Pi (π) shown (in Figure 1.3), providing superior electrostatic control, better suppression of short-channel effects, and improved performance for future nanoscale electronics compared to traditional single-gate devices, making it ideal for high-density integrated circuits [55]. This MOS device structure provides better gate-channel coupling reduces leakage currents than conventional MOSFETs and enhances immunity to short-channel effects like drain-induced barrier lowering (DIBL).

## 1.3 Junctionless Multigate FETs

### 1.3.1 Junctionless Transistors Design

In conventional MOSFETs, the source and drain regions are doped with opposite types of impurities (n-type or p-type) relative to the channel. A p-n junction forms between these regions and the channel, and the transistor operates by using the gate voltage to modulate the carrier concentration in the channel, allowing or blocking current flow between the source and drain. Unlike traditional MOSFETs, where the channel, source, and drain are doped differently, a junctionless transistor features a uniformly doped semiconductor channel. The entire channel is either n-type or p-type, depending on the desired type of carrier (electrons or holes). As the name suggests, junctionless transistors do not have p-n junctions (Figure 1.6). This concept began to emerge in the late 1990s and early 2000s. Researchers theorized that

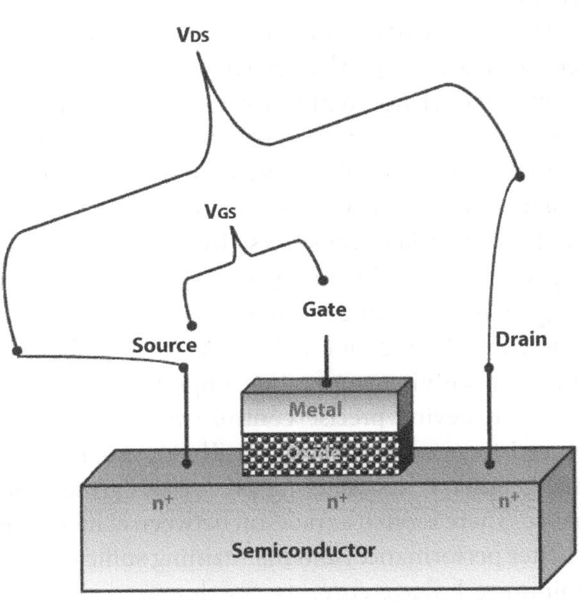


**Figure 1.6** Junctionless FET.

eliminating the p-n junctions in transistors could simplify the manufacturing process and reduce some of the issues associated with traditional transistors. This contrasts with conventional MOSFETs, where the gate modulates the carrier density across a junction. In 2009, researchers led by Jean-Pierre Colinge at the Tyndall Institute presented the first experimental demonstrations of junctionless transistors. This was a major breakthrough, proving that junctionless transistors could function effectively and offer certain advantages over conventional transistors [56–58]. The initial junctionless transistors were fabricated using silicon nanowires, which allowed for precise control of the channel thickness and doping concentration. The nanowire structure also made it easier to achieve full depletion of the channel.

### 1.3.2 Basic Principle of Junctionless Transistors

The junctionless transistor mainly functions in depletion mode. Without any gate voltage, the channel remains conductive, permitting current flow. When voltage is applied to the gate, it depletes the carriers in the channel, thereby reducing or halting the current flow. Essentially, the gate voltage controls the degree of depletion in the uniformly doped channel. When the gate voltage is low or negative (for n-type junctionless transistors), the channel remains conductive and current flows in the channel. When the gate voltage is increased (positive for n-type), it depletes the majority carriers in the channel, effectively "pinching off" the current flow, turning the transistor off. The absence of p–n junctions simplifies the doping process, as there is no need to create sharply defined regions with different doping types. This reduces manufacturing complexity, especially at nanoscale dimensions where precise junction formation becomes challenging. Additionally, the uniform doping profile makes junctionless transistors more scalable as device dimensions shrink, offering potential advantages in producing smaller, more densely packed transistors. The gate's capacity to control the uniformly doped channel helps mitigate short-channel effects, which are prevalent in aggressively scaled conventional MOSFETs [59–62]. With better electrostatic control, junctionless transistors can exhibit lower leakage currents and higher on/off current ratios, improving overall device performance. However, achieving precise control over the threshold voltage in junctionless transistors can be challenging. The threshold voltage is affected by the channel geometry, necessitating precise design to maintain consistent performance. There is often a trade-off between achieving low threshold voltages (for better performance) and maintaining sufficient depletion in the off-state to minimize leakage currents. So, despite their advantages, junctionless transistors are not yet widely adopted in commercial semiconductor manufacturing. They are primarily used in niche applications where their specific benefits outweigh the challenges, such as in certain low-power and high-reliability devices.

### 1.3.3 Evolution to Junctionless Multigate FETs

The loss gate control over channel charges observed in very-short channel transistors (due to the proximity of the source and drain regions) renders the use of planar MOS devices quite challenging in transistors of ultra scaled dimensions. For this reason, alternative technologies, such as the multigate architecture, have gained interest in the literature. The presence of the gate electrode on more than one side of the device effectively improves the electrostatic control over the channel, reducing short-channel effects. The narrow multigate transistors with fin width ($W_{fin}$ < 50–60 nm) have particularly good immunity against short-channel effects, promising to continue the scaling of silicon complimentary MOS beyond the limits established for planar devices.

With conventional MOS devices, when the channel length of classical multigate MOS transistors (e.g., FinFETs, trigate, and gate-all-around (GAA) devices) is scaled down to extreme dimensions, the formation of ultra sharp junctions between source/drain and channel becomes complex since the doping concentration has to vary by several orders of magnitudes over a distance of a few nanometers. Therefore, highly accurate doping techniques and ultrafast dopant activation processes are required to avoid the lateral diffusion of source and drain impurities into the channel region. A solution to this problem has recently been proposed in the form of a novel structure called the "junctionless (JL) multiple gate transistor", which exhibits an unconventional architecture that presents neither source/drain junctions nor doping concentration gradients.

The junctionless concept was naturally extended to multigate FETs, where multiple gates are used. Multigate structures, such as FinFETs and

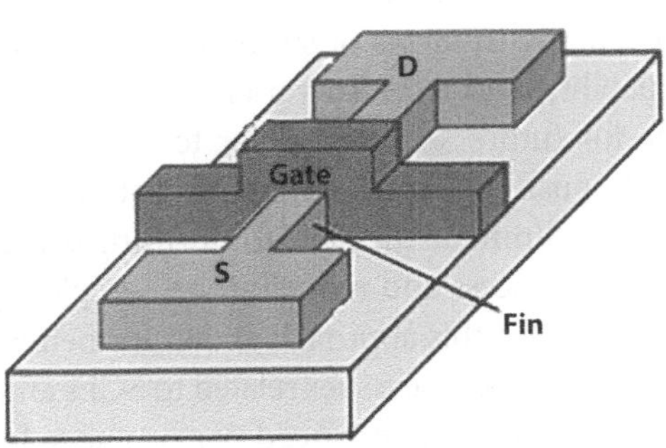


**Figure 1.7** Junctionless FET longitudinal cross section [74].

GAA FETs, were already known for their superior electrostatic control, and combining these with the junctionless approach further enhanced performance. One of the first implementations was the junctionless FinFET (Figure 1.7), where a fin-like channel is controlled by gates on three sides. This structure provided excellent electrostatic control, reduced short-channel effects, and simplified the fabrication process by eliminating the need for precise junction formation.

Similarly, the junctionless concept was applied to Gate-All-Around FETs, where the gate surrounds the channel completely [63–65], offering even greater control and reducing leakage currents to near zero levels. Therefore, combining the junctionless transistor concept with multigate structures has led to a novel class of transistors that leverage the strengths of both approaches. A junctionless multigate FET combines the uniform doping of the junctionless transistor with the superior electrostatic control of multigate architectures. In this structure, the channel is uniformly doped, and the multiple gates work together to provide a stronger and more uniform electric field that can more precisely deplete the channel. This leads to better control over the on/off states of the transistor, reducing leakage currents and enhancing the on-state current.

Many structures have been introduced, in a junctionless FinFET, the channel takes the form of a thin fin, with gates placed on three sides. The gates deplete the uniformly doped fin to control the current, taking advantage of both the simplified fabrication of junctionless transistors and the strong electrostatic control of the FinFET design. In a junctionless GAA FET, the gate fully surrounds the channel, providing the ultimate level of control. This design minimizes short-channel effects even further, making it highly effective for nanoscale devices [66–68].

### 1.3.4 Advantages of Junctionless Multigate FET

Junctionless multigate FETs offer numerous advantages over traditional transistor designs, particularly in terms of simplified fabrication, enhanced control, improved scalability, and reduced variability. These benefits make them a strong candidate for future semiconductor technologies, especially as the industry continues to push the limits of miniaturization and performance. So, junctionless transistors eliminate the need for forming source/drain junctions, simplifying the doping process and reducing the complexity of the manufacturing process. Additionally, this approach allows for better scaling to smaller technology nodes, as it reduces the issues related to SCEs and threshold voltage changeability. And, it leads to an enhanced performance by offering low off-state leakage currents and high drive currents [66–69].

### 1.3.5 Challenges and Limitations

While junctionless multigate FETs offer several advantages, they also face several challenges that need to be addressed for their successful implementation in advanced semiconductor applications. These challenges include threshold voltage control, process variability, short-channel effects, self-heating, scaling difficulties, manufacturing complexity, integration with existing technologies, and reliability concerns. Overcoming these challenges requires ongoing research and development, as well as advancements in materials, fabrication techniques, and design tools. Despite these challenges, the potential benefits of junctionless multigate FETs make them a promising technology for the future of semiconductor devices, particularly as the industry continues to push the limits of miniaturization and performance. Their unique advantages, particularly in terms of simplified fabrication and improved performance, position them as a key technology in the continued evolution of electronic devices.

The simplified structure and excellent control make these transistors candidates for use in emerging computing paradigms, such as neuromorphic computing (which mimics the human brain) and quantum computing.

### 1.3.6 Junctionless GAA FET Numerical Simulations

Numerical simulation of junctionless GAA FETs is a powerful tool that provides accurate device modeling, enables design optimization, reduces development costs, and supports innovation. It helps in predicting device behavior, optimizing performance, and addressing scaling challenges. So, to model complex physical phenomena and provide valuable insights into the performance characteristics particularly in terms of power efficiency, switching speed and scaling potential of the investigated JL GAA FETs,

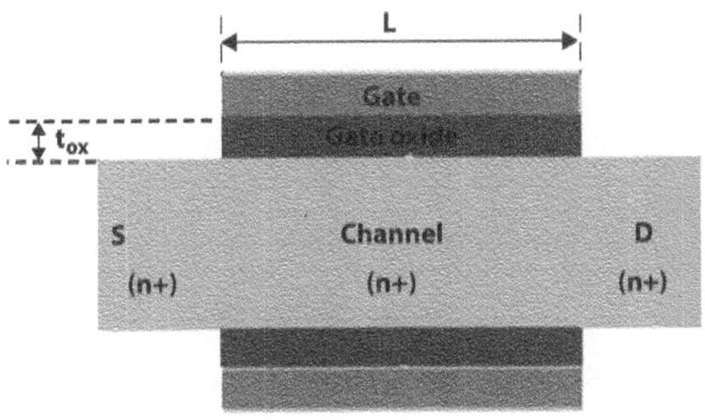


**Figure 1.8** Two-dimensional GAA JL MOSFET cross-sectional views [75].

simulations using ATLAS TCAD simulator were performed [70–73]. As shown in Figure 1.8, L is the gate length of the device channel, R is the radius (R = 2.5 nm), and $SiO_2$ is an interface layer with tox denoting its thickness (tox =1nm). These simulations are essential for understanding how the swing factor (SS) and off-state current ($I_{OFF}$) vary with different channel lengths (L from 10 to 60 nm). In this context, the device is considered with JL structure for which the channel region is considered n-type uniformly doped with concentration of $10^{19}$ $cm^{-3}$. Simulating the impact of channel length (L) variation on the SS, $I_{OFF}$, and the $I_{ON}/I_{OFF}$ current ratio is essential for understanding the behavior and performance of the model. These simulations show that the SS, which indicates the efficiency of the transistor switching from off to on state; does not vary significantly with changes in channel length (Figure 1.9). Similarly, Figure 1.10 shows a consistent $I_{OFF}$; $I_{OFF}$ represents the small amount of current that flows when the transistor is in the off state, An invariant SS indicates good control over the channel and consistent switching behavior. While, a consistent $I_{OFF}$ current is an indicator of good device design, strong electrostatic control, and low static power consumption., but it does not guarantee that the $I_{ON}/I_{OFF}$ ratio will remain invariant (Figure 1.11). The $I_{ON}/I_{OFF}$ ratio can vary due to factors like leakage currents, threshold voltage shifts, and changes in device characteristics with scaling. Therefore, while invariant SS is a good sign of robust device design, additional considerations are needed to ensure stable $I_{ON}/I_{OFF}$ ratio, Simulating these parameters is essential for

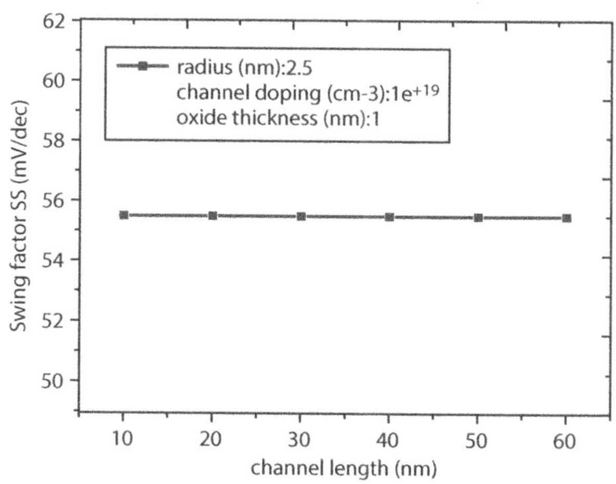


**Figure 1.9** Swing factor (SS) versus channel length (L) with R = 2.5 nm, $N_d = 10^{19} cm^{-3}$ and $t_{ox} = 1$ nm.

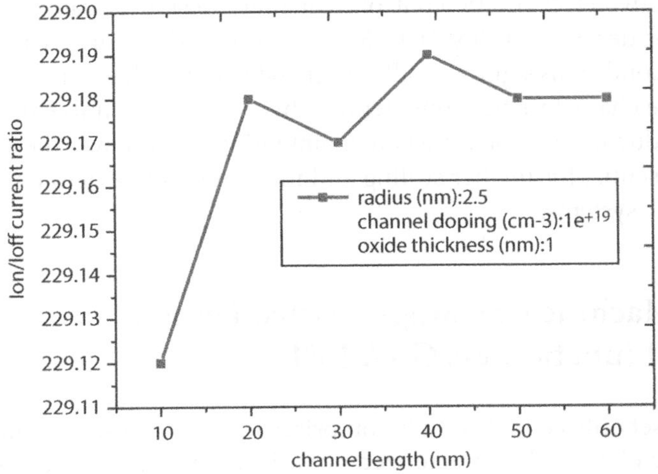


**Figure 1.10** Versus channel length (L) with R=2,5 nm, $N_d=10^{19}cm^{-3}$, and $t_{ox}$=1 nm.

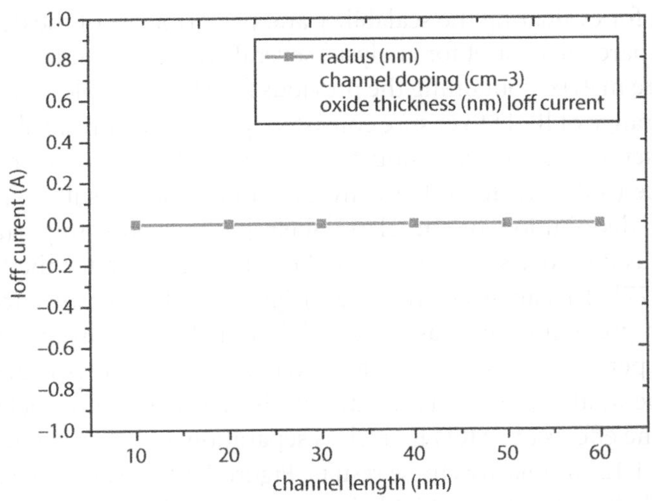


**Figure 1.11** $I_{OFF}$ versus channel length (L) with R=2,5 nm, Nd=1019cm$^{-3}$, and tox=1 nm.

designing next-generation transistors that meet the performance and efficiency requirements of modern electronics, including low-power, high-speed, and high-density applications.

In summary, when SS does not vary with channel length in junctionless GAA FET simulations, it generally indicates robust device performance and effective electrostatic control, which are desirable traits for modern transistor designs. An invariant SS, $I_{OFF}$, and $I_{ON}/I_{OFF}$ current ratio across varying conditions signify a well-designed transistor that maintains its performance characteristics consistently. This is highly desirable in advanced semiconductor technology, where maintaining control over these parameters is essential for further scaling and integration of devices into modern electronic systems.

## 1.4 Machine Learning-Assisted Design of Junctionless GAA FET

The dataset utilized in this study comprises 147 observations of junctionless multigate FETs, each characterized by variables including oxide permittivity, channel doping, oxide thickness, and channel radius. The Ioff is also recorded for each observation and categorized into three classes—Good, Medium, and Worst—based on predefined thresholds. This categorization is crucial for evaluating the scalability and performance of the devices.

To prepare the dataset for analysis, several steps were undertaken. First, the feature matrix, containing the previous four key variables, was normalized to a range of [0, 1] to ensure consistency across features. Subsequently, the dataset was balanced using the Synthetic Minority Over-sampling Technique (SMOTE) to address any class imbalance, resulting in a more equitable distribution of the class labels [76]. Feature importance was assessed using a decision tree model both before and after the balancing process [77]. This analysis provided insights into the significance of each feature in predicting the class labels. Additionally, dimensionality reduction was performed using t-sne to visualize the data in two dimensions [78]. t-sne scatter plots were created both before and after balancing to observe the effects of SMOTE on class separation and clustering.

Figure 1.12 summarizes these results. Figure 1.12 (a) shows the class proportions before balancing, highlighting an initial high imbalance. Figure 1.12 (b) depicts the adjusted class proportions after balancing. Figure 1.12 (c) presents the feature importance before balancing. Figure 1.12 (d) illustrates the feature importance post-balancing. Figure 1.12 (e) provides a t-SNE scatter plot of the data before balancing; and Figure 1.12 (f) displays the t-sne scatter plot after balancing. These visualizations and analyses collectively demonstrate the impact of data preparation on the scaling capability analysis of junctionless multigate FETs. As a result, the feature importance analysis reveals that the channel radius exhibits the highest significance, followed by

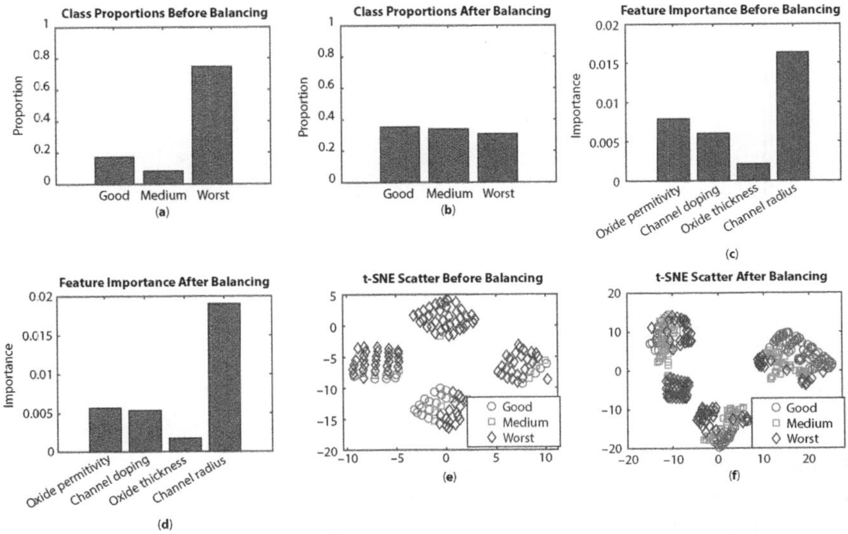


**Figure 1.12** Analysis of class proportions, feature importance, and t-SNE scatter plots: (a) Class proportions before balancing; (b) Class proportions after balancing; (c) Feature importance before balancing; (d) Feature importance after balancing; (e) t-SNE scatter plot before balancing; (f) t-SNE scatter plot after balancing.

oxide permittivity in the second stage. This indicates that the channel radius plays a crucial role in predicting the class labels of junctionless multigate FETs, while oxide permittivity also significantly contributes to the predictive model. This insight underscores the relative importance of these features in assessing the scaling capability and performance of the devices. Additionally, the data scatter plots reveal a nonlinear distribution of class labels both before and after balancing. This observation suggests that nonlinear modeling techniques may be more appropriate for future machine learning analysis to effectively capture the complexities in the data distributions.

Accordingly, this work utilized a Gated Recurrent Unit with a projected layer (PGRU) to effectively model and analyze the nonlinear patterns and complexities observed in the data [79]. The GRU was selected for its ability to capture temporal dependencies, while the projected layer enhances its capacity to model complex, high-dimensional relationships. Hyperparameters were tuned and optimized using a Bayesian optimization approach, ensuring robust performance and efficient convergence [80]. The model was evaluated under various criteria, including Accuracy, F1 score, Precision, and Recall, allowing for a comprehensive assessment of its performance and effectiveness in handling the dataset.

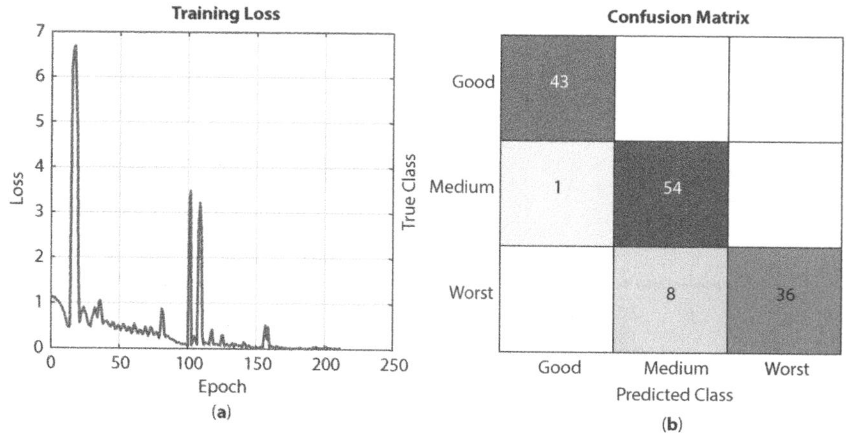


**Figure 1.13** PGRU model performances: (a) Plot of the training loss over epochs for the PGRU model, showing the model's learning progression. (b) Visualization of the test classification performance, with true vs. predicted class labels.

In Figure 1.13, the training loss curve (Figure 1.13 (a)) reveals a strong convergence of the PGRU model, with only minor fluctuations and a few spikes, reflecting effective learning and stable performance improvements. The confusion matrix (Figure 1.13 (b)) illustrates that the model performs excellently overall, with only 8 samples misclassified from 'Worst' to 'Medium' and 1 sample from 'Medium' to 'Good.' This indicates that, although the model is highly accurate, there is potential for further refinement in distinguishing between 'Worst' and 'Medium' categories, as well as in improving the classification of 'Good' and 'Medium' labels.

Table 1.1 presents the performance metrics for both the training and testing phases of the PGRU model. The F1 score, used as the objective function during the hyperparameter optimization process, was chosen because it captures the balance between precision and recall, providing a comprehensive measure of the model's classification performance. The training accuracy is exceptionally high at approximately 99.53%, indicating that the model performs very well on the training data. Precision, recall, and F1 scores are also impressive, with values of 0.99, 1, and 0.99, respectively. These metrics reflect the model's effectiveness in correctly identifying and classifying instances during training. For testing, the accuracy is slightly lower at around 93.66%, suggesting that the model generalizes well but with a minor decrease in performance compared to training. The test precision, recall, and F1 scores are strong as well, at 1, 0.98, and 0.99, respectively, demonstrating that the model

**Table 1.1** Performance metrics for training and testing.

| Metric | Training value | Testing value |
|---|---|---|
| Accuracy | 0.9953 | 0.9366 |
| Precision | 0.9878 | 1.0000 |
| Recall | 1.0000 | 0.9773 |
| F1 | 0.9939 | 0.9885 |
| Training Time | 1.9359 seconds | - |

maintains high performance on unseen data. The training time is approximately 1.94 seconds reflecting the efficiency of the training and optimization processes. Overall, the metrics indicate that the model is robust and performs effectively across both training and testing datasets.

## 1.5 Conclusion

In this work, we examined the shift from single-gate to MG-FETs to enhance device performance and address miniaturization challenges. MG-FETs offer improvements in ultra-low power consumption, energy efficiency, and low voltage operation, although they may not yet achieve the drive currents and switching speeds of traditional MOSFETs. JLMG-FETs were explored as a more advanced alternative to conventional MG-FETs. By eliminating junctions and using a uniformly doped structure, JLMG-FETs improve scalability and reduce fabrication costs, offering better electrostatic control. Machine learning techniques were applied alongside 2D numerical simulations to optimize these devices. The obtained F1 score of 0.9885 highlights the effectiveness of these models in evaluating transistor performance. This approach provides valuable insights into the scaling potential of JL-DG and GAA MOSFETs, supporting the development of advanced, cost-efficient semiconductor technologies.

**AI disclosure statement:** The author acknowledges the use of AI-based tools for minor language editing or grammatical correction. However, all intellectual contributions, analyses, and conclusions presented in this chapter are entirely the author's own.

## References


1. Chen, Q. and Meindl, J.D., Nanoscale metal–oxide–semiconductor field-effect transistors: Scaling limits and opportunities. *Nanotechnology*, 15, 8549–8555, 2004.
2. Bendib, T. and Djeffal, F., Electrical performance optimization of nanoscale double-gate MOSFETs using multiobjective genetic algorithms. *IEEE Trans. Electron Devices*, 58, 3743–3750, 2011.
3. Deshpande, A.A., Semwal, S., Raskin, J.-P., Kranti, A., Insights Into Parasitic Capacitance and Reconfigurable FET Architecture for Enhancing Analog/RF Metrics. *IEEE Trans. Electron Devices*, 70, 5983–5990, 2023.
4. Rai, S., *et al.*, Designing efficient circuits based on runtime-reconfigurable field-effect transistors. *IEEE Trans. Very Large Scale Integr. VLSI Syst.*, 27, 560–572, 2019.
5. Bendib, T., Djeffal, F., Arar, D., Subthreshold behavior optimization of nanoscale Graded Channel Gate Stack Double Gate (GCGSDG) MOSFET using multi-objective genetic algorithms. *J. Comput. Electron.*, 10, 210–215, 2011.
6. Bentrcia, T., Djeffal, F., Abdi, M.A., Chahdi, M., Boukhennoufa, N., An accurate two dimensional threshold voltage model for nanoscale GCGS DG MOSFET including traps effects, *2009 3rd International Conference on Signals, Circuits and Systems (SCS)*, pp. 1–6, Medenine, Tunisia, 2009, doi: 10.1109/ICSCS.2009.5412561.
7. Liu, L.N., Tang, W.M., Lai, P.T., Advances in La-Based High-k Dielectrics for MOS Applications. *Coatings*, 9, 4, 217, 2019. https://doi.org/10.3390/coatings9040217
8. Moore, G.E., Progress in Digital Integrated Electronics, Digest of the 1975 International Electron Devices Meeting, 1975.
9. Committee, I.R., *et al.*, International Technology Roadmap for Semiconductors, 2007.
10. Bentrcia, T., Djeffal, F., Arar, D., Dibi, Z., Gate-engineering-based approach to improve the nanoscale DG MOSFET behavior against interfacial trap effects. *Phys. Status Solidi C*, 11, 77–80, 2014.
11. Bentrcia, T., Djeffal, F., Benhaya, A.H., Continuous analytic I—V model for GS DG MOSFETs including hot-carrier degradation effects. *J. Semicond.*, 33, 014001, 2012.
12. Djeffal, F., Bentrcia, T., Abdi, M.A., Bendib, T., Drain current model for undoped Gate Stack Double Gate (GSDG) MOSFETs including the hot-carrier degradation effects. *Microelectron. Reliab.*, 51, 550–555, 2011.
13. Pandit, S., Nanoscale MOSFET: MOS Transistor as Basic Building Block, in: *Introduction to Nano*, Engineering Materials, A. Sengupta and C. Sarkar (Eds.), Springer, Berlin, Heidelberg, 2015.
14. Ferain, I., Colinge, C.A., Colinge, J.-P., Multigate transistors as the future of classical metal–oxide–semiconductor field-effect transistors. *Nature*, 479, 310–316, 2011.

15. Colinge, J.-P., Multigate transistors: Pushing Moore's law to the limit. *2014 International Conference on Simulation of Semiconductor Processes and Devices (SISPAD)*, Yokohama, Japan, pp. 313–316, 2014.
16. Meguellati, M. and Djeffal, F., New Dual-Dielectric Gate All Around (DDGAA) RADFET dosimeter design to improve the radiation sensitivity. *Nucl. Instrum. Methods Phys. Res., Sect. A*, 683, 24–28, 2012.
17. Djeffal, F., Ghoggali, Z., Dibi, Z., Lakhdar, N., Analytical analysis of nanoscale multiple gate MOSFETs including effects of hot-carrier induced interface charges. *J. Microelectron. Reliab.*, 49, 4, 377–381, 2009.
18. Abdi, M.A., Djeffal, F., Bentrcia, T., Arar, D., An explicit Continuous analytical model for Gate All Around (GAA) MOSFETs including the hot-carrier degradation effects. *J. Nanosci. Nanotechnol.*, 11, 9316–9320, 2011.
19. Abdi, M.A., Djeffal, F., Arar, D., Hafiane, M.L., Numerical analysis of Double Gate and Gate All Around MOSFETs with bulk trap states. *J. Mater. Sci.: Mater. Electron.*, 19 (Suppl 1), 248–253, 2008.
20. Bentrcia, T., Djeffal, F., Arar, D., Chebaki, E., Improved Reliability Performance of Junctionless Nanoscale DG MOSFET with Graded Channel Doping Engineering. *Phys. Status Solidi C*, 14, 146–152, 2017.
21. Bentrcia, T. and Djeffal, F., Compact modeling of multi-gate MOSFET including hot-carrier effects, in: *CMOS Technology: Electrical Engineering Developments Series 1*, 2011.
22. Roy, S., Jana, G., Chanda, M., Analysis of Sub-Threshold Adiabatic Logic Model Using Junctionless MOSFET for Low Power Application. *Silicon*, 14, 903–911, 2022.
23. Linienfeld, J.S.E., Method and apparatus for controlling electric currents, uS Patent 1,745,175, Jan. 28 1930.
24. Kahng, D. and Atalla, M., Silicon-silicon dioxide field induced surface devices, in: *IRE Solid-State Device Research Conference*, 1960.
25. Liou, J.J., Ortiz-Conde, A., Garcia-Sanchez, F., *Analysis and design of MOSFETs: modeling, simulation, and parameter extraction*, Springer, New York, NY, USA, 1998.
26. Moore, G.E., *et al.*, Cramming more components onto integrated circuits, 1965.
27. Skotnicki, T., Merckel, G., Pedron, T., The voltage-doping transformation: a new approach to the modeling of MOSFET short-channel effects. *Electron Device Lett. IEEE*, 9, 3, 109–112, 1988.
28. Gupta, S., Ghosh, B., Rahi, SB., Compact analytical model of double gate junction-less field effect transistor comprising quantum-mechanical effect. *J. Semiconduct.*, 36, 2, 024001, 2015.
29. Suwa, T., Iwasaki, T., Sato, K., Kato, H., Makino, T., Ogura, M., Takeuchi, D., Yamasaki, S., Hatano, M., Normally-Off Diamond Junction Field-Effect Transistors With Submicrometer Channel. *IEEE Electron Device Letters*, 37, 2, 209–211, Feb. 2016, doi: 10.1109/LED.2015.2513074.

30. Song, Y.S., Thoutam, L.R., Tayal, S., Rahi, S.B., Samuel, T.S.A., *Handbook of Emerging Materials for Semiconductor Industry*, Springer, 2024.
31. Sekigawa, T., Hayashia, Y., Ishii, K., Feasibility of very-short-channel mos transistors with double-gate structure. *Electron. Commun. Jpn. (Part II Electronics)*, 76, 10, 39–48, 1993.
32. Acovic, A., La Rosa, G., Sun, Y.C., A review of hot-carrier degradation mechanisms in MOSFETs. *Microelectron. Reliab.*, 36, 7–8, 845–869, 1996.
33. Frank, D.J., Dennard, R.H., Nowak, E., Solomon, P.M., Taur, Y., Wong, H.-S.P., Device scaling limits of Si MOSFETs and their application dependencies. *Proc. IEEE*, 89, 3, 259–288, 2001.
34. Frank, D.J., Power-constrained CMOS scaling limits. *IBM J. Res. Dev.*, 46, 2.3, 235–244, 2002.
35. Wilk, G.D., Wallace, R.M., Anthony, J., High-gate dielectrics: Current status and materials properties considerations. *J. Appl. Phys.*, 89, 10, 5243–5275, 2001.
36. Abdelmalek, N., Djeffal, F., Meguellati, M., Bendib, T., Numerical analysis of nanoscale junctionless MOSFET including effects of hot-carrier induced interface charges. *Adv. Mater. Res. J.*, 856, 137–141, 2014.
37. Chang, L., Choi, Y.-K., Ha, D., Ranade, P., Xiong, S., Bokor, J., Hu, C., King, T.-J., Extremely scaled silicon nano-CMOS devices. *Proc. IEEE*, 91, 11, 1860–1873, 2003.
38. Ferhati, H. and Djeffal, F., Novel high-performance SOI junctionless FET-based phototransistor using channel doping engineering: Numerical investigation and sensitivity analysis. *Optik-Int. J. Light Electron Opt.*, 138, 119–126, 2017.
39. Huang, X., Lee, W.-C., Kuo, C., Hisamoto, D., Chang, L., Kedzierski, J., Anderson, E., Takeuchi, H., Choi, Y.-K., Asano, K., *et al.*, Sub 50-nm FinFET: PMOS, in: *Electron Devices Meeting, 1999, IEDM'99, Technical Digest. International*, IEEE, pp. 67–70, 1999.
40. Woo, D.-S., Choi, B.Y., Choi, W.Y., Lee, M.W., Lee, J.D., Park, B.-G., 30 nm self-aligned FinFET with large source/drain fan-out structure. *Electron. Lett.*, 39, 15, 1154–1155, 2003.
41. Colinge, J.-P. and Colinge, J., *Silicon-on-insulator technology: materials to VLSI*, vol. 3, Springer, New York, NY, USA, 2004.
42. Djeffal., F., Arar, A., Abdelmalek, N., Abdi, M.A., Mahamdi, R., Errachid, A., A junctionless-multigate design to improve the electrical performances for deep submicron ISFET-based sensors. *Sens. Lett.*, 9, 6, 2309–2311, Dec. 2011, doi: 10.1166/sl.2011.1778, (Indexed by: Scopus/ISI Thomson/IF= 1.6).
43. Taur, Y., An analytical solution to a double-gate MOSFET with undoped body. *Electron Device Lett. IEEE*, 21, 5, 245–247, 2000.
44. Hisamoto, D., Lee, W.-C., Kedzierski, J., Takeuchi, H., Asano, K., Kuo, C., Anderson, E., King, T.-J., Bokor, J., Hu, C., FinFET-a self-aligned double-gate MOSFET scalable to 20 nm. *IEEE Trans. Electron Devices*, 47, 12, 2320–2325, 2000.

45. Kedzierski, J., High-performance symmetric-gate and CMOS-compatible Vt asymmetric-gate Fin-FET devices. *IEDM Tech. Dig., 2001*, 2001.
46. Liu, Y., Masahara, M., Ishii, K., Tsutsumi, T., Sekigawa, T., Takashima, H., Yamauchi, H., Suzuki, E., Flexible threshold voltage FinFETs with independent double gates and an ideal rectangular cross-section Si-Fin channel, in: *Electron Devices Meeting, 2003, IEDM'03 Technical Digest, IEEE International*, IEEE, pp. 18–8, 2003.
47. Doyle, B., Datta, S., Doczy, M., Hareland, S., Jin, B., Kavalieros, J., Linton, T., Murthy, A., Rios, R., Chau, R., High performance fully-depleted tri-gate CMOS transistors. *Electron Device Lett. IEEE*, 24, 4, 263–265, 2003.
48. Doyle, B., Boyanov, B., Datta, S., Doczy, M., Hareland, S., Jin, Tri-Gate fully-depleted CMOS transistors: fabrication, design and layout, in: *VLSI Technology, 2003, Digest of Technical Papers, 2003 Symposium on*, IEEE, pp. 133–134, 2003.
49. Bendib, T., Djeffal, F., Meguellati, M., An optimized junctionless GAA MOSFET design based on multi-objective computation for high-performance ultra-low power devices. *J. Semicond.*, 35, 074002-1–074002-6, 2014.
50. Kalavade, P. and Saraswat, K.C., Lateral gate-all-around (GAA) poly-Si transistors, in: *SOI Conference, 2001 IEEE International*, IEEE, pp. 109–110, 2001.
51. Cui, Y., Zhong, Z., Wang, D., Wang, W.U., Lieber, C.M., High performance silicon nanowire field effect transistors. *Nano Lett.*, 3, 2, 149–152, 2003.
52. Tian, Y., Huang, R., Wang, Y., Zhuge, J., Wang, R., Liu, J., Zhang, X., Wang, Y., *et al.*, New self-aligned silicon nanowire transistors on bulk substrate fabricated by epi-free compatible CMOS technology: Process integration, experimental characterization of carrier transport and low frequency noise, 2013.
53. Djeffal, F., Lakhdar, N., Yousfi, A., An optimized design of 10-nm-scale dual-material surrounded gate MOSFETs for digital circuit applications. *Physica E*, 44, 339–344, 2011.
54. Yang, F.-L., Chen, H.-Y., Chen, F.-C., Huang, C.-C., Chang, C.-Y., Chiu, H.-K., Lee, C.-C., Chen, C.-C., Huang, H.-T., Chen, C.-J., *et al.*, 25 nm CMOS omega FETs, in: *Electron Devices Meeting, 2002, IEDM'02, International*, IEEE, pp. 255–258, 2002.
55. Park, J.-T., Colinge, J., Diaz, C.H., Pi-gate SOI MOSFET. *Electron Device Lett. IEEE*, 22, 8, 405–406, 2001.
56. Ferhati, H. and Djeffal, F., Graded channel doping junctionless MOSFET: a potential high performance and low power leakage device for nanoelectronic applications. *J. Comput. Electron.*, 17, 129–137, 2018.
57. Ferhati, H. and Djeffal, F., Reliability Investigation of Nanoscale Junctionless GAA MOSFET Against Degradation-Related Ageing Effects. *J. Nanoelectron. Optoelectron.*, 13, 1106–1113, 2018.
58. Ferhati, H. and Djeffal, F., Role of Graded Channel Doping Engineering in Improving Junctionless GAA MOSFET Performance for Ultra Low-Leakage Power Applications. *J. Nanoelectron. Optoelectron.*, 13, 521–530, 2018.

59. Abdi, M.A., Djeffal, F., Meguellati, M., Arar, D., Two-dimensional analytical threshold voltage model for nanoscale graded channel gate stack DG MOSFETs, *2009 16th IEEE International Conference on Electronics, Circuits and Systems - (ICECS 2009)*, pp. 892–895, Yasmine Hammamet, Tunisia, 2009, doi: 10.1109/ICECS.2009.5410801.
60. Bendib, T., Djeffal, F., Benhaya, A., A PSO-based approach to study the nanoscale Double Gate- MOSFETs. *Rev. Sci. Technol.*, 1, 2, 14–22, 2010.
61. Djeffal, F. and Bendib, T., Multi-objective genetic algorithms based approach to optimize the electrical performances of the gate stack double gate (GSDG) MOSFET. *Microelectron. J.*, 42, 661–666, 2011.
62. Ram, M.K., Tiwari, N., Abdi, D.B., Saurabh, S., Effect of Drain Induced Barrier Enhancement on Subthreshold Swing and OFF-State Current of Short Channel MOSFETs: A TCAD Study. *IEEE Access*, 9, 141321–141328, 2021.
63. Lim, H.-R., *et al.*, Impact of bottom-gate biasing on implant-free junctionless Ge-on-insulator n-MOSFETs. *IEEE Electron Device Lett.*, 40, 9, 1362–1365, 2019.
64. Ganguli, T., Chanda, M., Sarkar, A., Impact of Aspect Ratio and Interface Trap Charge on the Performances of Junctionless MOSFET-Based Adiabatic Logic Circuit. *IEEE Trans. Electron Devices*, 70, 12, 6157–6162, 2023.
65. Chiang, T., A scaling theory for fully-depleted, surrounding-gate MOSFETs: including effective conducting path effect. *Microelectron. Eng.*, 77, 2, 175–183, 2005.
66. Djeffal, F., Lakhdar, N., Benhaya, A., An accurate two-dimensional analytical subthreshold swing model to study the scaling capability of the surrounding-gate MOSFET. *Int. Rev. Phys.*, 2, 5, 306–311, 2008.
67. Ghoggali, Z., Djeffal, F., Lakhdar, N., Analytical analysis of nanoscale Double-Gate MOSFETs including the hot-carrier degradation effects. *Int. J. Electron.*, 97, 2, 119–127, 2010.
68. Abdi, M.A., Djeffal, F., Dibi, Z., Arar, D., A two-dimensional analytical subthreshold behavior analysis including hot-carrier effect for nanoscale Gate Stack Gate All Around (GASGAA) MOSFETs. *J. Comput. Electron.*, 10, 179–185, 2011.
69. Bendib, T., Djeffal, F., Arar, D., Subthreshold behavior optimization of nanoscale Graded Channel Gate Stack Double Gate (GCGSDG) MOSFET using multi-objective genetic algorithms. *J. Comput. Electron.*, 10, 210–215, 2011.
70. Atlas User's manual, SILVACO TCAD, 2012.
71. Dhanaselvam, P.S. and Nithya Ananthi, A., Analytical Modeling of Electric Field Distribution in Dual Material Junctionless Surrounding Gate MOSFETs. *Int. J. VLSI Des. Commun. Syst.*, 5, 3, 83, 2014.
72. Manikandan, S., Suveetha Dhanaselvam, P., Karthigai Pandian, M., A 2D unified subthreshold drain current investigation for junctionless cylindrical surrounding gate (JCSG) silicon nanowire transistor. *Silicon*, 14, 10, 5311–5318, 2022.
73. Manikandan, S., Suveetha Dhanaselvam, P., Karthigai Pandian, M., A quasi 2-D electrostatic potential and threshold voltage model for junctionless triple

material cylindrical surrounding gate si nanowire transistor. *J. Nanoelectron. Optoelectron.*, 16, 2, 318–323, 2021.

74. Tocci, G., Performance estimation and Variability from Random Dopant Fluctuations in Multi-Gate Field Effect Transistors: a Simulation Study, Engineering, Physics, 2010.
75. Chebaki, E., Djeffal, F., Bentrcia, T., ANFIS-based Approach to Predict the Degradation-related Ageing of Junctionless GAA MOSFET. ScienceDirect, *Mater. Today Proc.*, 5, 8, Part 2, 15949–15958, 2018.
76. Chawla, N.V., Bowyer, K.W., Hall, L.O., Kegelmeyer, W.P., SMOTE: Synthetic minority over-sampling technique. *J. Artif. Intell. Res.*, 16, 2, 321–357, 2002, https://doi.org/10.1613/jair.953.
77. Musolf, A.M., Holzinger, E.R., Malley, J.D., *et al.*, What makes a good prediction? Feature importance and beginning to open the black box of machine learning in genetics. *Hum. Genet.*, 141, 1515–1528, 2022, https://doi.org/10.1007/s00439-021-02402-z.
78. Van der Maaten, L. and Hinton, G.E., Visualizing Data using t-SNE. *J. Mach. Learn. Res.*, 9, 2579–2605, 2008, https://api.semanticscholar.org/CorpusID:5855042.
79. Cheng, G., Povey, D., Huang, L., Xu, J., Khudanpur, S., Yan, Y., Output-Gate Projected Gated Recurrent Unit for Speech Recognition. *Interspeech 2018*, 2018-Septe, pp. 1793–1797, 2018, https://doi.org/10.21437/Interspeech.2018-1403.
80. Wu, J., Chen, X.Y., Zhang, H., Xiong, L.D., Lei, H., Deng, S.H., Hyper parameter optimization for machine learning models based on Bayesian optimization. *J. Electron. Sci. Technol.*, 17, 1, 26–40, 2019, https://doi.org/10.11989/JEST.1674-862X.80904120.
81. Ho, B., Sun, X., Shin, C., Liu, T.-J.K., Design optimization of multigate bulk MOSFETs. *IEEE Transactions on Electron Devices*, 60, 1, 28–33, Jan. 2013, doi: 10.1109/TED.2012.2224870.
82. Goel, E., Pandey, A., Rahi, S. B., Samuel, A. (Eds.), *Handbook of Advanced Semiconductor Field Effect Transistors*. John Wiley & Sons, 2025.
83. Nikonov, D.E. and Young, I.A., Overview of Beyond-CMOS Devices and a Uniform Methodology for their Benchmarking. *Proc. IEEE*, 101, 12, 2498–2533, 2013.
84. Tyagi, M. S., *Introduction to semiconductor materials and devices*. John Wiley & Sons, 2008.
85. Mishra, U. K. and Singh, J., *Semiconductor device physics and design*. vol. 83, Springer, Dordrecht, 2008.
86. Sze, S. M., *Semiconductor devices: physics and technology*. John Wiley & Sons, 2008.
87. Sze, S. M., Li, Y., Ng, K. K., *Physics of semiconductor devices*. John Wiley & Sons, 2021.
88. Colinge, J. P. and Colinge, C. A., *Physics of semiconductor devices*. Springer US, Boston, MA, 2002.

89. Neamen, D., *An introduction to semiconductor devices*. McGraw-Hill, Inc., 2005.
90. Ramesh, R., Influence of gate and channel engineering on multigate MOSFETs—A review, *Microelectronics*, 66, 136–154, 2017.
91. Moore, G., Progress in Digital Integrated Electronics. *IEEE Int'l Electron Devices Meeting Tech. Dig.*, 11–13, 1975.
92. Dennard, R. *et al.*, Design of Ion- Implanted MOSFETs with Very Small Physical Dimensions. *IEEE J. Solid State Circuits*, 9, 5, 256–268, 1974.
93. Avci, U.E., *et al.*, Energy Efficiency Comparison of Nanowire Heterojunction TFET and Si MOSFET at Lg 5 13 nm, Including P-TFET and Variation Considerations. *IEEE Int'l Electron Devices Meeting Tech. Dig.*, 33–36, 2013.

2

# Heterojunction Field Effect Transistor

**Shahnaz Kossar[1*], Asif Rasool[2], R. Amiruddin[3], Syed Abbas Ali[4], Mohammad Ayaz Ahmad[5], Nasharath Mapjan[6], Kasim Sakran Abass[7] and Mir Waqas Alam[8†]**

[1]*School of Natural Sciences, GNA University, Sri Hargobindgarh, Phagwara, Punjab, India*
[2]*Department of Applied Sciences, Maulana Mukhtar Ahmad Nadvi Technical Campus, Malegaon, India*
[3]*Department of Physics, B.S. Abdur Rahman Crescent Institute of Science and Technology, Chennai, India*
[4]*Department of Mechanical Engineering, SECAB Institute of Engineering and Technology, Vijaypur, India*
[5]*Department of Mathematics, Physics & Statistics, Faculty of Natural Sciences, University of Guyana, Greater Georgetown, Guyana, South America*
[6]*Department of Chemistry, Faculty of Natural Sciences, University of Guyana, Greater Georgetown, South America*
[7]*Department of Physiology, Biochemistry, and Pharmacology; College of Veterinary Medicine; University of Kirkuk, Kirkuk, Iraq*
[8]*Department of Physics, College of Science, King Faisal University, Al-Ahsa, Saudi Arabia*

## Abstract

Due to high power and high speed applications, high electron mobility transistors (HEMT) have been the area of focus for many research groups in recent years. Compared to conventional field effect transistors (FET), HEMT devices have demonstrated superior performance considering the operational frequency, output power density, and power added efficiency. High electron mobility transistor (HEMT) technology, considerable performance improvements have been brought to the fields of power electronics, radio frequency applications, and high-speed digital circuits. GaN-based complementary metal oxide semiconductor (CMOS) technology combines the wide-band gap characteristics of GaN with the low static

**Corresponding author*: shazia.rajput0505@gmail.com
[†]*Corresponding author*: wmir@kfu.edu.sa

P. Suveetha Dhanaselvam, Naima Guenifi, Shiromani Balmukund Rahi and J. Ajayan (eds.) Field-Effect Transistor Devices: Post-CMOS Theory and Applications, (33–52) © 2026 Scrivener Publishing LLC

power consumption advantages of CMOS, showing significant potential in promoting the development of integrated circuits (ICs). In recent years, numerous research organizations have focused on high electron mobility transistors due to their high power and speed applications, which provide a wide range of applications in electronics, optical and thermal properties. A comparison of field effect transistors with conventional HEMTs reveals that field effect transistors enhance power added efficiency, power density, and operating frequency. The vander Waals interlayer interaction allows you to exfoliate and reassemble various 2D materials into randomly and vertically layered heterostructures. The recently reported vapor phase development of 2D materials offers the path for directly constructing vertical and lateral heterojunction. This study provides insights into layered 2D heterostructures, as well as a brief introduction to 2D material and heterostructure preparation methods. These remarkable 2D heterostructures offer a wide range of possible applications.

***Keywords:*** Heterojunctions, ferroelectric, two dimensional, transistors

## 2.1 Introduction

The field-effect transistor (FET) is a semiconductor device in which current conduction is controlled by flow by applying an electric field inside the on applying voltage on gate terminal. Basically, field-effect transistors are classified into two categories: junction field-effect transistor known as JFET and metal-oxide semiconductor field-effect transistor popularly known as MOSFET. These field-effect transistors basically having three terminals: source, gate, and drain. In both cases, current flow inside device is controlled by modifying electric field by applying the voltage on the gate terminal. FET is essentially a semiconductor current path whose conductance is controlled by applying an electric field perpendicular to the current. This field and the carrier density in the semiconductor channel are changed by a voltage imposed on a metal gate. The more electrons in the channel, the higher its conductivity and better performances are obtained from the transistor. The flow of channel current occurs *via* the drain and source ohmic contacts placed on either side of, and parallel to, the gate metallization. For low values of drain-to-source bias, a current flows from the drain to the source through the electron gas. In this case, the average velocity of carriers is linearly proportional to the field strength. This is written as Equation 2.1:

$$v = \mu E \tag{2.1}$$

In Equation 2.1, $E$ is the electric field strength and $\mu$ is the low field electron mobility. This mobility is usually a strong function of doping density. For high electric field values, when the drain to source bias levels are increased, the steady-state carrier velocity becomes limited and the current levels saturate [1].

Due to high power and high speed applications, high electron mobility transistors (HEMT) have been the area of focus for many research groups in recent years. Compared to conventional field effect transistors (FET), HEMT devices have demonstrated superior performance considering the operational frequency, output power density, and power added efficiency. The more electrons in the channel, the higher its conductivity and better performances are obtained from the transistor. Moreover, with the development of heterostructure science and technology, heterojunction FETs now offer potential advantages in microwave, millimeter-wave, and high-speed digital integrated circuit applications over the former homojunction devices. Heterojunction field effect transistors (HFET) are more mature semiconductor components of the new generation of III-V transistors, as shown in Figure 2.1. Different names are used for these devices signifying the underlying physical mechanism or structural configuration. Examples are the high electron mobility transistor (HEMT), the selectively-doped heterojunction transistor (SDHT), the modulation-doped FET (MODFET), and the two-dimensional electron gas FET (TEGFET). Heterojunction-based field effect transistors (HFETs) are an advanced type of transistor that achieves higher electrical performance by forming a junction between two materials with distinct band gaps. The advancements in mobile communication applications, TV broadcasting, and satellite communications, microwave transistors and wide band gap materials have been the area of focus for much research. These devices have become critical components in current semiconductor technologies, with broad use in high-frequency, high-power, and optoelectronic applications. In comparison to traditional field-effect transistors (FETs), the heterojunction structure provides more control over carrier mobility, band alignment, and overall device efficiency [2]. A high output power per unit width and high temperature operation require a semiconductor material with higher band gap. This leads to fabricating compact devices that are easily matched when integrated with other circuit elements and the need for extra cooling will be either omitted or reduced. For using a device in high power mode, a high voltage operation is necessary for the device and this is fulfilled by using a material with high breakdown field. This will either reduce or completely eliminate the need for a step down converter in devices. A high gain and high velocity are the two enabling features of the demand for a low noise device generally used in all types of receivers with a high dynamic range. This chapter having a thorough introduction to HFETs, including their fundamental concepts, design benefits, material choices, operational features, and applications.

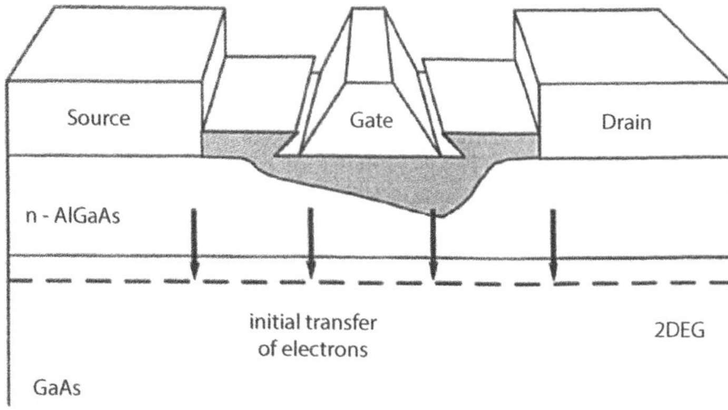


**Figure 2.1** Schematic diagram AlGaAs/GaAs HFET.

## 2.2 Basic History and Challenges

Physicist Julius Edgar Lilienfeld invented the FET in 1925, followed by Oskar Heil in 1934. However, neither was able to create a viable, working semiconductor device. In 1947, John Bardeen and Walter Houser Brattain, working at Bell Labs under William Shockley, identified and described the transistor effect, which marked a breakthrough in transistor technology. Shockley first sought to develop a functioning FET by altering a semiconductor's conductivity, but he encountered substantial problems, including surface states, dangling bonds, and the materials employed, such as germanium and copper compounds. This prompted Bardeen and Brattain to create the point-contact transistor in 1947, followed by Shockley's discovery of the bipolar junction transistor in 1948 [3].

Heinrich Welker invented the first effective FET device, the junction field-effect transistor (JFET), in 1945. Jun-ichi Nishizawa and Y. Watanabe, Japanese engineers, produced a more sophisticated version known as the static induction transistor (SIT) in 1950. Following Shockley's theoretical analysis of the JFET in 1952, George C. Dacey and Ian M. Ross built a working JFET in 1953 [4]. However, junction transistors, like the JFET, had manufacturing issues and remained hefty devices, restricting their usage in large-scale applications. The insulated-gate field-effect transistor (IGFET) was considered a viable replacement, but early attempts to construct one failed owing to the "surface state barrier," which prevented the electric field from permeating the material. By the mid-1950s, scholars had moved their emphasis to the promising bipolar junction transistor (BJT) technology [5].

William Shockley, John Bardeen, and Walter Brattain laid the basis for MOSFET (metal-oxide-semiconductor field effect transistor) technology.

Shockley independently envisioned the FET in 1945, but surface states slowed its development. Bardeen extended on this hypothesis, describing how electrons trapped in surface states prevented the external field from reaching the semiconductor. His work on surface physics, which developed the notion of a "inversion layer," paved the way for MOSFETs [6]. In 1948, Bardeen developed an insulated-gate FET (IGFET) with an inversion layer, which became the foundation of contemporary CMOS (complementary metal-oxide-semiconductor) technology. Shockley later hailed Bardeen's concept on surface states as a watershed moment in semiconductor development [7].

Despite these developments, Bardeen, Brattain, and other researchers failed to overcome the effects of surface states in FETs. In 1947, Robert Gibney and Brattain attempted to reduce surface states by sandwiching an electrolyte between the metal and semiconductor, but the results were unsatisfactory. Bardeen's subsequent research focused on the conductivity of the inversion layer, ultimately replacing the electrolyte with a solid oxide layer. However, the oxide was wiped off during research, resulting in the unexpected discovery of the point-contact transistor rather than a successful FET. According to historian Lillian Hoddeson, if Bardeen and Brattain had worked with silicon rather of germanium, they may have created a working FET considerably sooner [8].

By the 1950s, it was clear that surface states were a major challenge for FETs. Researchers discovered two types of surface states: rapid surface states connected with the semiconductor/oxide interface and bulk, and slow surface states linked with the oxide layer's adsorption of atoms, molecules, and ions.

Carl Frosch and Lincoln Derrick made a significant discovery in 1955 when they unexpectedly found that a coating of silicon dioxide may prevent some dopants from entering a silicon wafer. This "passivating effect" of oxidation became a critical approach in semiconductor production, allowing the formation of microscopic pores in the oxide layer for dopant migration [9].

Their 1957 publication cleared the door for the manufacturing of MOSFET devices. The significance of Frosch's discovery was rapidly recognized at Bell Labs, and it impacted researchers such as Jean Hoerni of Shockley Semiconductor. In the same year, Ian Munro Ross invented a ferroelectric FET (FeFET) or MFSFET, which use ferroelectric material as a dielectric/insulator rather than oxide. His invention, albeit ahead of its time, was comparable to current MOSFETs [10].

In the same year, Ian Munro Ross invented a ferroelectric FET (FeFET) or MFSFET, which use ferroelectric material as a dielectric/insulator rather than oxide. His invention, while ahead of its time, was comparable to current MOSFETs and was intended for use in memory systems. Around the same time, John Wallmark submitted a patent for a FET with germanium monoxide as a gate dielectric, but did not explore the concept further. Ernesto Labate, a physicist at Bell Labs, designed a device similar to a MOSFET in 1957, however it did not specifically employ silicon dioxide as an insulator. These contributions influenced the development of FET technology, eventually leading to the successful invention of the MOSFET and its widespread application in modern electronics [11].

Julius Edgar Lilienfeld first suggested the notion of field-effect transistors (FETs) in 1925. However, it wasn't until 1947 that John Bardeen, Walter Brattain, and William Shockley, working at Bell Labs, developed the first functioning transistors, beginning with the point-contact transistor. In 1948, Shockley presented the improved bipolar junction transistor (BJT), which went into production in the early 1950s and became widely used, ushering in the transistor age in electronics [12].

The metal-oxide-semiconductor field-effect transistor (MOSFET) was invented at Bell Labs between 1955 and 1960, after Carl Frosch and Lincoln Derick's discovery of silicon dioxide surface passivation. This innovation enabled the mass manufacture of MOS transistors, which laid the groundwork for current CPUs and solid-state memory. Today, the MOSFET is the most commonly manufactured electronic device in history.

### 2.2.1 Overview of Transistor Technology

Transistors are the fundamental components of contemporary electronics, and their growth has been characterized by substantial advances in both materials and device design. Traditional field-effect transistors (FETs), such as metal-oxide-semiconductor FETs (MOSFETs), have long dominated integrated circuits due to their fast switching speed, low power consumption, and scalability. However, when device dimensions decrease and performance demands rise, restrictions in mobility, power efficiency, and leakage current emerge [13].

To solve these issues, novel designs such as heterojunction-based FETs have been created. HFETs, also known as heterostructure FETs or modulation-doped FETs (MODFETs), rely on heterojunctions, which are interfaces between two materials with distinct band gaps. This design provides for greater control over the charge carriers, resulting in faster and more energy-efficient functioning than homojunction FETs, which employ a single material.

When attempting to develop an early FET, the fundamental issue was the vast number of electrons (or holes) required to inject into a crystal of any considerable size. This made the device less useful as an amplifier since it required a high injection current to start. The rationale behind utilizing a crystal diode was that the crystal could deliver electrons over extremely short distances, particularly in the depletion area. The important discovery was to place the input and output contacts close together on the crystal's surface, on each side of this area [14].

As the research team, led by Walter Brattain, worked on developing such a device, they noticed intermittent amplification effects that appeared and then disappeared abruptly. In one case, a nonfunctional setup became operational after being immersed in water. They discovered that electrons in the crystal moved according to neighboring charges, whereas those in the emitter or "holes" in the collector preferred to gather at the crystal surface, where they could interact with opposing charges from the surrounding air or water. By applying a modest charge, these carriers might be pushed off the surface, eliminating the requirement for a vast supply of injected electrons. A tiny number of carriers, put at the right area on the crystal, might do the same [15].

## 2.3 Heterojunction-Based Field-Effect Transistors

Heterojunction-based field-effect transistors are a significant innovation in semiconductor technology, allowing high-speed and high-efficiency operation in a variety of applications. This chapter investigates the application of semiconductor physics to HFET operation, with an emphasis on the fundamental concepts that enable their performance. Important subjects include heterojunction formation, band alignment, the two-dimensional electron gas (2DEG), carrier transport processes, and the impact of material qualities on HFET characteristics.

When two different semiconductor materials are combined, they generate a heterojunction, which has electronic characteristics that are distinct from the separate components. (i) This is critical for HFET functioning since it enables: Bandgap engineering is the process of tailoring the band structure to enhance performance. (ii) Enhanced Carrier Mobility: Achieving high-speed operation by forming a 2DEG and (iii) Improved Device Characteristics: Increased on-state current and decreased off-state leakage [1].

### 2.3.1 Principles of Heterojunctions

When two semiconductor materials with differing band gaps combine, they produce a heterojunction. At this contact, the materials' conduction and valence bands line in a certain way, resulting in an energy discontinuity. This discontinuity causes the production of a 2DEG or a hole gas, depending on the kind of doping and materials employed. This 2DEG is critical to the operation of HFETs because it allows for high electron mobility and reduced scattering, resulting in increased transistor performance. In comparison to homojunction transistors, which confine charge carriers inside the bulk material, heterojunctions enable superior carrier confinement due to energy barriers at the interface. This provides for improved control of carrier flow and helps to decrease power losses and leakage current.

Extremely high-performance modulation-doped field-effect transistors (FETs) using AlGaAs/GaAs and AlInAs/GaInAs heterojunctions have been developed. However, alternative materials may provide considerable benefits over these classic methods. Recent research has focused on both double- and single-barrier tunneling in the AlGaSb/InAs material combination. Other researchers have investigated hot-electron transistors with the GaSb/AlSb/InAs arrangement [3].

### 2.3.2 Types of Heterojunction

The concept heterojunction has become important semiconductor concept for modern technology. The heterojunction FETs are formed between semiconductors of different compositions, lattice constants, and bandgaps, such as AlGaN/GaN and InGaAs/InP. A carrier (electron/hole) barrier may be introduced into a semiconductor heterojunction due to the presence of a band offset, which is impacted by the bandgap and work function of each semiconductor in the semiconductor heterojunction. The carrier barrier regulates carrier injection and recombination, enabling the semiconductor heterojunction to possess various charge transport mechanisms. Up to now, semiconductor heterojunction have been employed in the structures of many important devices such as optoelectronic devices and HFMT. The concept of heterojunction is also very useful, a known steep slope FET used in ultra low applications, known as tunnel FET. Based on the sign of the conduction band offset ($\Delta E_C$) and valence band offset ($\Delta E_V$), a semiconductor heterojunction is often called a type I, type II, or type III semiconductor heterojunction as described below.

#### *2.3.2.1 Type I Heterojunction*

In type I, two semiconductor materials with identical band alignments allow for efficient carrier injection and transport. As shown in Figure 2.2

a type I semiconductor heterojunction can be also called a straddling gap heterojunction, for which the signs of $\Delta E_C$ and $\Delta E_V$ are opposite. Charge carriers (electrons and holes) in a type I semiconductor heterojunction are easily confined in the narrow bandgap semiconductor of the semiconductor heterojunction, facilitating efficient radiative electron-hole recombination. The conduction and valence bands in both materials are aligned, allowing carriers to readily travel between them. Charge carriers easily travel across the junction, resulting in high current conduction [16]. Type I heterojunction are commonly employed in low-voltage applications that need low barrier heights. The popular pairings include GaAs/InGaAs and AlGaAs/GaAs, which are commonly used in optoelectronics.

#### *2.3.2.2 Type II Heterojunction*

The signs of $\Delta E_C$ and $\Delta E_V$ are the same in a type II or type III semiconductor heterojunction. For a type II semiconductor heterojunction as shown Figure 2.2 the forbidden bands of two component semiconductors are partially overlapped. A type II semiconductor heterojunction can be also called a staggered gap heterojunction, for which electrons and holes may be confined in different semiconductors. Due to the overlap of the wave functions of electrons and holes near the interface of two-component semiconductors, optical matrix elements are reduced in a type II semiconductor heterojunction. Compared with a type I semiconductor heterojunction, a type II semiconductor heterojunction has a longer radiation lifetime of excitons and a smaller excitons binding energy. A type II semiconductor heterojunction is good for the separation of electrons and holes. Such type of heterojunction has staggered band edges between two semiconductor materials, resulting in charge separation at the contact. This arrangement may be useful for some applications, such as photodetectors and light generating devices. The staggered band alignment enables the separation of electrons and holes, which might improve certain capabilities. Limited Carrier Transport: While charge separation is helpful, it might impede carrier transport efficiency, leading in increased on-state resistance [6]. Type II heterojunction are commonly employed in devices requiring charge separation, such as solar cells and photodetectors. Quantum Well Structures: These heterojunction may be used to build quantum well structures, which confine carriers in two dimensions and improve optical characteristics. Advantages: Improved performance in light-absorbing and emitting applications. In type I1, useful for constructing charge separation-based devices like sensors and solar cells.

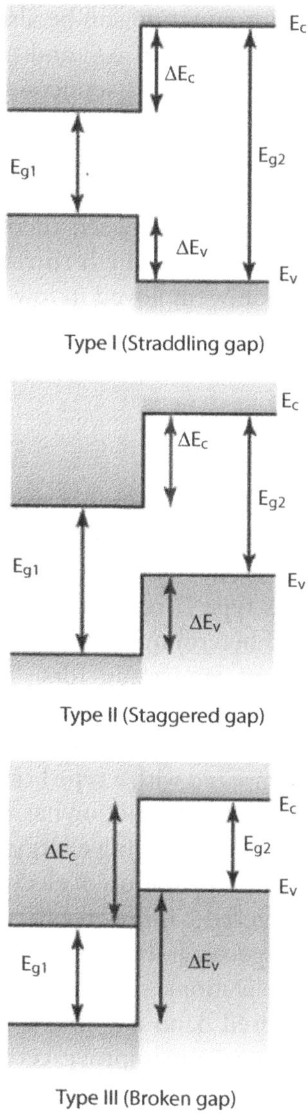


**Figure 2.2** Classification of heterojunction in semiconductors.

### 2.3.2.3 *Type III Heterojunction*

In a type III semiconductor heterojunction, the forbidden bands of two component semiconductors are fully staggered shown in Figure 2.2. A type III semiconductor heterojunction can be also called a broken gap heterojunction. The spontaneous charge transfer from the valence band of one semiconductor to the conduction band of the other semiconductor in a type III semiconductor heterojunction leads to a semimetallic behavior, as exemplified in the semiconductor heterojunction of GaSb/InAs. Please note that semiconductor heterojunctions are formed as long as the component semiconductors have different bandgaps, which may be altered by phase transition, crystallinity, strain, composition, quantum confinement, edge states [17]. Type III heterojunction occur, when one semiconductor material has a greater bandgap than the other. This design may provide a possible barrier that confines carriers to a certain location, hence enhancing device performance. The wider bandgap material can form a barrier that confines carriers (electrons or holes) in the smaller bandgap material, hence increasing device efficiency. Type III heterojunction frequently have better thermal stability and durability, making them ideal for high-power applications [17]. The AlGaN/gallium nitride (GaN), which is commonly, utilized in high-power RF applications due to GaN's high breakdown voltage and thermal conductivity [18].

## 2.4 Selection of Material for HFETs

The selection of materials for HFETs is important to their performance. The most widely utilized materials are III-V semiconductors such as gallium arsenide (GaAs), aluminum gallium arsenide (AlGaAs), and indium phosphide (InP). These materials have strong electron mobility and excellent band gap engineering properties, which are required for high-speed and high-frequency operation. One of the most common heterojunction couples is GaAs/AlGaAs [19]. In Table 2.1 some popular material property has included. The material GaN has higher mobility having unique property for HEMT applications. In Table 2.2 is a FETs comparison for relative rank of IC technology.

**Table 2.1** Properties of competing semiconductor materials.

| Material | Mobility | Dielectric constant | Bandgap | Breakdown field | Tmax |
|---|---|---|---|---|---|
| Si | 1300 | 11.4 | 1.12 | $0.3*10^6$ | 300 °C |
| GaAs | 5000 | 13.1 | 1.42 | $0.4*10^6$ | 300 °C |
| 4H-SiC | 260 | 9.7 | 3.2 | $3.5*10^6$ | 600 °C |
| GaN | 1500 | 9.5 | 3.4 | $2.0*10^6$ | 700 °C |

**Table 2.2** FETs comparison.

| Parameters | HFET | GaAs MESFET | Vertical FET | HBT | Silicon MOSFET | BJT |
|---|---|---|---|---|---|---|
| Speed | 1 | 3 | 1 | 2 | 5 | 4 |
| Power-delay product | 1 | 2 | 2 | 4 | 1 | 4 |
| Lithography requirements | 2 | 3 | 5 | 1 | 4 | 1 |
| Threshold control | 4 | 4 | 4 | 1 | 2 | 1 |
| Processing complexity | 2 | 1 | 4 | 5 | 3 | 5 |
| Materials problem | 4 | 3 | 5 | 4 | 1 | 2 |
| Score | 14 | 16 | 21 | 14 | 16 | 17 |

Relative Rank of IC technology (1: Excellent, 5: poor) Information source: Drury, Robert. "The physical modelling of heterojunction field effect transistors." PhD diss., University of Leeds, 1994.

In this design, GaAs serves as the channel material for the 2DEG, while AlGaAs serves as the barrier material for modulation doping. The difference in band gaps between GaAs and AlGaAs produces the circumstances required for the creation of the 2DEG, resulting in increased carrier mobility and lower noise. Other material systems, including silicon-germanium (SiGe) and GaN, have been investigated for HFET applications. GaN-based HFETs, in particular, have grown in favor for high-power and high-temperature applications because to their wide band gap, high breakdown voltage, and thermal stability [20].

Gallium nitride (GaN) and related wide bandgap semiconducting alloys have been regarded as the candidate materials for the next-generation of high-power and high-frequency electronic devices. Compared with Si counterparts, the higher breakdown field and drift velocity of GaN based devices can produce lower losses at high voltage and temperature as shown in Table 2.1. The superior performances are not only originated from these inherent material properties of GaN itself, but also related to the device technology based on AlGaN/GaN heterostructures. Owing to the high mobility of 2D electron gas (2DEG) formed at the AlGaN/GaN interface, a lateral device called heterojunction field-effect transistor (HFET), also known as high electron mobility transistor (HEMT), offers tremendous potential to the power switching and radio frequency (RF) applications.

Cascode devices of GaN heterojunction field effect transistors (HFETs) plus Si MOSFETs have recently attracted much attention and are now commercially available. The typical layout consists of a series connection of a high voltage depletion-mode (D-mode) GaN device and a low voltage enhancement-mode (E-mode) Si device. The GaN plus Si hybrid cascode device not only enables normally-OFF operation, but the cascode structure also offers the advantage of mitigation of the Miller effect, leading to an improved switching speed and reduced switching losses. This makes cascode devices strong candidates for high power and high frequency switching applications. Due to high power and high speed applications, high electron mobility transistors (HEMT) have been the area of focus for many research groups in recent years. Compared to conventional field effect transistors (FET), HEMT devices have demonstrated superior performance considering the operational frequency, output power density, and power added efficiency.

### 2.4.1 Band Structure in HFETs

The band structure at the heterojunction interface is critical to understanding HFET functioning. The orientation of the conduction and valence bands affects carrier dynamics and transport properties: Conduction Band Alignment: Indicates how quickly electrons may travel from one substance to another. Valence Band Alignment affects hole transport and total device conductivity. When two distinct semiconductor materials create a junction, the difference in work functions causes band bending at the contact. The applied electric field in HFET causes band bending, which influence movement of charge carriers. In depletion region, the area around the junction where charge carriers are exhausted, which influences device turn-on properties. While in accumulation layer in some instances, electrons can collect at the junction, increasing conductivity and on-state performance [9].

### 2.4.2 Effect of Band Gap on Device Performance

The band gap difference ($\Delta E_g$) in a HEMT enables the formation of a high-performance channel. A wider band gap material (doped) donates electrons, which then transfer to a narrow band gap material (undoped), creating a 2D electron gas (2DEG) in a quantum. This 2DEG channel has high electron mobility because electrons move with fewer impurities due to the undoped nature of the narrow band gap layer. Wide band gap materials also provide high breakdown fields, allowing for higher operating voltages and power, and often have better thermal conductivity for heat dissipation.

### 2.4.3 Formation of 2DEG Features

In a HEMT (High Electron Mobility Transistor), "2DEG" refers to a two dimensional electron gas which is a sheet of electrons confined to move only in two dimensions at the interface between two different semiconductor layers. This 2DEG acts as the channel through which current flows in the transistor and is key to the high-speed and high-power performance of HEMTs, as the electrons in it have high mobility due to being in an intentionally undoped channel [3, 18, 24–29]. Figure 2.3 represent the formation of 2DEG in HEMT having AlGaN/GaN semiconductor materials. In Figure 2.3, $E_C$, $E_V$ and $E_f$ denotes conduction band, valance band and Fermi energy level.

### 2.4.4 Mechanisms for 2DEG Formation

The polarization effect of an AlGaN layer grown on a GaN buffer induces positive polarization charge at the AlGaN/GaN interface and negative polarization charge at the top of the AlGaN layer. Thus an electric field is formed within the AlGaN layer. It is reasonable to regard the AlGaN surface and the AlGaN/GaN interface as two infinite charged planes,

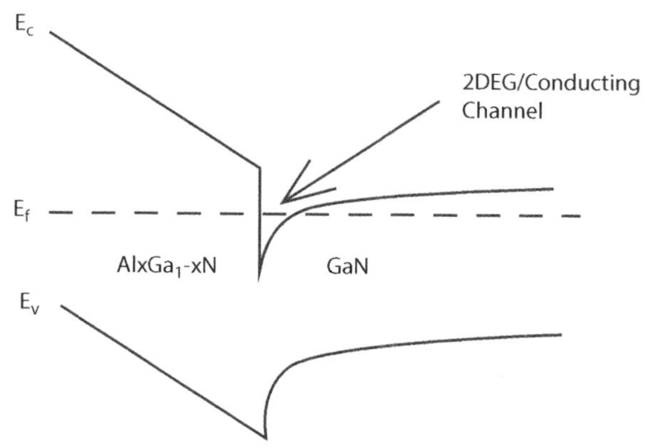


**Figure 2.3** Formation of 2DEG in HEMT.

as the thickness of the whole epilayer is just several μm, which is very small in comparison to the lateral size of the substrate. The formation of 2DEG of an AlGaN/GaN heterostructure with an n-doped AlGaN layer is easy to understand. In the AlGaN layer, the positive polarization charge at the AlGaN/GaN interface and the negative polarization charge at the top of the AlGaN layer induce an electric field and make the energy band (along with the Fermi level) tilt toward the interface. This situation is equal to applying a voltage to a piece of thin free standing AlGaN film, in other words, a planar plate capacitor with AlGaN dielectric medium. Under the force of the electric field, conducting electrons (note that the AlGaN layer is n-doped) in AlGaN move to the positive electrode and accumulate there, leaving positive space charges in AlGaN, bending the energy band and making the Fermi level flat again. Also, the built-in electric field will reduce the whole electric field in the AlGaN layer and thus make the inclined energy band flatter. Once contacting GaN, electrons on the AlGaN side will flow into the GaN and form 2DEG, because the Fermi level of GaN is lower than that of AlGaN. This process will continue until the Fermi levels of AlGaN. The total polarization charge at the AlGaN/GaN interface can be divided into two parts. One part comes from AlGaN, and the other comes from GaN. The AlGaN part will have no effect on the GaN side because the negative polarization charge at the top of the AlGaN layer and the positive polarization charge at the interface forms a planar plate capacitor, which generates no electric field outside its two planes. Similarly, the GaN part, together with its counterpart at the bottom of the GaN layer, generates a

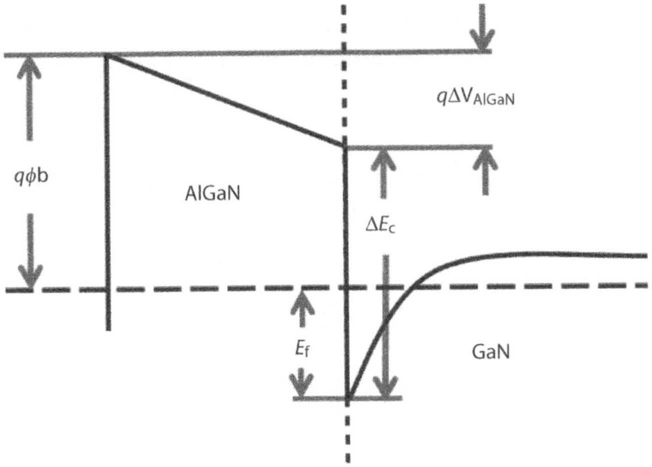


**Figure 2.4** Conductive band diagram of an AlGaN/GaN heterostructure.

uniform electric field pointing from the bottom of the GaN layer toward the AlGaN/GaN interface. This electric field will reduce the net electron density at the AlGaN/GaN interface, let alone form 2DEG. Second, the 2DEG cannot come from the body of the AlGaN barrier because there are few conductive electrons in undoped AlGaN. Figure 2.4 the conductive band diagram of an AlGaN/ GaN heterostructure for polarization-induced charge. In AlGaN polarization-induced charge and the 2DEG shown in Equation 2.2.

$$\frac{\Delta V_{AlGaN}}{d} = \frac{\sigma_{AlGaN}}{\varepsilon} - \frac{qn_s}{\in} \tag{2.2}$$

where $\Delta V_{AlGaN}$ is the potential drop across the conductive band of the AlGaN surface and the AlGaN/GaN interface. The variable $d$ is the thickness of the AlGaN layer, $\sigma_{AlGaN}$ is the sheet charge density that is induced by AlGaN polarization, ns is the sheet density of 2DEG, $q$ is the electron charge, and $\varepsilon$ is the dielectric constant of AlGaN.

In energy band diagram, the potential drop $\Delta V_{AlGaN}$ can be expressed as Equation 2.3:

$$q\Delta V_{AlGaN} = q\varnothing_B + E_f - \Delta E_c \tag{2.3}$$

In Equation 2.3, $\phi_b$ is the surface barrier height, $E_f$ is the Fermi level position with respect to the GaN conduction-band edge at the AlGaN/GaN interface, and $\Delta E_c$ is the conduction band discontinuity between GaN and AlGaN. The sheet density of 2DEG can be expressed as Equation 2.4. The conduction band edge discontinuity in HEMT is equal to difference in electron affinities of semiconductors used. In Equations 2.4–2.5 is the mole fraction.

$$n_s = \frac{\sigma_{AlGaN}}{q} - \frac{\in}{q^2 d}\left(q\theta + E_f - \Delta E_c\right) \tag{2.4}$$

$$\Delta E_c = \chi_B - \chi_B \tag{2.5}$$

### 2.4.5 Implications for HFET Operation

The inclusion of the 2DEG increases the HFET's on-state current while decreasing the threshold voltage. The increased electron density allows for larger current flow, making HFETs excellent for applications that need high speed and efficiency, such as RF and microwave devices [21–23]. Table 2.2 shows the FETs comparison.

## 2.5 Conclusion

Heterojunction-based field effect transistors offer a big step forward in semiconductor technology, integrating the distinct features of many materials to improve device performance. Their capacity to generate a 2DEG at the heterojunction interface leads to increased electron mobility, faster switching rates, and lower power consumption, making them excellent for high-frequency and high-power applications. The variety of material choices, such as the use of wide-bandgap semiconductors like GaN and InP, enables HFETs to be adapted to particular operating needs. This versatility not only broadens the spectrum of applications—from telecommunications to power electronics—but also improves integration with other semiconductor devices, opening the door for smaller and more efficient electronic systems.

Furthermore, continued research and innovation in HFET technology promise even greater performance and efficiency gains. Exploration of innovative materials and enhanced production processes may result in the creation of next-generation HFETs capable of meeting the difficulties provided by developing technologies and demanding applications. In conclusion, HFETs are a critical technology that improves the capabilities of electronic devices by combining speed, efficiency, and adaptability. Their continuing progress is anticipated to play an important role in influencing the future of electronic systems across several sectors.

**AI disclosure statement:** The author acknowledges the use of AI-based tools for minor language editing or grammatical correction. However, all intellectual contributions, analyses, and conclusions presented in this chapter are entirely the author's own.

## References


1. Solomon, P.M. and Morkoc, H., Modulation-doped GaAs/AlGaAs heterojunction field-effect transistors (MODFET's), ultrahigh-speed device for supercomputers. *IEEE Transac. Electron Devices,* 31, 8, 1015–1027, Aug. 1984, doi: 10.1109/T-ED.1984.21654.
2. Chen, Y., Yu, S., Sun, Y., Liang, Z., Phase engineering in quasi-2D Ruddlesden–Popper perovskites. *J. Phys. Chem. Lett.*, 9, 10, 2627–31, 2018 Apr 30.
3. Jeen, H., Choi, W.S., Freeland, J.W., Ohta, H., Jung, C.U., Lee, H.N., Topotactic phase transformation of the brownmillerite SrCoO2. 5 to the perovskite $SrCoO_{3-\delta}$. *Adv. Mater.*, 25, 27, 3651–6, 2013 Jul 19.
4. Kossar, S., Rasool, A., Khan, R., Amiruddin, R., Koser, K. Bhat, A.A., Investigations of lanthanum-doped bismuth ferrite thin films for resistive switching random access memory (ReRAM) applications. *J. Mater. Sci.: Mater. Electron.,* 36, 10, 1–13, 2005.
5. Gamal, A., Alruqi, M., Rabia, M., $CsPbI_3$ lead and $CsSnI_3$ lead-free perovskite materials for solar cell device. *Int. J. Energy Res.*, 2022 Sep 13.
6. Liang, J., Wang, C., Wang, Y., Xu, Z., Lu, Z., Ma, Y., Zhu, H., Hu, Y., Xiao, C., Yi, X., Zhu, G., All-inorganic perovskite solar cells. *J. Am. Chem. Soc.*, 138, 49, 15829–32, 2016 Dec 14.
7. Zhang, W., Ge, M., Li, Y., Tan, S., Yu, C., Chen, D., TCAD Simulation of an E-Mode Heterojunction Bipolar p-FET with Imax> 240 mA/mm. *Electronics,* 13, 23, 4752, 2024.
8. Yin, W.J., Weng, B., Ge, J., Sun, Q., Li, Z., Yan, Y., Oxide perovskites, double perovskites and derivatives for electrocatalysis, photocatalysis, and photovoltaics. *Energy Environ. Sci.*, 12, 2, 442–62, 2019.
9. Mizoguchi, H., Bhuvanesh, N.S., Kim, Y.I., Ohara, S., Woodward, P.M., Hydrothermal crystal growth and structure determination of double hydroxides LiSb (OH) 6, BaSn (OH) 6, and SrSn (OH) 6. *Inorg. Chem.*, 53, 19, 10570–7, 2014 Oct 6.
10. Jiang, S., *et al.*, All-GaN-Integrated Cascode Heterojunction Field Effect Transistors. *IEEE Trans. Power Electron.*, 32, 11, 8743–8750, Nov. 2017, doi: 10.1109/TPEL.2016.2643499.
11. Tian, W., Zhou, H., Li, L., Hybrid organic–inorganic perovskite photodetectors. *Small,* 13, 41, 1702107, 2017 Nov.
12. Jadhav, V.V., Mane, R.S., Shinde, P.V., Jadhav, V.V., Mane, R.S., Shinde, P.V., Basics of Ferrites: Structures and Properties, in: *Bismuth-Ferrite-Based Electrochemical Supercapacitors*, pp. 37–45, 2020.
13. Yao, L., Majumdar, S., Äkäslompolo, L., Inkinen, S., Qin, Q.H., van Dijken, S., Electron-Beam-Induced Perovskite–Brownmillerite–Perovskite Structural Phase Transitions in Epitaxial $La_{2/3}Sr_{1/3}MnO_3$ Films. *Adv. Mater.*, 26, 18, 2789–93, 2014 May.

14. Lebedinskaya, A.R. and Kupriyanov, M.F., Phenomena of structural order/disorder in perovskites with properties of ferroelectrics-relaxers, in: *II National crystallochemical conference. Summaries of reports*, 2000.
15. Jung, H.S. and Park, N.G., Perovskite solar cells: from materials to devices. *Small*, 11, 1, 10–25, 2015 Jan.
16. Ha, W., Shinohara, K., Brar, B., Enhancement-Mode Metamorphic HEMT on GaAs Substrate With 2 S/mm gmg_{m} and 490 GHz fTf_{T} in: *IEEE Electron Device Letters*, 29, 5, 419–421, May 2008, doi: 10.1109/LED.2008.920283.
17. Juarez-Perez, E.J., Sanchez, R.S., Badia, L., Garcia-Belmonte, G., Kang, Y.S., Mora-Sero, I., Bisquert, J., Photoinduced giant dielectric constant in lead halide perovskite solar cells. *J. Phys. Chem. Lett.*, 5, 13, 2390–4, 2014 Jul 3.
18. Song, Y. S., Thoutam, L.R., Tayal, S., Rahi, S.B., Samuel, T.S.A. (Eds.), *Handbook of Emerging Materials for Semiconductor Industry*, Springer, 2024.
19. Avval, A.G., Larique, E., El-Ghazaly, S.M., Heterojunction field effect transistors, in: *Reference Module in Materials Science and Materials Engineering*, Elsevier, New York, 2018.
20. Parker, M., A gallium nitride HEMT that enhances. *Nat. Electron.*, 4, 12, 858, 2021.
21. Yao, Y., Huang, S., Cao, R., Zhaofu Zhang, Xinhua Wang, Qimeng Jiang, Jingyuan Shi, Chenrui Zhang, Jiaolong Liu, Ke Wei, Yi Pei, Hui Zhang, Hongtu Qian, Fuqiang Guo, Guoping Li, Ning Tang, Jun-Wei Luo, Weikun Ge, Xinyu Liu, Bo Shen , Kevin J. Chen, Dislocation-assisted electron and hole transport in GaN epitaxial layers. *Nat. Commun.*, 16, 6448, 2025,. https://doi.org/10.1038/s41467-025-61510-w
22. He, J., Cheng, W.-C., Wang, Q., Cheng, K., Yu, H., Chai, Y., Recent advances in GaN-based power HEMT devices. *Adv. Electron. Mater.* 7, 4, 2001045, 2021.
23. Sun, T., *et al.*, High-Voltage Polarization-Superjunction GaN HEMT with Built-In SBD for Low Reverse Conduction Loss, in: *IEEE Journal of the Electron Devices Society*, vol. 10, pp. 808–812, 2022, doi: 10.1109/JEDS.2022.3208731.
24. Wang, X. *et al.*, Effect of GaN Channel Layer Thickness on DC and RF Performance of GaN HEMTs With Composite AlGaN/GaN Buffer, in: *IEEE Transactions on Electron Devices*, vol. 61, no. 5, pp. 1341–1346, May 2014, doi: 10.1109/TED.2014.2312232.
25. Goel, E., Pandey, A., Rahi, S. B., Samuel, A. (Eds.), *Handbook of Advanced Semiconductor Field Effect Transistors*. John Wiley & Sons, 2025.
26. He, J., *et al.*, Normally-OFF AlGaN/GaN MIS-HEMTs With Low RON and Vth Hysteresis by Functioning In-situ SiNx in Regrowth Process, in: *IEEE Electron Device Letters*, vol. 43, no. 4, pp. 529–532, April 2022, doi: 10.1109/LED.2022.3149943.

27. Wang, W., *et al.*, Suppression of Dynamic Resistance Degradation in 1200-V GaN-on-Sapphire E-Mode GaN HEMTs by Drain-Side Thin p-GaN Design, in: *IEEE Transactions on Electron Devices*, vol. 72, no. 3, pp. 1537–1540, March 2025, doi: 10.1109/TED.2025.3534746.
28. Wang, W.-K., *et al.*, Performance enhancement by using the n/sup +/-GaN cap layer and gate recess technology on the AlGaN-GaN HEMT fabrication, in: *IEEE Electron Device Letters*, vol. 26, no. 1, pp. 5–7, Jan. 2005, doi: 10.1109/LED.2004.840395.
29. Hwang, I., *et al.*, Source-Connected p-GaN Gate HEMTs for Increased Threshold Voltage, in: *IEEE Electron Device Letters*, vol. 34, no. 5, pp. 605–607, May 2013, doi: 10.1109/LED.2013.2249038.

3

# Nanosheet Field Effect Transistor

**P. Suveetha Dhanaselvam[1], B. Karthikeyan[1*], K. Kavitha[1], S. Nagarajan[2], P. Karthikeyan[1] and G. Pradeep Kumar[1]**

[1]*Department of Electronics and Communication Engineering, Velammal College of Engineering and Technology, Madurai, Tamilnadu, India*
[2]*Department of Electronics and Communication Engineering, Easwari Engineering College, Chennai, Tamilnadu, India*

## *Abstract*
Nanosheet field-effect transistors (FETs) represent a transformative advancement in semiconductor technology, addressing the challenges of traditional scaling limits while delivering enhanced performance for next-generation electronics. This chapter provides a comprehensive overview of nanosheet FETs, focusing on their structural innovations, fabrication methodologies, and operational principles. The discussion begins with the evolution of transistor architectures, leading to the development of nanosheet designs that leverage horizontal nanostructures for superior electrostatic control. Key fabrication techniques, including lithography, epitaxy, and selective etching, are examined, emphasizing their role in achieving the precise dimensional control required for nanosheet production. The chapter also delves into the advantages of nanosheet FETs, such as reduced power consumption, higher current drive, and scalability beyond FinFET technologies. Furthermore, challenges related to process complexity, material compatibility, and device variability are critically analyzed. By exploring the latest research trends, practical applications, and potential future directions, this chapter aims to serve as an essential resource for researchers, engineers, and students working in the field of advanced semiconductor devices.

***Keywords*:** Nanosheet FETs, 2D materials, semiconductor devices, carrier mobility, short-channel effects, device scaling, high-performance electronics

**Corresponding author*: om.bkarthikeyan@gmail.com

P. Suveetha Dhanaselvam, Naima Guenifi, Shiromani Balmukund Rahi and J. Ajayan (eds.) Field-Effect Transistor Devices: Post-CMOS Theory and Applications, (53–68) © 2026 Scrivener Publishing LLC

## 3.1 Introduction to Nanosheet FETs

The evolution of transistors forms the cornerstone of modern semiconductor technology. Since the invention of the first point-contact transistor in 1947, the field has witnessed rapid advancements, driving the miniaturization and performance improvement of electronic devices. The transition from planar transistors to three-dimensional structures, such as FinFETs [1, 2], marked a significant milestone in overcoming scaling challenges. Each technological iteration has been driven by Moore's Law, which predicted the doubling of transistor density on integrated circuits approximately every two years. While planar metal-oxide-semiconductor field-effect transistors (MOSFETs) dominated for decades, their limitations in terms of leakage current, power efficiency, and scalability necessitated the transition to FinFETs, which offered superior electrostatic control. However, as technology nodes approach the sub-3nm regime, FinFETs encounter physical and process-related constraints, setting the stage for the introduction of nanosheet FETs as the next-generation solution [3, 4].

With changing technology nodes, the semiconductor industry is however not eager to switch to other transistor architectures. Some companies might even decide to stay at certain nodes longer. But still, there are applications such as machine learning, big data analysis and data center servers that will require the latest universal CMOS solutions. With such a universal CMOS solution, one and the same transistor architecture in one and the same technology node can be used to perform all functionalities on the chip. To increase the number of field-effect transistors (FETs) in the same area, alternatives to the typically planar channel architecture have been proposed including FinFETs and nanosheet FETs that can increase the amount of drive current per channel area. Core MOSFETs are developed from planar transistors and FinFETs to the latest Stacked NanoSheet/NanoWire Gate-All-Around FETs (GAAFETs), and now to the cutting-edge vertical transistor 3D stacking in CFET/3DS-FET or vertical-channel transistor structure with the potential 2DM, CNT, or AOS materials for the future [25–31]. Figure 3.1: development of MOSFET technology.

The relentless pursuit of higher transistor density has exposed inherent limitations in conventional scaling. The planar MOSFET architecture, which relies on a single gate structure, suffers from poor control over the channel, especially at smaller gate lengths. This leads to increased leakage currents and degraded device performance. FinFETs addressed these issues by introducing a three-dimensional fin-shaped channel surrounded by gates on three sides, enhancing electrostatic control. However, as gate lengths

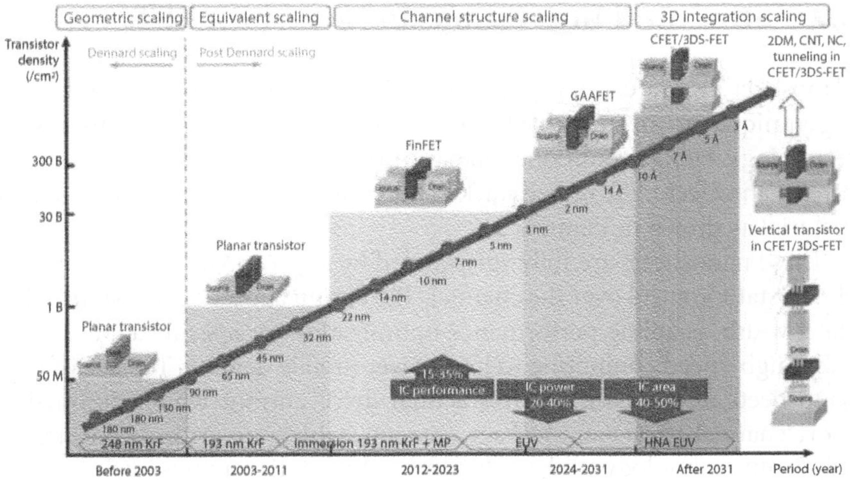


**Figure 3.1** Development of MOSFET technology.

shrink further, even FinFETs struggle to provide adequate control, resulting in challenges such as increased short-channel effects, higher variability, and difficulties in manufacturing. Furthermore, the high aspect ratio of FinFET structures introduces complexities in lithography, etching, and doping processes. These limitations highlight the urgent need for alternative transistor architectures that can maintain scalability, improve energy efficiency, and ensure reliable performance in advanced technology nodes.

Nanosheet FETs have emerged as a promising solution to the challenges posed by conventional transistor designs [5–7]. These devices build upon the strengths of FinFETs while introducing a horizontal nanostructure-based architecture that allows for superior gate control and scalability. The key innovation lies in their multi-bridge channel configuration, where thin nanosheets are stacked horizontally and fully surrounded by the gate [8]. This gate-all-around (GAA) design significantly reduces leakage currents, enhances channel control, and improves drive current per unit area [9]. Unlike FinFETs, where performance depends on the fin height and width, nanosheet FETs allow flexible width modulation of the nanosheets, enabling precise performance tuning for various applications. As a result, nanosheet FETs are considered the ideal candidate for scaling to sub-3nm technology nodes, meeting the stringent demands of high-performance computing, mobile devices, and low-power applications. Their emergence represents a paradigm shift in transistor design, marking a new chapter in the evolution of semiconductor technology.

## 3.2 Structural Innovations in Nanosheet FETs

Nanosheet FETs are at the forefront of advanced transistor design, featuring a unique architecture that builds upon the principles of gate-all-around technology. Unlike traditional planar or three-dimensional FinFET structures, nanosheet FETs utilize stacked horizontal nanosheets as the channel material as shown in Figure 3.2.

These nanosheets are fully surrounded by a gate, providing exceptional electrostatic control over the channel. The design allows for a customizable sheet width, enabling performance optimization for specific applications, from high-speed computing to low-power mobile devices. The use of thin nanosheets as channels ensures a high drive current while minimizing short-channel effects and leakage currents, which are critical concerns at sub-5 nm technology nodes. This innovative design represents a significant departure from earlier transistor architectures, laying the groundwork for continued scaling and performance improvements in semiconductor devices.

The defining feature of nanosheet FETs is their horizontal nanostructure configuration, which plays a crucial role in enhancing electrostatic control. The horizontal stacking of nanosheets provides a larger effective channel width, improving current drive compared to FinFETs. The gate-all-around structure ensures uniform control of the channel from all sides, significantly reducing short-channel effects. This is especially important as technology nodes shrink and transistor dimensions approach the atomic scale. The precise control offered by the horizontal nanostructures also leads to improved subthreshold slope, reduced off-state leakage, and better overall energy efficiency. By allowing for independent tuning of nanosheet thickness and width, the architecture enables manufacturers to balance performance and power consumption effectively, addressing the diverse demands of modern electronics.

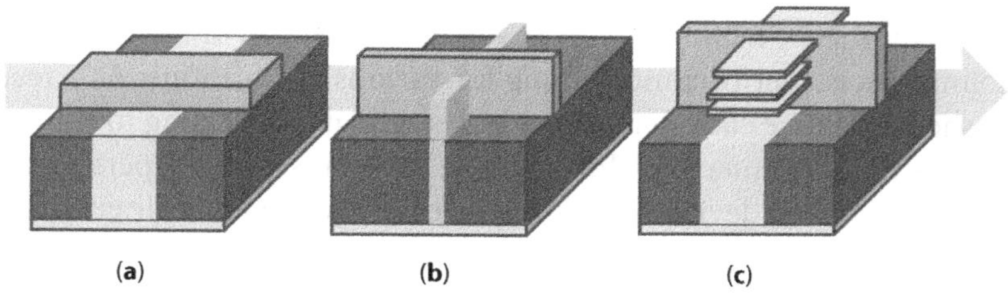


**Figure 3.2** Evolution of FETs. (a) Planar FET, (b) FinFET, (c) nanosheet FET.

The structure of the Si-based transistors would be changed from the fin field-effect transistor (FinFET) to the cutting-edge Stacked NanoSheet/NanoWire Gate-All-Around FETs (GAAFETs) at 3 nm node. Therefore, they introduce up-to-date GAAFET integration process method and the recent technical progress in research institutions and IC industries. Furthermore, the key challenges in fabricating the GAAFETs are illustrated detailly, including high-quality GeSi/Si superlattice periodic epitaxy, channel release, inner spacer module, SD-selective epitaxial defects, parasitic sub-fin channel leakage, and HKMG filling, low hole mobility in the (100) orientation, high voltage (HV) and input/output (IO) integration, and high parasitic capacitance during AC operation. Some innovations in GAA devices, e.g. Forksheet FETs, Tree FET, Fishbone FET, CombFET [22–31].

Nanosheet FETs outperform FinFETs and other advanced transistor designs in several critical aspects. While FinFETs offer improved electrostatic control by surrounding the channel with the gate on three sides, their performance is limited by the height and width of the fin. Nanosheet FETs, on the other hand, eliminate this limitation by using horizontally stacked sheets, which provide greater flexibility in optimizing the channel dimensions. Furthermore, the gate-all-around structure of nanosheet FETs delivers superior short-channel effect mitigation and leakage current reduction compared to the tri-gate design of FinFETs. Compared to planar MOSFETs, nanosheet FETs offer exponentially better performance at smaller nodes, as planar designs suffer from poor electrostatic control and high leakage currents. When juxtaposed with other emerging technologies, such as nanowire FETs, nanosheet FETs provide a higher drive current due to the wider channel area. This advantage, combined with their compatibility with existing FinFET fabrication processes, makes nanosheet FETs a more practical and scalable choice for mass production. As a result, nanosheet FETs are rapidly gaining traction as the go-to solution for scaling beyond the limits of FinFET technology, ensuring continued innovation in the semiconductor industry.

## 3.3 Fabrication Techniques for Nanosheet FETs

Fabricating nanosheet FETs requires the use of advanced lithography techniques to achieve the precision necessary for sub-5nm nodes [5]. Figure 3.3 illustrates the fabrication methods employed in the production of nanosheet FETs. Extreme ultraviolet (EUV) lithography is the primary enabler for defining the nanosheet patterns with high resolution and accuracy. EUV provides

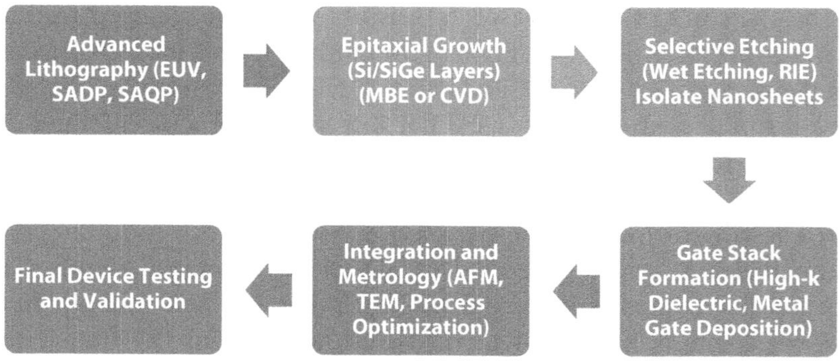


**Figure 3.3** Fabrication process flow for nanosheet FETs.

a shorter wavelength of light (13.5 nm), allowing for the creation of finer features and tighter pitches. The lithographic process begins with the patterning of the nanosheet stack, which involves a series of alternating layers of semiconducting and sacrificial materials. These layers are deposited on the wafer and patterned to define the horizontal nanosheet structure. Multipatterning techniques, such as self-aligned double patterning (SADP) and self-aligned quadruple patterning (SAQP) [10, 11], are often employed to further refine the features and ensure uniformity across the wafer. The precision offered by these lithographic methods is critical for maintaining the structural integrity and performance of nanosheet FETs at advanced technology nodes.

Epitaxial growth is a fundamental step in the fabrication of nanosheet FETs, enabling the deposition of high-quality semiconductor layers with controlled thickness and composition. During this process, alternating layers of silicon (Si) and silicon-germanium (SiGe) are grown on a substrate. The SiGe layers serve as sacrificial materials, while the Si layers form the active nanosheet channels. Molecular beam epitaxy (MBE) and chemical vapor deposition (CVD) are commonly used for this purpose, as they provide atomic-level precision and uniformity. The epitaxial process ensures that the nanosheets have the desired electrical properties, such as high carrier mobility and low defect density. Furthermore, the ability to tune the composition and strain within the nanosheets through epitaxy enables enhanced device performance, such as increased drive current and reduced leakage.

Selective etching is a critical step for isolating the nanosheets from the sacrificial SiGe layers, creating the horizontal channel structures. This process requires a high degree of precision to avoid damaging the active Si nanosheets while completely removing the sacrificial material. Wet etching methods, such as the use of hydrochloric acid (HCl) solutions, are typically employed due to their high selectivity for SiGe over Si. Dry etching techniques, such as reactive ion etching (RIE), may also be used to refine the nanosheet dimensions and ensure uniformity. After etching, the gate stack is formed by depositing a high-k dielectric material, such as hafnium oxide ($HfO_2$), followed by a metal gate material. The selective etching process is instrumental in maintaining the structural integrity of the nanosheets, ensuring consistent performance across all devices on the wafer.

The fabrication of nanosheet FETs introduces several integration challenges, including alignment accuracy, process complexity, and material compatibility. The stacking of multiple nanosheets requires precise control over the thickness, uniformity, and spacing of each layer. Any deviation in these parameters can lead to performance variability and reduced yield. Additionally, the transition from FinFET to nanosheet FET technology demands significant modifications to existing manufacturing processes, including the adoption of new materials and etching techniques. To address these challenges, advanced metrology tools, such as atomic force microscopy (AFM) and transmission electron microscopy (TEM), are used to monitor and ensure dimensional accuracy during fabrication. Process simulations and machine learning-based optimization techniques are also being employed to refine manufacturing workflows and reduce variability. Collaborative efforts between semiconductor manufacturers and equipment providers have led to the development of specialized tools and processes tailored for nanosheet FET production. By overcoming these integration challenges, the industry is paving the way for the widespread adoption of nanosheet FETs in high-performance and low-power applications.

## 3.4 Performance Advantages of Nanosheet FETs

Nanosheet FETs deliver exceptional power efficiency and current drive, making them ideal for advanced semiconductor applications. The horizontal stacking of nanosheets provides a larger effective channel width compared to FinFETs, resulting in a higher drive current per unit area. This increased current drive translates to faster switching speeds and improved performance in high-frequency applications. The gate-all-around (GAA) architecture reduces power consumption by minimizing leakage currents and enabling precise control over the channel. This is particularly beneficial for low-power devices, such as those used in mobile and IoT applications, where energy efficiency is paramount. By combining high current drive with low power dissipation, nanosheet FETs meet the demands of both performance-intensive and energy-sensitive technologies.

The GAA structure of nanosheet FETs provides superior electrostatic control over the channel, a critical factor in mitigating short-channel effects at smaller technology nodes. Unlike FinFETs, where the gate wraps around three sides of the fin, nanosheet FETs feature a gate that completely surrounds the channel from all sides. This configuration enhances gate control, reducing the susceptibility to variations in threshold voltage and improving device reliability. Moreover, the improved gate control leads to a steeper subthreshold slope, allowing for lower operating voltages without compromising performance [12]. This not only reduces dynamic power consumption but also minimizes off-state leakage currents, which are a significant source of power loss in modern transistors. By addressing these challenges, nanosheet FETs ensure robust performance even as devices scale to sub-3nm nodes.

As semiconductor technology approaches the physical and process limitations of FinFETs, nanosheet FETs emerge as the ideal solution for continued scaling. The flexible design of nanosheet FETs allows for precise tuning of nanosheet width and spacing, enabling seamless scaling to smaller nodes while maintaining performance and power efficiency. Furthermore, the architecture is compatible with existing FinFET manufacturing processes, reducing the cost and complexity of transitioning to nanosheet technology. The modular nature of the nanosheet design also allows for easy integration with emerging materials, such as high-mobility channel materials and new dielectric layers, further extending the scalability of the technology. By combining enhanced electrostatic control, high current drive, and flexibility in scaling, nanosheet FETs provide a clear pathway for overcoming the challenges of advanced technology nodes. Their ability to deliver superior performance while meeting the stringent power and scalability requirements of modern electronics ensures their position as a key enabler of next-generation semiconductor devices.

## 3.5 Challenges in Nanosheet FET Development

The fabrication of nanosheet FETs involves a multi-step process that introduces significant complexity and variability. The precise stacking of nanosheets, along with the removal of sacrificial layers, requires advanced lithography, epitaxy, and etching techniques [13, 14]. Even minor deviations in these processes can lead to variations in nanosheet dimensions, spacing, and alignment, resulting in performance inconsistencies across devices on the wafer. Moreover, the introduction of new process steps, such as gate-all-around (GAA) formation and high-k metal gate deposition, demands tighter control over manufacturing tolerances. As device dimensions shrink to sub-3nm nodes, process-induced variability, such as line edge roughness and critical dimension (CD) variations, becomes more pronounced, adversely affecting transistor performance and yield. To address these challenges, manufacturers are leveraging advanced metrology tools, process simulations, and artificial intelligence-based optimization techniques to monitor and minimize variability throughout the fabrication process. Despite these advancements, achieving consistent results at such small scales remains a significant hurdle in nanosheet FET development.

Material selection plays a crucial role in the performance and manufacturability of nanosheet FETs, but integrating different materials poses significant challenges. The use of silicon-germanium (SiGe) as a sacrificial layer introduces stress and strain during fabrication, which can affect the structural integrity of the nanosheets [15]. Additionally, the deposition of high-k dielectrics and metal gates requires materials that are thermally stable and compatible with the nanosheet structure. As the industry explores high-mobility materials, such as III-V compounds (e.g., InGaAs) or two-dimensional materials (e.g., $MoS_2$), for improved performance, issues related to material compatibility with existing processes and substrates become more pronounced. These challenges include lattice mismatch, thermal expansion coefficient differences, and interfacial defects, all of which can degrade device performance and reliability. To overcome these issues, researchers are investigating innovative solutions such as new bonding techniques, interfacial engineering, and alternative substrate materials.

Ensuring the reliability and long-term performance of nanosheet FETs is another critical challenge, particularly as device dimensions shrink and operating conditions become more demanding. Nanosheet FETs are susceptible to various reliability concerns, including bias temperature instability (BTI), hot carrier injection (HCI), and time-dependent dielectric breakdown (TDDB). These phenomena can lead to shifts in threshold voltage, increased leakage currents, and eventual device failure over time. The integration of new materials, such as high-k dielectrics and advanced metals, introduces additional reliability risks, as these materials may exhibit unique degradation mechanisms under stress. Moreover, the reduced dimensions of nanosheets increase their sensitivity to process-induced defects and variability, further complicating reliability assurance.

Addressing these challenges requires a combination of rigorous testing protocols, advanced reliability modeling, and the development of novel materials with improved stability. Efforts are also being directed toward the design of self-healing or fault-tolerant architectures that can mitigate the impact of reliability issues in nanosheet FETs. Despite these advancements, achieving the required levels of reliability and long-term performance for widespread adoption remains a significant challenge for the semiconductor industry.

## 3.6 Applications of Nanosheet FETs

Nanosheet FETs are poised to revolutionize consumer electronics by enabling more powerful, energy-efficient, and compact devices [16]. The superior performance of nanosheet FETs, characterized by their high drive current and low power consumption, makes them ideal for smartphones, tablets, laptops, and wearable devices [17]. The enhanced electrostatic control provided by the gate-all-around (GAA) architecture allows for higher processing speeds and longer battery life, addressing key demands in consumer electronics. The scalability of nanosheet FETs to sub-3nm nodes supports the integration of more transistors per chip, enabling advancements in features such as real-time image processing, augmented reality (AR), and artificial intelligence (AI) capabilities. Moreover, their low power dissipation ensures that thermal management challenges are minimized, a crucial factor in compact devices. By combining performance and efficiency, nanosheet FETs are driving the next generation of smart, connected, and portable electronic devices.

The high current drive and superior scalability of nanosheet FETs make them a key enabler for high-performance computing (HPC) applications [18]. Data centers, supercomputers, and servers demand processors that can deliver exceptional performance while maintaining energy efficiency, and nanosheet FETs meet these requirements by reducing leakage currents and enhancing transistor density. Their ability to operate reliably at smaller nodes ensures higher computational throughput and enables more complex workloads, such as large-scale data analytics, machine learning, and scientific simulations. Also, nanosheet FETs support the development of specialized accelerators, such as graphics processing units (GPUs) and tensor processing units (TPUs), which are critical for AI and deep learning applications. By addressing the dual challenges of performance and power efficiency, nanosheet FETs are transforming the HPC landscape.

The Internet of Things (IoT) relies on a vast network of interconnected devices that require low power consumption and efficient operation. Nanosheet FETs are well-suited for IoT applications due to their energy efficiency and ability to operate at ultra-low power levels [19]. Their superior electrostatic control minimizes leakage currents, enabling devices to function effectively in power-constrained environments. IoT devices often need to balance computational power with energy efficiency, as they are typically battery-operated or energy-harvesting systems. The tunable design of nanosheet FETs allows for optimization based on specific application needs, whether it is processing sensor data, transmitting information, or executing AI algorithms at the edge. Furthermore, the scalability of nanosheet FETs supports the integration of multiple functionalities on a single chip, reducing size and cost, which is essential for IoT ecosystems. From smart homes and industrial automation to healthcare monitoring and environmental sensing, nanosheet FETs are playing a pivotal role in enabling the next wave of IoT innovation. Their performance, efficiency, and versatility position them as a foundational technology for the rapidly expanding IoT landscape.

While nanosheet FETs are driving the current generation of semiconductor advancements, scaling beyond the nanosheet era is an active area of research. Potential successors include stacked forksheet transistors and complementary FETs (CFETs), which aim to further enhance transistor density and performance [20–22]. Stacked CFETs combine n-type and p-type transistors in a vertical configuration, significantly reducing the area required for device integration. Researchers are also exploring advanced 2D materials and vertical transport mechanisms to overcome limitations associated with conventional scaling. The push toward sub-2nm technology nodes will demand innovative lithographic methods, novel device architectures, and reimagined interconnect technologies to sustain progress. The block diagram in Figure 3.4 illustrates the applications and

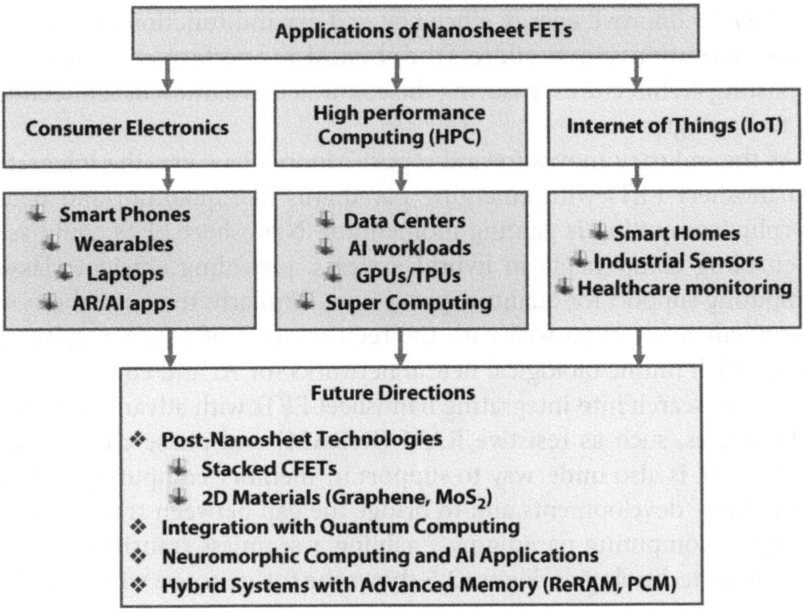


**Figure 3.4** Application of nanosheet FETs.

future directions of nanosheet FETs by categorizing their impact into three primary sectors and projecting potential advancements. In Consumer Electronics, nanosheet FETs drive the development of portable, energy-efficient devices such as smartphones, wearables, and AR/AI-enabled gadgets. For High-Performance Computing (HPC), they enable the efficient execution of AI workloads, data center operations, and supercomputing tasks through high transistor density and enhanced performance. Within the Internet of Things (IoT) domain, their low power consumption and scalability support smart homes, industrial automation, and healthcare monitoring applications. The future directions section highlights innovations like stacked CFETs and 2D materials for even higher transistor density and performance. Nanosheet FETs are poised to integrate seamlessly into quantum computing and neuromorphic systems, paving the way for groundbreaking advancements in computing architectures.

The integration of new materials is essential for enhancing the performance of nanosheet FETs and their successors. High-mobility materials, such as III-V compounds (e.g., InGaAs) and two-dimensional materials (e.g., graphene, $MoS_2$), are being investigated to replace silicon as the channel material. These materials offer superior electron or hole mobility, enabling faster switching speeds and lower power consumption. Beyond materials, researchers are exploring alternative device structures, such as nanowire FETs, negative capacitance FETs (NCFETs), and tunnel FETs (TFETs), to improve energy efficiency and expand functionality [23, 24]. These innovations aim to address the physical and performance constraints of existing architectures, ensuring the continued evolution of semiconductor technologies.

As the industry moves toward a post-Moore's Law era, the integration of nanosheet FETs with emerging paradigms like quantum and neuromorphic computing is gaining momentum. Nanosheet FETs could serve as enabling components in hybrid systems, providing efficient classical computing support for quantum processors. Similarly, their scalability and energy efficiency align well with the requirements of neuromorphic systems, which mimic biological neural networks for AI and cognitive applications. Research into integrating nanosheet FETs with advanced memory technologies, such as resistive RAM (ReRAM) and phase-change memory (PCM), is also underway to support in-memory computing architectures. These developments aim to bridge the gap between traditional and emerging computing paradigms, enabling a seamless transition to next-generation technologies. Figure 3.5 shows the future scope with nanosheet FETs.

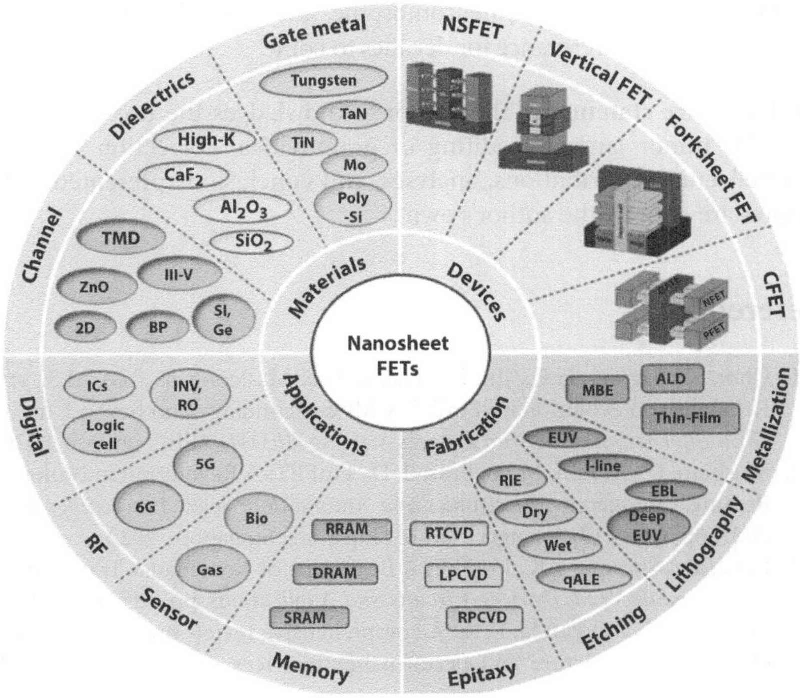


**Figure 3.5** Future scope with nanosheet FETs.

## 3.7 Conclusion

Nanosheet FETs represent a transformative advancement in transistor technology, addressing the limitations of FinFETs and enabling continued scaling to smaller nodes. Their innovative gate-all-around architecture, combined with superior electrostatic control, high current drive, and energy efficiency, positions them as a cornerstone for modern semiconductor applications. Despite challenges in fabrication, material integration, and reliability, nanosheet FETs are already making significant contributions to consumer electronics, high-performance computing, and IoT devices. The adoption of nanosheet FETs marks a pivotal moment for the semiconductor industry, ensuring the continuation of Moore's Law in the near term and setting the stage for future innovations. By overcoming current limitations and exploring new materials, architectures, and computing paradigms, the industry can unlock unprecedented levels of performance and efficiency. Nanosheet FETs not only address immediate technological needs but also pave the way for advancements in quantum, neuromorphic, and beyond-CMOS technologies, ensuring their relevance for decades to come.

**AI disclosure statement:** The author acknowledges the use of AI-based tools for minor language editing or grammatical correction. However, all intellectual contributions, analyses, and conclusions presented in this chapter are entirely the author's own.

## References

1. Choi, K., Sagong, H.C., Jin, M., Hai, J., Lee, M., Jeong, T., Yeo, M.S., *et al.*, Reliability on evolutionary FinFET CMOS technology and beyond, in: *2020 IEEE International Electron Devices Meeting (IEDM)*, IEEE, pp. 9–3, 2020.
2. Mushtaq, U., Akram, M.W., Prasad, D., FinFET: A Revolution in Nanometer Regime, in: *Emerging Electronics and Automation: Select Proceedings of E2A 2021*, pp. 403–417, Springer Nature Singapore, Singapore, 2022.
3. Kajal, and Sharma, V.K., FinFET: A Beginning of Non-Planar Transistor Era, in: *Nanoscale VLSI: Devices, Circuits and Applications*, pp. 139–159, 2020.
4. Reddy, M.N. and Panda, D.K., A comprehensive review on FinFET in terms of its device structure and performance matrices. *Silicon*, 14, 18, 12015–12030, 2022.
5. Yakimets, D., Bhuwalka, K.K., Wu, H., Rzepa, G., Karner, M., Liu, C., Inflection Points in GAA NS-FET to C-FET Scaling Considering Impact of DTCO Boosters, in: *IEEE Transactions on Electron Devices*, vol. 71, no. 4, pp. 2309–2314, April 2024, doi: 10.1109/TED.2024.3368380.
6. Mukesh, S. and Zhang, J., A review of the gate-all-around nanosheet FET process opportunities. *Electronics*, 11, 21, 3589, 2022.
7. Kumari, N.A., Bharath Sreenivasulu, V., Vijayvargiya, V., Upadhyay, A.K., Ajayan, J., Uma, M., Performance comparison of nanosheet FET, CombFET, and TreeFET: device and circuit perspective. *IEEE Access*, 12, 9563–95715, 2024.
8. Kumari, N.A. and Prithvi, P., Performance evaluation of GAA nanosheet FET with varied geometrical and process parameters. *Silicon*, 14, 15, 9821–9831, 2022.
9. Wang, D., Sun, X., Liu, T., Chen, K., Yang, J., Wu, C., Xu, M., Zhang, W., Investigation of Source/Drain Recess Engineering and Its Impacts on FinFET and GAA Nanosheet FET at 5 nm Node. *Electronics*, 12, 3, 770, 2023.
10. Samavedam, S.B., Ryckaert, J., Beyne, E., Ronse, K., Horiguchi, N., Tokei, Z., Radu, I., *et al.*, Future logic scaling: Towards atomic channels and deconstructed chips, in: *2020 IEEE International Electron Devices Meeting (IEDM)*, IEEE, pp. 1-1, 2020.

11. Horiguchi, N., Milenin, A.P., Tao, Z., Hubert, H., Altamirano-Sanchez, E., Veloso, A., Witters, L., *et al.*, Patterning challenges in advanced device architectures: FinFETs to nanowires, in: *Advanced Etch Technology for Nanopatterning V*, vol. 9782, SPIE, p. 978209, 2016.
12. Dhanaselvam, P.S., Sriram Kumar, D., Karthikeyan, B., Vanitha, P., Basic Science and Development of Subthreshold Swing Technology, in: *Tunneling Field Effect Transistors*, pp. 17–27, CRC Press, 2023.
13. Peng, W., *et al.*, Process Development and Optimization of Flip FET (FFET) for Stacked Transistors Technology, in: *IEEE Transactions on Electron Devices*, 2025. doi: 10.1109/TED.2025.3609731.
14. Park, K.-W., *et al.*, A Gate-All-Around Oxide Semiconductor FETs With Selectively Crystallized InGaOx Channel for Performance and Reliability Improvement, in: *IEEE Transactions on Electron Devices*, doi: 10.1109/TED.2025.3605574.
15. Müller, J., Lecestre, A., Demoulin, R., Cristiano, F., Hartmann, J.M., Larrieu, G., Engineering of dense arrays of Vertical Si1-x Ge x nanostructures. *Nanotechnology*, 34, 10, 105303, 2022.
16. Chen, J., Sun, M.-Y., Wang, Z.-H., Zhang, Z., Zhang, K., Wang, S., Zhang, Y., *et al.*, Performance Limits and Advancements in Single 2D Transition Metal Dichalcogenide Transistor. *Nano Micro Lett.*, 16, 1, 264, 2024.
17. Sreenivasulu, V.B. and Narendar, V., Design and temperature assessment of junctionless nanosheet FET for nanoscale applications. *Silicon*, 14, 8, 3823–3834, 2022.
18. Mii, Y.-J., Semiconductor innovations, from device to system, in: *2022 IEEE Symposium on VLSI Technology and Circuits (VLSI Technology and Circuits)*, IEEE, pp. 276–281, 2022.
19. Devnath, A., Lee, G., Ji, H., Alimkhanuly, B., Patil, S., Kadyrov, A., Lee, S., IoT-enabled novel heterostructure FET-based hybrid sensor for real-time arsenic detection. *Sens. Actuators, B*, 136146, 2024.
20. Ritzenthaler, R., Mertens, H., Eneman, G., Simoen, E., Bury, E., Eyben, P., Bufler, F.M., *et al.*, Comparison of electrical performance of co-integrated forksheets and nanosheets transistors for the 2nm technological node and beyond, in: *2021 IEEE International Electron Devices Meeting (IEDM)*, IEEE, pp. 26–2, 2021.
21. Dhanaselvam, P.S., Srinivasa Rao, K., Rahi, S.B., Yadav, D.S. (Eds.), *Field Effect Transistors*, John Wiley & Sons, 2025.
22. Ryckaert, J., Schuddinck, P., Weckx, P., Bouche, G., Vincent, B., Smith, J., Sherazi, Y., *et al.*, The Complementary FET (CFET) for CMOS scaling beyond N3, in: *2018 IEEE Symposium on Vlsi Technology*, IEEE, pp. 141–142, 2018.
23. Tian, L., *et al.*, Study of Scattering Effect on AC Reliability of Gate-all-Around Nanosheet FETs From Quantum Transport Perspective, in: *IEEE Transactions on Device and Materials Reliability.*
24. Goel, A., Rawat, A., Rawat, B., Finding a Promising CMOS Inverter Architecture With Silicon Nanosheet for Future Technology Node, in: *IEEE*

*Transactions on Electron Devices,* vol. 72, no. 4, pp. 1574–1581, April 2025, doi: 10.1109/TED.2025.3540040.
25. Hung, L.-C., Stanojević, Z., Schanovsky, F., Kosina, H., Karner, M., On-Current Degradation in Ultra-Scaled Nanosheet FETs with S/D Underlap Doping, in: *IEEE Electron Device Letters,* doi: 10.1109/LED.2025.3614369.
26. Zhang, S., *et al.*, Design Considerations for Si- and Ge-Stacked Nanosheet pMOSFETs Based on Quantum Transport Simulations, in: *IEEE Transactions on Electron Devices,* vol. 67, no. 1, pp. 26–32, Jan. 2020, doi: 10.1109/TED.2019.2954308.
27. Lee, C.-C., Hsieh, D.-R., Li, S.-W., Kuo, Y.-S., Chao, T.-S., Hysteresis-Free Gate-All-Around Stacked Poly-Si Nanosheet Channel Ferroelectric HfxZr1-xO2 Negative Capacitance FETs With Internal Metal Gate and NH3 Plasma Nitridation, in: *IEEE Transactions on Electron Devices,* vol. 69, no. 3, pp. 1512–1518, March 2022, doi: 10.1109/TED.2022.3147445.
28 Smaani, B., Paras, N., Rahi, S.B., Song, Y.S., Yadav, R., Tayal, S., Impact of the self-heating effect on nanosheet field effect transistor performance. *ECS J. Solid State Sci. Technol.,* 12, 2, 021005, 2023.
30. Yoon, J.-S., Jeong, J., Lee, S., Baek, R.-H., Systematic DC/AC Performance Benchmarking of Sub-7-nm Node FinFETs and Nanosheet FETs, in: *IEEE Journal of the Electron Devices Society,* vol. 6, pp. 942–947, 2018, doi: 10.1109/JEDS.2018.2866026.
31. Parker, M. Nanosheet FETs with low leakage. *Nat. Electron.,* 8, 555, 2025. https://doi.org/10.1038/s41928-025-01428-9.

4

# Post CMOS Semiconductor Field Effect Transistor

**Vishnudevi N.[1,2*], Thenmozhi T.[2], Vijay M.[3] and P. Suveetha Dhanaselvam[1]**

[1]*Department of Electronics Engineering (VLSI Design and Technology), Velammal College of Engineering and Technology, Madurai, India*
[2]*Department of Electronics and Communication Engineering, Velammal College of Engineering and Technology, Madurai, India*
[3]*Department of Electrical and Electronics Engineering, Aditya University, Surampalem, Andhra Pradesh, India*

***Abstract***
Semiconductor devices have undergone significant advancements, propelling innovation across sectors such as computing, telecommunications, automotive, and healthcare. Key developments in material science, fabrication techniques, and device architecture have led to improvements in performance, miniaturization, and energy efficiency. Transitioning from traditional silicon to materials like gallium nitride (GaN) and silicon carbide (SiC) has enabled the creation of devices that excel in high-power and high-frequency applications, offering greater efficiency in power electronics and faster communication systems. The introduction of three-dimensional (3D) semiconductor architectures, including Fin FETs and gate-all-around (GAA) transistors, has successfully tackled the challenges of scaling down devices, resulting in better performance and reduced power consumption. Furthermore, the integration of artificial intelligence (AI) and machine learning (ML) in the design and optimization processes of semiconductor devices has accelerated innovation, allowing for the development of more complex and intelligent electronic systems. Emerging fields like quantum computing and neuromorphic engineering are also driving the evolution of semiconductor technology, pushing the boundaries of conventional computing and enabling new possibilities. As the need for higher performance, lower energy consumption, and greater functionality continues to grow, ongoing research in semiconductor technologies is expected to lead the next generation of technological advancements, shaping the future of electronics. In summary,

**Corresponding author*: vichu0408@gmail.com

P. Suveetha Dhanaselvam, Naima Guenifi, Shiromani Balmukund Rahi and J. Ajayan (eds.) Field-Effect Transistor Devices: Post-CMOS Theory and Applications, (69–98) © 2026 Scrivener Publishing LLC

the progress in semiconductor devices is central to the rapid advancement of modern technology, resulting in more powerful, efficient, and intelligent systems that will continue to meet the increasing demands of the digital age.

***Keywords:*** Semiconductor devices, artificial intelligence, machine learning

# 4.1 Introduction

## 4.1.1 Overview of Semiconductor Devices

Miniaturization of semiconductor devices influence the production of electronic gadgets like smartphones and high-speed computers. Semiconductor devices are fundamental components in modern electronics, encompassing diodes, transistors, LEDs, solar cells and integrated circuits. Conventional materials like silicon and gallium arsenide have evolved significantly with the emergence of wide bandgap semiconductors such as gallium nitride (GaN) and silicon carbide (SiC), which offer superior performance in high-power and high-temperature environments. Recent research is increasingly focused on advanced materials, including two-dimensional (2D) materials like graphene and molybdenum disulfide ($MoS_2$), which promise ultra-thin, flexible, and high-speed electronic applications. In addition, quantum devices based on semiconductor quantum dots and spin qubits are laying the groundwork for quantum computing technologies. Another key area is neuromorphic computing, where novel devices like memristors are being developed to emulate the brain's synaptic functions, aiming for more efficient artificial intelligence systems. Meanwhile, the transistor landscape is transitioning from FinFETs to Gate-All-Around (GAA) architectures as researchers push beyond the limitations of Moore's Law. Optoelectronic integration is also gaining momentum, enabling faster data transmission through photonic circuits. Furthermore, the rise of flexible and wearable electronics is driving innovation in organic semiconductors. In renewable energy, developments in perovskite solar cells and tandem solar technologies are enhancing efficiency while reducing costs. These advancements are being supported by artificial intelligence and machine learning techniques to accelerate device design and material discovery. Overall, the current research landscape in semiconductor devices is vibrant and multidisciplinary, targeting enhanced performance, miniaturization, energy efficiency, and sustainability [1].

In sonducemictor sector, very-large-scale (VLSI) is the technical practice of combining billions of transistors to create an integrated circuit. According to Moore's law, the number of transistors doubles in every 2 years because of transistor's size is scaled down. Though the scaling of MOSFET has been the driving force towards the technological advancement, but due to continuous scaling various secondary effect which include; short channel effects, high leakage current, excessive process variation and reliability issue degrades the device performance. At every new technology generation, chipmakers have been able to scale transistor specs by 0.7x, delivering a 15% performance boost, a 50% area gain, a 40% power reduction and a 35% cost decrease at device level.

However, the demand for ever smaller, more efficient circuits is now pushing the limits of CMOS. The post-CMOS refers to the possible future digital logic technologies beyond the CMOS scaling limits. The demand for ever smaller and portable electronic devices has driven metal oxide semiconductor-based (CMOS) technology to its physical limit with the smallest possible feature sizes. This presents various size-related problems such as high power leakage, low-reliability, and thermal effects, and is a limit on further miniaturization. To enable even smaller electronics, various nano-FET devices including carbon nanotube transistors, graphene transistors, and tunnel transistors, junctionless FETs, FeFETs, NCFETs, collectively called post-CMOS devices are emerging that could replace the traditional and ubiquitous silicon MOSFET technology.

The last 50 years was based on Moore's law scaling and predominantly the Complementary Metal Oxide Semiconductor (CMOS) technology. The aggressive conventional CMOS technology has reached the end of Moore's law or approaching it in the near future. As conventional CMOS technology scaling results in smaller and smaller transistor sizes, existing CMOS technology starts to encounter certain fundamental physical limitations. These limitations can potentially result in drastic reduction in performance and energy efficiencies of these transistors, which could make it untenable to continue scaling them. In this scenario, experts have been developed several alternatives at the device level. However, each of these devices comes with varying energy-performance tradeoffs and optimal design points in comparison to CMOS technology. It is also used in various applications, such as sensors, image processing, and wireless communications. The conventional CMOS technology faces physical limits (short-channel effects, power density) as transistors shrink, requiring new physics and materials to continue performance improvements beyond traditional scaling [55–59].

The post-CMOS technologies explore alternatives and complements to traditional silicon CMOS for future computing. It focuses on new materials (2D materials, nanotubes), devices (TFETs, spintronics, memristors), and integration (photonic, neuromorphic, 3D stacking) to overcome CMOS scaling limits, enabling lower power, higher speed, and novel functions like in-memory computing or quantum systems. The key areas include Tunnel FETs, Graphene/CNTs, Spintronics, Photonics, Topological Materials, and Molecular Electronics, often used in hybrid systems with CMOS. Various beyond-CMOS device technologies including tunneling FET (TFET), hybrid phase transition FET (HyperFET), carbon nanotube FET (CNTFET), silicon nanowire FET (SiNWFET), symmetrical tunneling FET (SymFET), phase-change memory (PCM), spin-transfer torque magnetic tunnel junction (STT-MTJ), resistive random access memory (RRAM) have been considered [55–59].

### 4.1.2 Historical Background

The discovery of semiconductor materials brought forth the birth of semiconductor devices near the beginning of the twentieth century. The transistor originated in 1947 when John Bardeen together with Walter Brattain and William Shockley invented it at Bell Laboratories. The electronics industry experienced a transformation because of this innovation which delivered small-sized devices with advanced capabilities. The 1960s brought revolutionary progress to electronics through the invention of integrated circuits (ICs) because these electronic chips enabled simultaneous production of multiple transistor components as well as other electronic components [2].

Solid state electronic revolution began from semiconductor surface exploration. Surface states provided the inspiration that resulted in the transistor discovery during 1947. The reactive Ge surfaces caused new devices to become unstable along with being non-reproducible. An unpredictable threat to semiconductor electronics stimulated one of the biggest research efforts in scientific and technological history aimed toward controlling semiconductor surfaces as shown in Figure 4.1. The enormous worldwide research on semiconductor surfaces from that time has generated all present-day capabilities and principles which engineers and scientists use in surface science today. The birth of solid-state electronics on semiconductor surfaces developed into the contemporary scientific field that researches and controls any kind of surface.

The silicon material stands as the primary substance used for producing modern semiconductor devices. Tools based on submicron compound semiconductor structures featuring gallium arsenide and indium phosphide perform effectively for microwave and ultra-fast digital circuit

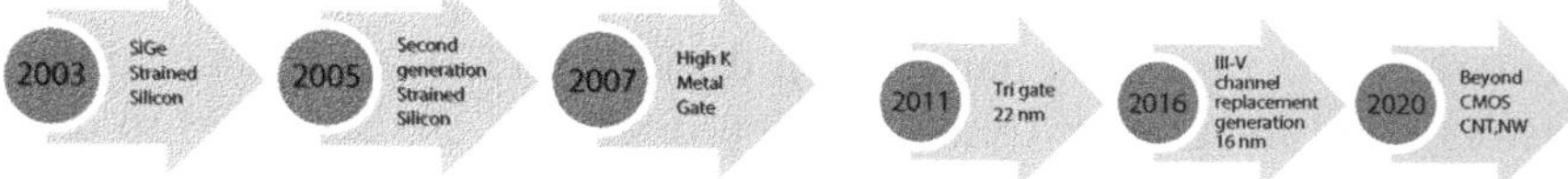


**Figure 4.1** Timeline of semiconductor growth.

purposes. The salt mercury cadmium telluride alongside other semiconductors serve the detection of infrared frequencies. The materials silicon carbide aluminium nitride together with gallium nitride show potential for power devices to function at high temperatures and in extreme conditions. Several display and photovoltaics applications utilize the hydrogenated amorphous silicon material and its related compounds. The electronic technology benefits from three important substances: ternary compounds, quaternary compounds as well as heterostructure materials. The electronic industry faces a prolonged challenge to develop new semiconductor materials alongside their unique property implementation.

## 4.2 CMOS Technology

CMOS technology has revolutionized the electronics industry by balancing performance, energy efficiency, and affordability. The System-on-a-Chip (SoC) paradigm has allowed a general-purpose approach to drive increasingly complex systems, integrating an ever-larger number of transistors on a single chip. This also enabled high-volume and low-cost production, boosting affordability of electronics. In the dennard scaling era of MOS transistors, extraordinary improvements in terms of switching speed, device density, functionality and cost have been achieved by the additive application of several technology. The boosters such as substrate engineering, strain, multigate, high-k/metal gate stacks and high-mobility channel materials are very common for semiconductor experts. However, the concept of a metal-oxide-semiconductor field-effect transistor (MOSFET) remained unchanged. Recently, aggressive scaling of the gate length dimensions down to few tens of nanometers is facing major challenges in terms of process variability, high leakage power, and unsalable voltage supply and degraded current switching ratios. CMOS technology advancements focus on transistor scaling and architecture (like 3D FinFETs), materials innovation (such as 2D materials and high-k/metal gates), and integration with other technologies (like photonic-CMOS and quantum dots) to improve performance, efficiency, and functionality. These developments enable greater integration on System-on-a-Chip (SoC) devices, leading to higher speeds, lower power consumption, and new applications in areas like AI and scientific imaging.

Modern electronic devices that include digital integrated circuits build their design using CMOS (Complementary Metal-Oxide semiconductor) technology. This technology serves microprocessors and microcontrollers and memory chips along with different digital logic circuits. CMOS technology remains popular because it features a versatile design along with superior noise resistance along with power efficiency capabilities. CMOS technology demonstrates no better performance than its competition when it comes to these characteristics. Modern electronic systems derive their 'gates' from CMOS devices due to reasons which this text soon explains. Technologically advanced devices like smartphones rely on CMOS components because of their small size along with their ability to grow in number easily and their broad capabilities and their ability to operate on small amounts of power which lets them work in portable electronic devices and 'always-on' wearables.

Although the modern semiconductor industry faces difficulty in CMOS dimensional scaling. Due to this semiconductor CMOS industry are promising for alternative directions for sustained improvement in IC performance and functionality. Integrating new high-speed, high-density, and low-power memory technologies onto the CMOS platform may break the memory limits, improve system performance, and enable new computing paradigms (e.g., in-memory compute). Novel switches based on new materials, alternative mechanisms, and non-charge state variables (e.g., steep-slope transistors, phase-transition devices, nonvolatile switches) may achieve significantly better energy efficiency than CMOS transistors based on thermionic emission over the channel energy barrier. New information processing technologies substantially beyond the capabilities of CMOS may originate from an innovative combination of new devices, interconnect, and architecture for extending CMOS, which may eventually lead to new platform technologies. Ultimately scaled CMOS as a platform technology may also extend into new domain of functionalities and applications (e.g., security, sensing, communication), a general direction known as "more than Moore". For all these promising directions beyond CMOS and dimensional scaling, CMOS scaling has driven a broad spectrum of applications through increased performance and complexity. As dimensional scaling of CMOS eventually approaches fundamental limits, new information processing devices and architectures have been explored extensively in both academia and industry. The Beyond CMOS chapter in the International Roadmap for Devices and Systems (IRDS) surveys, catalogs, and assesses viable emerging devices and novel architectures for their long-range potential, and identifies the scientific and technological challenges gating their acceptance by the semiconductor industry.

### 4.2.1 Historical Background

During the 1960s Complementary Metal-Oxide-Semiconductor technology introduced as CMOS for the first time thus starting its path towards the electronics revolution of today. The innovative technology gained popularity because of its minimal power requirements that exceeded those of traditional logic forms. The initial integrated circuit functioned using TTL technology though it delivered sufficient speed alongside low power use but semiconductor development outran its capability. The installation of transistors in semiconductor chips experienced a significant rise in both manufacturing efficiency and device speed and energy efficiency through improvements in production techniques. Modern technology based on nanometer dimensions creates both high-performance microprocessors along with elaborate integrated circuits that contain thousands of millions of CMOS transistors.

The semiconductor sector now considers FinFET (Fin Field-Effect Transistor) technology as their leading technology because it extends CMOS's capabilities for performance and efficiency. Modern electronic devices use CMOS technology to develop future generation systems which include AI processors along with high-performance computing equipment and 5G communication infrastructure.

CMOS demonstrates better power efficiency and broader scalability and superior performance than all other semiconductor technologies. CMOS technology represents the most suitable semiconductor choice because it enables dependable operation while using less power while minimizing heat generation when compared to TTL and BJT systems. The long-term operation without charging stands as a crucial benefit for mobile devices like smartphones and digital cameras which need to remain functional throughout an average day. Judge BJT devices for their speed capabilities but CMOS achieved superior metrics for large-scale integration which makes it the present foundation of LSI circuits [3].

CMOS technology maintains its position as an economical production method that provides efficient scaling for consumer electronics mass manufacturing even when compared to GaAs (Gallium Arsenide) which shows better electron mobility and high-frequency characteristics. Most consumer electronics now use CMOS-Based Mix-SigTech as their key analog and digital solution even though FinFET and SOI (Silicon-on-Insulator) emerged with improved traditional CMOS capabilities that operate below the microscopic scale. The ongoing significance of CMOS-Based Mix-SigTech stems from its basic structure combined with easy use and market support together with many possible applications covering digital and both analog and mixed-signal domains.

Very intensive research targeting the reduction of power and energy consumption of complementary metal-oxide-semiconductor (CMOS) electronics circuits and systems has been being conducted during the last decade. The need for ultra-low-power and energy-efficient circuits is due to different reasons. The impossibility of further increasing the system complexity because of power density problem of CMOS and the boom of portable devices working from batteries are the main ones. More recently, the arising of IoT or implantable medical devices which eventually use energy harvesting has boosted the activity in the area. The reduction of power and energy has been addressed from different domains, including both software and hardware approximations. Concerning the latter, methodologies and techniques at distinct abstraction levels are possible. At the circuit level, it is well known that reducing the supply voltage is a very powerful way of reducing power. In fact, supply voltage reduction enabled by the scaling of CMOS technologies has significantly reduced power consumption in the past. However, the further scaling of supply voltages to reduce dynamic power while maintaining adequate speed is counterbalanced by the exponential growth of leakage currents. Keeping speed with lower voltage supply requires lowering the threshold voltage of the transistors, which, because of the 60 mV/decade minimum subthreshold slope (*SS*) of CMOS devices, produces unacceptable off-state leakage currents. This has motivated intensive research into devices with *SS* below the physical limit of CMOS, known as steep slope devices, to achieve better transistor performance at reduced supply voltages [21–31].

Along with size scaling, switching power has been controlled with reductions in the supply voltage $V_{DD}$. However, steady state standby power increases rapidly with the reductions in $V_{DD}$ due to the thermionic emission of carriers over the source-to-channel barrier, so that CMOS voltage scaling is approaching an impasse. The "end-of-the-roadmap" estimates of the MOS field-effect transistor (MOSFET) threshold voltage $V_{Th}$ to maintain sufficient ON-state-current ($I_{ON}$)-to-OFF-state current ($I_{OFF)}$ ratios are 0.19 and 0.37 V for low operating and standby power applications, respectively, or about three and six times the minimum possible subthreshold slope for MOSFETs ($\ln(10)kBT/q$) of 60 mV/decade at room temperature $T = 300$ K. The switching times $\tau_s$, followed by Equation 4.1.0 for short-channel MOSFETs to charge or discharge the load capacitance values $C_L$ of interconnects and following transistor gates (approximated here as constants for simplicity) are roughly given by where $g_m$ is the transconductance. Thus, fast switching requires the overdrive voltages ($V_{DD} - V_{Th}$) of the same scale as $V_{Th}$, leading to the end-of-the-roadmap values of $V_{DD}$ of 0.5 and 0.7 V, respectively [25–31]. Even for high-performance applications requiring smaller ON/OFF ratios but also more speed, the limiting supply voltage appears to be 0.7 V.

$$\tau_S \cong \frac{V_{DD}C_L}{I_{ON}} \cong \frac{V_{DD}C_L}{g_m(V_{DD} - V_{Th})} \tag{4.1}$$

Novel device concepts are being widely studied to overcome scaling limitations of traditional planar bulk CMOS technologies, such as short-channel effects and leakage currents. One major issue with conventional MOSFETs is that carrier injection at the MOSFET source-channel junction is bound by the exponential tail of Fermi statistics because it is governed by thermionic emission over a potential barrier. This implies that the MOSFET inverse subthreshold slope (SS) cannot be reduced below 60 mV/decade at room temperature. However, power-dissipation increase in next-generation digital systems must be contrasted by reducing the voltage supply $V_{DD}$ [22–28].

MOSFET transistor researchers are exploring device structure and channel material changes to enable further generations of MOSFET scaling. The MOSFET implemented with stacks of multiple horizontal nanowires is one option that, due to its superior electrostatics, could enable further gate-length scaling beyond what the FinFET can achieve. MOSFETs with III-V semiconductor channel materials are a promising option for realizing a higher-mobility channel than silicon. This higher mobility can be used either to provide higher drive current and higher performance or to allow the MOSFET to be operated at lower voltage for lower active power.8 Lowering the supply voltage of CMOS logic below about 0.5 V leads to a dilemma between logic having high performance and high static leakage current versus logic with lower performance and low leakage current. This is due to the choice of MOSFET threshold voltage and its electron "thermal tail" determined sub-threshold gate voltage swing of 60 mV/decade. One alternative transistor option that operates differently than a MOSFET (and as such could be classified as a *beyond-CMOS device*) is the Tunneling Field Effect Transistor. The TFET can achieve subthreshold swing smaller than 60 mV/decade (that is, steeper current turn-on) and can therefore operate at a lower power supply voltage than a MOSFET.

### 4.2.2 Principles of CMOS Technology

The CMOS transistor integrates two complimentary MOSFETs which are built as NMOS and PMOS gates on top of a silicon substrate. Both transistor layers included three different components resulting in gate and drain and source features. Silicon dioxide ($SiO_2$) forms a thin insulation around the gate to enable control of source-drain current without electrical gate current passage [4]. When voltage of positive charge is applied through the gate of an NMOS transistor it allows current to pass between the n-doped source and drain terminals. This current travels from a p-doped source through drain only when the gate receives negative voltage in PMOS transistors. The complementary operation of NMOS with PMOS transistors enables power reduction by turning the remaining transistor off after the other one activates. The integration enables logic gate and digital circuit BUILD before creating an efficient and worthy CMOS platform. The operational attributes of N-channel and P-channel form an architectural basis for CMOS technology. A positive gate signal in NMOS devices generates a gate electric field that pulls electrons through the device channel which enables drain-to-source conduction. Current from drain to source will pass because of this construction which makes the NMOS perform extremely well in logic circuits for output ground. Connections through PMOS transistors just occur under gate conditions with negative voltage which attracts holes.

Integrated circuit (IC) chips represent one of the most revolutionary inventions of the modern era, serving as the backbone of virtually every electronic device we use today. CMOS play an important role in IC sector development. From smartphones to complex systems such as in the auto-driving cars and space probes, IC chips have played an important role in making modern technology possible. These tiny but incredibly sophisticated components can solve many problems, from basic calculations to complex data processing, communication, and control, covering countless applications in consumer electronics, healthcare, automotive, and aerospace industries.

The process of IC manufacturing began in the 1960s, with the development of planar transistor technology, which made it possible to shrink electronic components to previously unimaginable sizes. This breakthrough boosted decades of explosive growth in the semiconductor industry, mainly because of the famous Moore's Law. This Law has predicted that the number of transistors that can fit on a computer chip doubles approximately every two years. What makes this trend so impressive is not only its accuracy, but also its consistency over a long period.

The shrinking of transistors to dimensions below 100 nm enables hundreds of millions transistors to be placed on a single chip. The increased functionality and reduced cost of large variety of integrated circuits and systems has brought its own benefit to the end users and above all the semiconductor industry. A low cost of manufacturing, increased speed of data transfer, computer processing power and the ability to accomplish multiple tasks simultaneously are some of the major advantages gained as a result of transistor scaling. The packing density of IC on microelectronics roughly estimated with the help of Moore's Law (an empirical compact mathematical formulation;

$$N = \frac{D^2 P_E}{F^2}.$$

In this empirical formula $N$ is the number of transistors per chip, $F$ is the minimum feature size, $D$ is the chip area, and $P_E$ is transistor packaging efficiency measured per minimum feature area). This empirical formula plays a crucial for observation on the rate of growth of semiconductor technology.

## 4.3 Advancements in Semiconductor Devices

Although the IC chip industry has traditionally relied on flat and two-dimensional chip designs, the whole industry is changing from two-dimensional (2D) technologies to three-dimensional (3D) integrated circuit technologies now. Instead of spreading components across a single flat surface, these new 3D technologies stack multiple layers of active components on top of each other, like building a skyscraper instead of a sprawling ranch house. Engineers can use this vertical approach to fit more transistors and other components into the same footprint and also reduce the distance that electrical signals need to travel, which can significantly improve performance semiconductor devices.

However, this transition to 3D technologies brings a series of relevant challenges. For example, producing these multi-layered structures requires an incredibly precise control system and the ability to work with different materials with properties and behaviors, respectively. Additionally, stacking components vertically causes new problems with heat management and power distribution.

### 4.3.1 Thermal Transistors

Semiconductor devices focus on innovation through the creation of heat-based transistors. Thermal transistors operate through temperature gradients instead of traditional electrical signal-based switching mechanisms of traditional transistors. The transduction method depends on the thermoelectric effect to generate electrical voltage when a substance experiences temperature gradients. The new generation of devices shows great potential to improve heat removal capabilities and could transform heating management in electronic systems especially in the fields of power electronics and high-performance computing systems. The use of heat for control purposes in thermal transistors enables potential achievements in energy-efficient systems. Figure 4.4 shows the developed thermal transistor.

Rapid thermal annealing (RTA) serves as a vital process for semiconductors due to its role in fulfilling International Technology Roadmap for Semiconductors (ITRS) future technology specifications. Rapid thermal annealing (RTA) finds its most important application in USJ formation to address fabrication needs of deep submicron CMOS devices in next-generation technology [5]. Quick thermal annealing processes work together to prevent dopant deactivation from heating and to protect junction widths from transient enhanced dopant diffusion.

The preferred manufacturing process includes low-energy beamline ion implantation that proceeds with RTA treatment. Mainstream processes have relied on this approach yet its limitations will prevent its use in producing the tiny junctions needed for emerging technological advances as shown in Figure 4.2. Cutting ion energy below 1 keV results in beam transport limitations caused by space charge effects that reduce factory production rates.

Tungsten–halogen lamp spike RTA functions as the current industry standard for annealing yet experts suggest that its standard 1-second heating period needs further improvement for ultra-shallow implant profile control. Various alternative implantation and ultra-rapid heat treatment approaches have been researched in recent times. Scientists now explore the potential of mixing plasma doping with flash lamp annealing as an effective solution.

Semiconductor device scientist has developed a technique to not only manage heat but help route it away from delicate devices. In this research approach, a nanoscale thermal transistor has been trying to develop for years as shown in Figure 4.3. The circuit symbol of the thermal transistor device in Figure 4.3. Previous thermal transistors proved too big, too slow and not sensitive enough for practical use. The challenge has been finding a nanoscale technology that could toggle on and off repeatedly, have a large hot-to-cool switching contrast and no moving parts. The thermal transistor design starts with a thin layer of molybdenum disulfide, a semiconducting crystal that is made up of layered sheets of atoms. Just 10 nm thick and effective at room

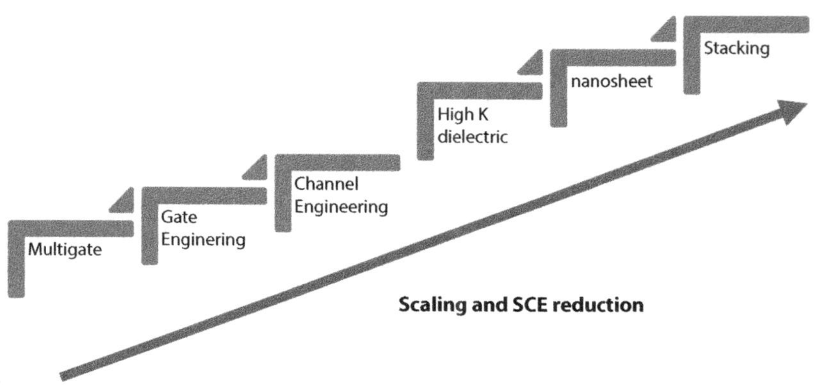


**Figure 4.2** Innovations in semiconductor devices.

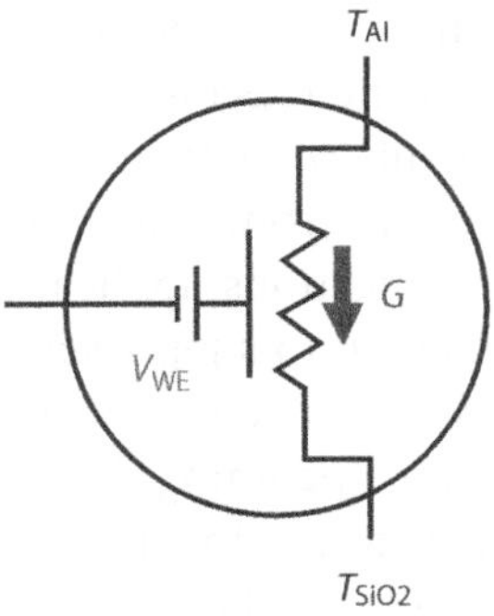


**Figure 4.3** Circuit symbol of the thermal transistor device: The gating voltage $V_{WE,}$ applied between Li and Al/$MoS_2$ electrodes, changes the thermal conductance G to heat flow between the Al transducer and $SiO_2$ substrate, which are at temperatures $T_{Al}$ and $T_{SiO2}$ respectively.

temperatures, this material could be integrated into ordinary semiconductors, a critical factor in making the technology practical.

Specifically, the working device is a 10-nm-thick layered $MoS_2$ crystal, prepared on $SiO_2$ (90 nm) on p-type Si. An 80-nm Al layer patterned on top of the $MoS_2$ serves as an electrical contact and as an opt thermal transducer for measurements. A solid Li pellet acts as the reference and counter electrodes, and 1.0 M $LiPF_6$ in ethylene carbonate/diethyl carbonate is the liquid electrolyte. Researchers were able to modulate thermal conductance reversibly in the device by a factor of nearly 10×, on time scales of minutes. The researchers envision that thermal transistors connected to computer chips would switch on and off to help limit the heat damage in sensitive electronic devices.

### 4.3.2 Advanced Materials

Semiconductor technology advances directly because of meaningful material development progress. Researchers discover great interest in graphene because this flat hexagonally arranged carbon atom layer has exceptional electrical and mechanical properties. Graphene performs well as an electrically conductive material because it demonstrates both outstanding strength and high flexibility.

Transistors using this material are positioned to deliver faster performance at higher energy efficiency. The semiconductor device functionality grows better because technologies like Gallium nitride (GaN) and silicon carbide (SiC) serve to boost device performance in applications involving enhanced power and frequency capabilities [6]. The materials enable devices to function effectively from elevated voltage and temperature conditions which suits power electronics and telecommunications industries.

The increasing requirements of current electronics demand advanced semiconductor materials for electronics as well as energy systems and scientific research and communications applications. New materials below silicon and germanium emerged to supply semiconductor manufacturers with properties that can drive advanced performance and boosted efficiency while creating capabilities for upcoming devices. The three fundamental compound semiconductor materials include the semiconductor combinations of gallium arsenide (GaAs) and gallium nitride (GaN) together with indium phosphide (InP). The various electrical properties including elevated electron movements and expanded bandgap edges of these materials make them exceptionally suited for high-speed high-frequency and high-power devices. The telecommunications industry utilizes GaAs primarily in microwave and RF applications including satellite communications whereas the power and light generation industry chooses GaN for its optimal performance in LED technology and power electronics.

The category of advanced materials also includes wide-bandgap semiconductors which include silicon carbide (SiC) and diamond. Elaborate materials offer superior capabilities to operate at intense voltages and elevated temperatures and high-frequency applications than silicon does so they enable crucial performance in electric vehicle setups (EVs) and power grid management and renewable energy deployment. SiC has entered power electronics systems for EV applications and delivers improved charging efficiency and accelerated charging speed. The field of electronics is emerging with promising new candidates consisting of 2D materials which include graphene alongside transition metal dichalcogenides (TMDs) [7].

The excellent electrical capability and elastic properties of graphene make it suitable for transforming sensor technology alongside transistor development and flexible electronic device production. Industrial innovations result from the active advancement and implementation of these superior materials which produces devices that are faster and smaller yet more energy efficient. The role of these materials remains crucial in developing advanced quantum computing systems because they also advance telecommunications-based technologies and renewable energy sectors to build the electronics landscape of tomorrow.

### 4.3.3 3D Integrated Circuits

Nowadays, the semiconductor industry is approaching a limit to the number of transistors that can be packed onto the surface of a computer chip.

So, chip manufacturers are looking to build up rather than out. Instead of squeezing ever-smaller transistors onto a single surface, the industry is aiming to stack multiple surfaces of transistors and semiconducting elements — akin to turning a ranch house into a high-rise. Such multilayered chips could handle exponentially more data and carry out many more complex functions than today's electronics [32–39].

In the landscape of integrated circuits (ICs) design science and technology, semiconductor industry has experiencing a profound transformation. With the physical and economic limits of conventional two-dimensional scaling, the industry is rapidly embracing three-dimensional integrated circuits (3D IC). In conventional ICs technology, the IC component exists as two-dimensional structures because they distribute their components throughout a single flat surface. The 3-D view of the thermal transistor is shown in Figure 4.4. The vertical design of 3D ICs combines various component layers to minimize component distances which results in superior performance. In 3D Integrated Circuits (ICs) technology stacks multiple chips vertically, connecting them with tiny wires like through-silicon vias (TSVs) to create a single, powerful device. The 3D Integrated Circuits (ICs) technology overcoming limitations of traditional 2D chips by shrinking footprint, boosting performance, reducing latency, and enabling heterogeneous integration (mixing different types of chips like logic and memory).

**Figure 4.4** 3-D view of developed thermal transistor.

This advanced packaging allows for smaller, faster, and more power-efficient electronics, crucial for AI, mobile devices, and data centers. The shorter interconnects mean lower latency, higher bandwidth, and faster speeds. The vertical stacking dramatically reduces the physical space of wafer needed for fabrication. The shorter signal paths require less energy to transmit data [32–39].

The 3D chip stacking offers numerous advantages, it also presents challenges such as thermal management, manufacturing complexity, and cost considerations. It helps managing heat dissipation in densely packed chips and ensuring reliable interconnects is among the key challenges. The adoption of 3D chip stacking is expected to continue growing as semiconductor manufacturers seek ways to improve performance, reduce power consumption, and meet the demands of emerging technologies such as artificial intelligence, 5G, and edge computing. Ongoing research and development efforts are focused on addressing existing challenges and further optimizing this technology for future applications [32–40].

### 4.3.4 Technological Trends

#### *4.3.4.1 Integration of Artificial Intelligence (AI)*

The incorporation of artificial intelligence capabilities into semiconductors drives forward numerous industrial sectors toward their future direction [8]. Current technology allows AI algorithms to operate directly in chips which enables instant data processing and decision sequences. The vehicle uses AI-enabled chips to evaluate and process significant sensor data in real-time thus ensuring safe navigation and efficient operation. AI applications spread throughout healthcare services along with industrial automation and consumer electronics enable better device capabilities as well as reduced power usage and expanded device capabilities.

##### 4.3.4.1.1 Semiconductor Fabrication

Semiconductor fabrication receives substantial impact from AI technology because precision alongside efficiency matter significantly at this stage [9]. Manipulations of photolithography and etching and deposition processes become more effective through AI algorithms which evaluate real-time information to change parameters and enhance productivity and reliability. The detection of microscopic anomalies and prediction of manufacturing defects form two core functions of deep learning models when applied to wafer testing. Semiconductor production costs decrease as the precision improves while device reliability levels increase.

##### 4.3.4.1.2 Performance Prediction and Process Control

Semiconductor materials undergo electrical and optical performance and thermal evaluation through AI models available across various examination scenarios [10]. AI accomplishes performance prediction through atomic-level simulation that shows how temperature changes as well as pressure and doping affect semiconductor material performance. Semiconductor optimization for applications such as high-frequency communication systems and power electronics along with quantum computing becomes possible through this method.

##### 4.3.4.1.3 AI-Driven Innovations in Device Integration

The deployment of advanced semiconductor devices requires the essential use of AI for both floor plan and advanced fabrication integration. Artificial intelligence algorithms maximize system-level performance while helping to create better integrated circuits and memory chips as well as processors. The AI-driven ASIC development platform speeds up the process of designing custom processors that specialize in AI and machine learning operations.

#### *4.3.4.2 Internet of Things (IoT)*

High growth in Internet of Things devices depends fundamentally on the utilization of semiconductor technology. Specialized low-power yet high-performance chips are essential for efficient operation of IoT devices that include smart home appliances together with industrial sensors [11]. The combination of microcontrollers and system-on-chips (SoCs) powers devices for mutual communication using wireless networks which results in forming linked ecosystems. The production of chips represents an innovative driving force behind the advancement of smart cities together with smart homes and smart industries which in turn bring enhanced operational efficiency and safety while delivering convenience.

#### *4.3.4.3 5G Technology*

High-frequency chip development closely follows the deployment of 5G networks because such networks necessitate chips able to transmit data at superior rates and through higher frequency ranges. The semiconductor industry works on developing GaN materials alongside unique chip designs to fulfill 5G technology requirements. 5G-compatible chips allow the next generation of technology to use virtual reality (VR), augmented reality (AR) and self-driving vehicles while enhancing both mobile communication systems and manufacturing along with healthcare and logistics sectors through their fast and dependable connections [12].

### 4.3.5 Manufacturing Techniques

#### *4.3.5.1 Advanced Packaging*

The industry requires new packaging methods to support semiconductor devices running complex operations at high performance rates. System-in-package (SiP) technology represents an innovation that combines several chips inside a single package [13]. The approach enables better component communication while saving power thus making it best suited for mobile devices as well as high-performance computing systems. Chip let design technology recently emerged as an effective approach because it allows manufacturers to combine chips into modular units with improved flexibility.

#### *4.3.5.2 In-House Chip Design*

Presently numerous technology firms switch to designing their own processing chips because it helps them direct their product development processes and establish company-specific product differences [14]. Three major technology companies Apple, Google and Amazon dedicate significant resources to design their proprietary chips which specifically cater to their unique application needs. Through in-house design organizations obtain the ability to create specialized high-performance chips while shortening product development timelines which reduces their time-to-market.

#### *4.3.5.3 Sustainable Manufacturing*

Sustainability has risen to prominence in the semiconductor sector since its manufacturers address the rising environmental challenges in their production methods. Manufacturers currently work on multiple efforts to cut down energy use and waste production during chip manufacturing. The manufacturing process relies on sustainable implementation of three methods: renewable energy consumption, material recycling and minimal water and chemical substance usage. Such critical initiatives maintain semiconductor industry durability while electronics markets continue their expansion.

## 4.4 Applications of Semiconductor Devices

### 4.4.1 Consumer Electronics

Semiconductor materials work well with adjustments because they are vital components in electronic devices. Consumer electronics have developed due to semiconductor technological advancements [15]. New generation smartphones together with tablets and laptops along with various electronic devices achieve superior performance attributes and extended battery utilization and additional features through advanced semiconductor methodologies. The essential role of semiconductors exists at factory-smart electronics because they offer powerful processors and AI-powered sensors with computing capacity and processing speeds needed for innovation. The increasing consumer need for advanced and slim electronic products drives semiconductor technology developers to innovate at its highest potential. The material exists across many commercial products including automobiles and electronic calculators. All industries ranging from computer production to aerospace utilize semiconductor devices as part of their operations. These industries also include communication businesses along with manufacturing facilities and agricultural and healthcare sectors. The technological revolution in music players and LED TVs along with DVD players and laptops and smartphones became possible due to semiconductor technology which produced small cost-effective rapid and dependable electronic devices. Semiconductors serve essential roles in scientific progress alongside their functions that sustain the operations of industrial businesses.

All active circuit components which include both diodes and transistors come from semiconductor materials in electronic devices. Pure semiconductor materials function as materials that do not behave either as electrically conducting or electrically insulating. The conducted level of electric current depends on how conductive a material works. The following section of this chapter introduces a model describing semiconductor current movement alongside an explanation of electrical conductivity principles. The electrical flow through conductor metals such as copper remains fast because they exhibit high conductivity levels at low voltage application. Electrical current may not occur at all during voltage applications when the insulator material mica demonstrates its low conductivity behaviour.

Because of their intermediate position regarding conductivity semiconductors serve as the preferred materials for creating semiconductor devices which engineers refer to as solid-state devices. The most essential semiconductor devices used in automotive electronics consist of diodes and transistors. Modern integrated circuits function as the collective platform created through these devices. Our examination centres on the discussion of these two device types. Semiconductor devices exist as silicon or germanium materials although manufacturers add gallium arsenide to select devices during production. Deliberately adding impurities to semiconductor materials enables alteration of their conductivity rates. Purified semiconductors can be controlled through planned introduction of precise impurity amounts.

### 4.4.2 Digital Applications

The basic components used in digital technology development rely on CMOS devices for their operation. ELECTRONIC systems in the present age utilize CMOS technology through its vital application [16]. Microprocessors mainly select CMOS devices because these devices offer great integration capabilities and minimize power usage. SRAM and DRAM memory technologies base their operation on these components since they exist at the core of modern computers as well as portable devices. The digital logic circuit design that implements fundamental tasks of addition and data processing uses CMOS technology for control system management functions. High-definition photos and videos from digital cameras and smartphones are possible because of CMOS sensors that operate on low power consumption. Communication devices harness this technology for power management because it supplies the required energy for transmitting data signals and processing electrical signals. The range of digital challenges undergo improvement through these devices which enable higher efficiency and support innovation for different modern devices.

### 4.4.3 Analog Applications

CMOS devices function as key components in various analog applications because they take analog signals effectively with low power usage. CMOS technology delivers accurate digital-to-analog signal coexistence in ADCs and DACs therefore making them vital components for audio-video-telecommunications systems [17]. Operational amplifiers constructed from CMOS materials perform various functions during signal correction and amplification tasks by delivering outstanding performance in reduced packages. CMOS technology serves as a fundamental element in mixed-signal circuit development because it makes sensor interfaces and RF transceivers function efficiently with analog and digital signals combined. Battery-powered portable medical devices together with wearable technology benefit from CMOS devices due to their low operating voltage specifications. The analog applications of CMOS technology improve the overall performance in a wide range of electronic systems dedicated for consumer and industrial usage.

### 4.4.4 Mixed Signal Applications

The integration of mixed-signal applications benefits from CMOS devices because they connect analog to digital operations within a single electronic device [18]. The capability of bridging digital analog signals enables processing engines to work together with analog components through CMOS technology thus making it critical for data converters and signal processing and communication system applications. In analog-to-digital converters (ADCs) and digital-to-analog converters (DACs), for instance, CMOS technology facilitates accurate signal conversion, critical for audio and video systems, as well as sensor-based devices. Through CMOS technology people can develop dense integrated systems that use phase-locked loops for frequency synthesis purposes in telecommunications. MOS devices known as complementary metal-oxide-semiconductors operate at low power consumption while achieving high integration density and this combination makes them ideal for current technology developments.

### 4.4.5 Power Management Applications

The key requirement for prolonging battery life depends on using energy efficiently and effectively. CMOS devices have applications in such situations. The power distribution system of electronic devices relies on CMOS technology which governs power management integration in PMICs [19]. The regulated circuits maintain power for numerous devices including industrial machines along with mobile gadgets which need correct voltage potentials. The application of CMOS power amplifiers and low-dropout regulators becomes standard when operating equipment with sensitive components which need precise voltage regulation without noise interference. The technology enables the development of circuits which harvest surrounding environmental energy to generate usable power thus enhancing total efficiency levels. CMOS technology enables low static power usage together with higher power capacity in devices and therefore improves both smart grid technology and renewable energy system functionality. The implementation yields sustainable energy practices that are followed. Power management achieved through CMOS technology ensures its wide application across electronic devices which makes them convenient to handle while extending their operational lifespan.

### 4.4.6 Automobile Sector

Semiconductor devices brought an extensive transformation to the automotive industry. Modern vehicles exist as mobile computing devices which contain various semiconductor-controlled systems that regulate the functioning of both motors and safety protection mechanisms. Electric vehicles and autonomy in automobiles increased the necessity for semiconductors to become critical components [20]. The electronic management of energy in electric vehicles depends on power electronics and AI-powered chips provide real-time sensor evaluation for autonomous driving alongside ADAS capabilities.

### 4.4.7 Healthcare

Healthcare technology continues to advance through the essential role that semiconductors provide in its improvement [19, 20]. The precise monitoring of human health and advanced diagnostic equipment both utilize semiconductor-based devices for providing specific healthcare solutions. Pakistan Medical Research Council initiated a project to screen pregnant women for possible vitamin D deficiency through an assessment of their dietary behaviour making selection tactics much more accurate. Semantic devices allow doctors to create AI medical systems that analyze substantial medical data to support diagnostic diagnosis and treatment development.

## 4.5 Future Directions

### 4.5.1 Research Trends in CMOS Technology

The main successful effort is in innovating new transistor designs such as Fin FETs and gate-all-around (GAA) structures to improve the electrostatic control for lower leakage current [20]. Two new materials families of interest include two dimensional materials substances, for example of metal (cobalt) sandwiched with another substance such as transition metal dichalcogenides to enhance device performance and enable further scaling beyond conventional silicon. This emerging trend is to use machine learning, artificial intelligence to optimize design of manufacturing processes, to optimize the constructed circuit performance and to reduce R&D duration. A prolonged study on energy efficiency leads to better power management methodologies and low power circuit techniques that are being prioritized for addressing power constraints. The new markets will emerge due to application of such devices and neuromorphic computing which requires specialized mixed signal designs and dedicated CMOS architectures for specific functionality. However, all these research trends together try to extend the CMOS technology limits so that it is effective in ever changing electronics domain.

With the ongoing demand for compact and energy-efficient electronic devices, traditional planar CMOS technology faces significant challenges due to increased power leakage and area constraints at nanoscale dimensions. The integration of FinFET-based CMOS technology into very large scale integration (VLSI) design has revolutionized the semiconductor industry by addressing power and area limitations. This research proposes a novel FinFET-based CMOS architecture employing multi-threshold voltage (MTCMOS) and dynamic voltage scaling (DVS) techniques to optimize power consumption and chip area [41–45].

The increasingly high standard of electronics, the semiconductor material science and semiconductor manufacturing have been booming in the last few decades, with massive data accumulated in both fields. If analyzed effectively, the data will be conducive to the discovery of new semiconductor materials and the development of semiconductor manufacturing. Fortunately, machine learning, as a fast-growing tool from computer science, is expected to significantly speed up the data analysis. Machine learning (ML) is poised to play an important part in advancing the predicting capability in semiconductor device compact modeling domain. One major advantage of ML-based compact modeling is its ability to capture complex relationships and patterns in large datasets [46–50].

Machine learning refers to algorithms and statistical models that allow computers to "learn" from data, identify patterns, and make predictions or decisions based on that data without being explicitly programmed. In the context of semiconductor materials, ML models can analyze large sets of data from experiments, simulations, and manufacturing processes to identify material properties, optimize material selection, and predict performance characteristics [41–50].

The integration of artificial intelligence (AI) and machine learning (ML) into Very Large Scale Integration (VLSI) design and manufacturing is revolutionizing the semiconductor industry. As VLSI circuits become increasingly complex, traditional design and verification methodologies face significant challenges. The AI and ML offer promising solutions to enhance design automation, optimize performance, reduce power consumption, and accelerate time-to-market. This article provides a comprehensive analysis of the opportunities, key algorithms, and emerging trends in applying AI and ML techniques within the VLSI domain [45–54].

### 4.5.2 Potential for New Applications

With frequent advances in technology and intensifying need for finding new solutions at multiple frontiers, the employability scope of CMOS devices is broad and growing. CMOS is also being used in AI and within machine learning, where CMOS is being used to make special purpose hardware which increases processing power and enables the real time analysis of data. As backbone of the specification of smart sensors and ultralow power communication devices in IoT, CMOS technology holds large use for seamless connectivity and transfer of data from smart homes to industrial processes. As ADAS and autonomous vehicles become increasingly common, the automotive industry is now heavily using CMOS devices while reliability and high efficiency are very important for handling massive amount of sensor data. Other emerging areas include biomedical engineering for example, CMOS system is being studied to use in wearable health monitoring devices and diagnostic tools partially because of low power requirement and high integration. Meanwhile, the CMOS technology innovation will open the door for a variety of avenues across the board ranging from consumer electronics to healthcare and beyond.

### 4.5.3 Environmental Impact and Sustainability

As the demand for electronics increases, environmental impact and sustainability issues of CMOS devices also become more important. Although CMOS technology is highly efficient energy-wise during operation, the manufacturing process may consume a lot of resources, and it is characterized by significant energy consumption and the use of hazardous material. By doing less harmful chemicals use and materials recycling, it improves sustainability because it promotes greener manufacturing practices. But mostly the designing the electronic product with all the energy efficiency decreases the carbon foot print of the electronic products most especially during the rise of their use in IoT and AI applications. Another important aspect in sustainability is the longevity and robustness of CMOS devices; in this case products are desired to have a higher lifetime making the US apparent smaller as less replacements are needed. However, the developments in semiconductor technologies, for instance, energy harvesting circuits and low power design, aim at improving the environmental benefits from CMOS devices. It is both welcomed, and asked for, the tech industry to transform into more environmentally sustainable practices, and for the promptly changing prospects for the still growing CMOS technology field, all in the sake of the environment's future and electronics' future innovation.

## 4.6 Conclusion

Semiconductor devices are used everywhere and the application range grows even larger whenever new technologies and opportunities come up for further development and solutions. CMOS is a key player in creating the specialized hardware requisite for the speed required in training and operation of processors in AI and machine learning to provide real time data processing. In the context of Internet of Things (IoT), CMOS technology facilitates the evolution of smart sensors and ultra-low power communication devices which forms the bedrock for the sensor to sensor and sensor to info system communication. In the automotive industry, CMOS devices are gaining more usage in Advanced Driver Assistance Systems (ADAS) and self-driving cars, where dependability and high efficiency is very important especially for the handling of large amount of sensor data. Outer applications such as wearable health monitoring or diagnostics are also likely to use CMOS technology in the field of biomedical engineering due to low power and high integration. Moreover, as the research further progresses, CMOS technology should enable further development for new solutions across a whole host of industries and consumer electronics as well as healthcare.

**AI disclosure statement:** The author acknowledges the use of AI-based tools for minor language editing or grammatical correction. However, all intellectual contributions, analyses, and conclusions presented in this chapter are entirely the author's own.

## References


1. Harada, S., *et al.*, Non-destructive analysis of threading mixed dislocations in SiC using x-ray topography and birefringence. *AIP Adv.*, 15, 2, 2025.
2. Rai, H., Singh, K.R., Natarajan, A., Pandey, S.S., Advances in field effect transistor based electronic devices integrated with CMOS technology for biosensing. *Talanta Open*, 100394, 2024.
3. Mukherjee, S., Pal, D., Bhattacharyya, A., Roy, S., 28 Future of the Semiconductor Industry, in: *Handbook of Semiconductors: Fundamentals to Emerging Applications*, p. 359, 2024.
4. Yeboah, L.A., Oppong, P.A., Malik, A.A., Acheampong, P.S., Morgan, J.A., Addo, R.A.A., Henyo, B.W., Exploring Innovations, Sustainability and Future Opportunities in Semiconductor Technologies. *Int. J. Adv. Nano Comput. Anal.*, 3, 2, 01–42, 2024.

5. Arooj, N., Muneer, I., Ali, D., Ahmad, S., Bashir, F., Hussain, T., Examining enhanced electrochemical performance through analysis of charge storage mechanisms in WO3/Bi2MoO6/rGO nanocomposites. *Mater. Sci. Eng., B*, 306, 117425, 2024.
6. Nam, D. and Choi, G., The identification of emerging technologies of automotive semiconductor. *KSII Trans. Internet Inf. Syst. (TIIS)*, 17, 2, 663–677, 2023.
7. Chen, J., Integrated circuit design based on CMOS technology principle and its application in GPU. *Theor. Nat. Sci., 12*, 141–146, 2023.
8. Lee, C. and Zhao, Y., Advanced CMOS Devices and Applications. *Electronics*, 13, 1, 134, 2023.
9. Hsu, C.Y., Rheima, A.M., sabri Abbas, Z., Faryad, M.U., Kadhim, M.M., Altimari, U.S., Kianfar, E., Nanowires properties and applications: a review study. *S. Afr. J. Chem. Eng.*, 2023.
10. Rafiq, J., Singh, H., Pandey, H., Srivastava, A., Awasthi, D., Design and Fabrication of Multifunctional, Portable and Economical Agriculture Machine. *J. Prod. Ind. Eng. (JPIE), 4*, 1, 2023.
11. Mathur, D., IEEE–Advancing Technology for Humanity, 2022.
12. Bespalov, V.A., Dyuzhev, N.A., Kireev, V.Y., Possibilities and limitations of CMOS Technology for the production of various Microelectronic systems and devices. *Nanobiotechnol. Rep., 17*, 1, 24–38, 2022.
13. Kim, H., Recent Trends in Copper Metallization. *Electronics*, 11, 2914, 2022.
14. Zhao, Y., Gobbi, M., Hueso, L.E., Samorì, P., Molecular approach to engineer two-dimensional devices for CMOS and beyond-CMOS applications. *Chem. Rev., 122*, 1, 50–131, 2021.
15. Vandana, B., Patro, B.S., Das, J.K., Mohapatra, S.K., Tripathi, S.L., Emerging Trends in Nanoscale Semiconductor Devices, in: *Advanced VLSI Design and Testability Issues*, pp. 111–127, 2020.
16. Fu, Q., Kokalj, D., Stangier, D., Kruis, F.E., Tillmann, W., Aerosol synthesis of titanium nitride nanoparticles by direct current arc discharge method. *Adv. Powder Technol., 31*, 9, 4119–4128, 2020.
17. Chaudhury, S. and Sinha, S.K., Carbon nanotube and nanowires for future semiconductor devices applications, in: *Nanoelectronics*, pp. 375–398, Elsevier, 2019.
18. Zhuang, H., Yang, N., Zhang, L., Fuchs, R., Jiang, X., Electrochemical properties and applications of nanocrystalline, microcrystalline, and epitaxial cubic silicon carbide films. *ACS Appl. Mater. Interfaces, 7*, 20, 10886–10895, 2015.
19. Uranga, A., Verd, J., Barniol, N., CMOS–MEMS resonators: From devices to applications. *Microelectron. Eng.*, 132, 58–73, 2015.
20. Kim, Y.B., Challenges for nanoscale MOSFETs and emerging nanoelectronics. *Trans. Electr. Electron. Mater.*, 11, 3, 93–105, 2010.
21. Nikonov, D.E. and Young, I.A., Overview of Beyond-CMOS Devices and a Uniform Methodology for their Benchmarking. *Proc. IEEE*, 101, 12, 2498–2533, 2013.

22. Tyagi, M.S., *Introduction to semiconductor materials and devices.* John Wiley & Sons, 2008.
23. Mishra, U.K. and Singh, J., *Semiconductor device physics and design,* vol. 83, Springer, Dordrecht, 2008.
24. Sze, S.M., *Semiconductor devices: physics and technology.* John Wiley & Sons, 2008.
25. Sze, S. M., Li, Y., Ng, K. K., *Physics of semiconductor devices.* John Wiley & Sons, 2021.
26. Colinge, J.P. and Colinge, C.A., *Physics of semiconductor devices.* Springer US, Boston, MA, 2002.
27. Neamen, D., *An introduction to semiconductor devices.* McGraw-Hill, Inc., 2005.
28. Colinge, J.P., Multi-gate soi mosfets. *Microelectronic Engineering,* 84, 9–10, 2071–2076, 2007.
29. Moore, G., Progress in Digital Integrated Electronics. *IEEE Int'l Electron Devices Meeting Tech. Dig.,* 11–13, 1975.
30. Dennard, R. *et al.,* Design of Ion-Implanted MOSFETs with Very Small Physical Dimensions. *IEEE J. Solid State Circuits,* 9, 5, 256–268, 1974.
31. Avci, U.E. *et al.,* Energy Efficiency Comparison of Nanowire Heterojunction TFET and Si MOSFET at Lg 5 13 nm, Including P-TFET and Variation Considerations. *IEEE Int'l Electron Devices Meeting Tech. Dig.,* 33–36, 2013.
32. Aoyagi, M. *et al.,* Wide bus chip-to-chip interconnection technology using fine pitch bump joint array for 3D LSI chip stacking, in: *2012 2nd IEEE CPMT Symposium Japan,* Kyoto, Japan, pp. 1–4, 2012, doi: 10.1109/ICSJ.2012.6523418.
33. Lu, J. *et al.,* Oxide-based CMOS Logic Circuits Enabled by Monolithically 3D Stacked SnOx p-FET and IGZO n-FET with Matched Threshold Voltage, in: *IEEE Electron Device Letters,* doi: 10.1109/LED.2025.3635160.
34. Rahi, S. B., 3D Stacked IC Chip Design: From CMOS to 2D Materials. In *Handbook of Emerging Materials for Semiconductor Industry,* pp. 839–849, Springer Nature Singapore, Singapore, 2024.
35. Jia, X. *et al.,* A Reliable ESD 3D-Integrated Design and Simulation (3D-IDS) Methodology for Wafer on Wafer Stacked DRAM, in: *IEEE Transactions on Computer-Aided Design of Integrated Circuits and Systems,* doi: 10.1109/TCAD.2025.3607152.
36. Liu, P. *et al.,* Leveraging Tensor Dataflow for Improved Thermal Performance on 3D-Stacked SRAM Architecture, in: *IEEE Transactions on Computer-Aided Design of Integrated Circuits and Systems,* doi: 10.1109/TCAD.2025.3607137.
37. Feng, H., Tang, G., Zhang, X., Lau, B. L., Jong, M. C., Two-phase Liquid Cooling of Vertically Stacked High-Power Chips with Backside-Embedded Micro-Pin Fins, in: *IEEE Transactions on Components, Packaging and Manufacturing Technology,* doi: 10.1109/TCPMT.2025.3631836.
38. Goel, E., Pandey, A., Rahi, S. B., Samuel, A. (Eds.), *Handbook of Advanced Semiconductor Field Effect Transistors.* John Wiley & Sons, 2025.

39. Khan, N. H., Alam, S. M., Hassoun, S., System-level comparison of power delivery design for 2D and 3D ICs, in: *2009 IEEE International Conference on 3D System Integration*, San Francisco, CA, USA, pp. 1-7, 2009, doi: 10.1109/3DIC.2009.5306539.
40. Kim, K.S., Seo, S., Kwon, J., Lee, D., Kim, C., Ryu, J.E., ... Kim, J., Growth-based monolithic 3D integration of single-crystal 2D semiconductors. *Nature*, 636, 8043, 615–621, 2024.
41. Wang, Y., Peng, Z., Wu, Z., Yuan, Z., Miao, X., Wang, X., Real-Time Optimization Design of GAAFETs Based on Machine Learning Method, *2023 International Symposium of Electronics Design Automation (ISEDA)*, Nanjing, China, 2023, pp. 414–417, doi: 10.1109/ISEDA59274.2023.10218718.
42. Butola, R., Li, Y., Kola, S.R., A Comprehensive Technique Based on Machine Learning for Device and Circuit Modeling of Gate-All-Around Nanosheet Transistors, in: *IEEE Open Journal of Nanotechnology*, vol. 4, pp. 181–194, 2023, doi: 10.1109/OJNANO.2023.3328425.
43. Gunter, D. and Wilton, S., Machine Learning VLSI CAD Experiments Should Consider Atomic Data Groups, *2024 ACM/IEEE 6th Symposium on Machine Learning for CAD (MLCAD)*, Salt Lake City (Snowbird), UT, USA, 2024, pp. 1–8, doi: 10.1109/MLCAD62225.2024.10740199.
44. Martins, R. and Lourenço, N., An Efficient Performance-Driven Analog IC Placement Optimizer *Via* Extremely Randomized Tree-Based Post-Layout Performance Regressors, *2024 IFIP/IEEE 32nd International Conference on Very Large Scale Integration (VLSI-SoC)*, Tanger, Morocco, 2024, pp. 1–4, doi: 10.1109/VLSI-SoC62099.2024.10869513.
45. Kumar, A., *et al.*, ML-based Fast On-Chip Transient Thermal Simulation for Heterogeneous 2.5D/3D IC Designs, *2022 International Symposium on VLSI Design, Automation and Test (VLSI-DAT)*, Hsinchu, Taiwan, 2022, pp. 1–8, doi: 10.1109/VLSI-DAT54769.2022.9768082.
46. Gambra, A., Rioja, U., Chatterjee, D., Armendariz, I., Batina, L., Machine Learning Fault Injection Detection in Clock Signals: An Analysis of Frequency Impact, *2025 IEEE Computer Society Annual Symposium on VLSI (ISVLSI)*, Kalamata, Greece, 2025, pp. 1–6, doi: 10.1109/ISVLSI65124.2025.11130208.
47. Ko, K., Lee, J.K., Kang, M., Jeon, J., Shin, H., Prediction of Process Variation Effect for Ultrascaled GAA Vertical FET Devices Using a Machine Learning Approach, in: *IEEE Transactions on Electron Devices*, vol. 66, no. 10, pp. 4474–4477, Oct. 2019, doi: 10.1109/TED.2019.2937786.
48. Card, H.C., McNeill, D.K., Schneider, C.R., Schneider, R.S., The impact of VLSI fabrication on neural learning, *Proceedings of ISCAS'95 - International Symposium on Circuits and Systems*, Seattle, WA, USA, 1995, pp. 985–988 vol.2, doi: 10.1109/ISCAS.1995.519931.
49. Fettes, Q., Clark, M., Bunescu, R., Karanth, A., Louri, A., Dynamic Voltage and Frequency Scaling in NoCs with Supervised and Reinforcement Learning Techniques, in: *IEEE Transactions on Computers*, vol. 68, no. 3, pp. 375–389, 1 March 2019, doi: 10.1109/TC.2018.2875476.

50. Raghavendra, N.V., Amuru, D., Abbas, Z., MetaCirc: A Meta-learning Approach for Statistical Leakage Estimation Improvement in Digital Circuits, *2024 IEEE International Symposium on Circuits and Systems (ISCAS)*, Singapore, Singapore, 2024, pp. 1–5, doi: 10.1109/ISCAS58744.2024.10558185.
51. Kumar, S.R.R., K.V., Reddy, U.S., Naidu, K.V., Enhancing VLSI Design and Manufacturing using AI Techniques in MOSFET, *2025 7th International Conference on Inventive Material Science and Applications (ICIMA)*, Namakkal, India, 2025, pp. 279–283, doi: 10.1109/ICIMA64861.2025.11074227.
52. Kundu, S., *et al.*, Special Session: Reliability Analysis for AI/ML Hardware, *2021 IEEE 39th VLSI Test Symposium (VTS)*, San Diego, CA, USA, 2021, pp. 1–10, doi: 10.1109/VTS50974.2021.9441050.
53. Kumar, R.V., C.A.A.C., A.D.V.M., Revolutionizing VLSI: Power and Area Reduction Using Novel FinFET-Based CMOS Techniques, *2025 Second International Conference on Cognitive Robotics and Intelligent Systems (ICC - ROBINS)*, Coimbatore, India, 2025, pp. 14–20, doi: 10.1109/ICC-ROBINS64345.2025.11086214.
54. Akhtar, M.R., Goswami, R.B., Abbas, Z., RelOps: Reliability Optimization in Standard Cells across PVT Variations in FinFET Digital Circuits, in: *IEEE Transactions on Computer-Aided Design.*
55. Li, M., Review of advanced CMOS technology for post-Moore era. *Science China Physics, Mechanics and Astronomy* 55, 12, 2316–2325, 2012.
56. Iwai, H., Future of nano CMOS technology, in: *28th Symposium on Microelectronics Technology and Devices (SBMicro 2013)*, pp. 1–10. IEEE, 2013.
57. Brewer, J.E., Zhirnov, V.V., Hutchby, J.A., Memory technology for post CMOS era. *IEEE Circuits and Devices Magazine* 21, 2, 13–20, 2005.
58. Ionescu, A.M., New functionality and ultra low power: Key opportunities for post-CMOS era, in: *2008 International Symposium on VLSI Technology, Systems and Applications (VLSI-TSA)*, pp. 72–73. IEEE, 2008.
59. Davari, B., CMOS technology: Present and future, in: *1999 Symposium on VLSI Circuits. Digest of Papers (IEEE Cat. No. 99CH36326)*, pp. 5–10. IEEE, 1999.

# 5

# Underlapped MOSFETs and Its Characteristics

**P. Suveetha Dhanaselvam*, B. Karthikeyan, K. Kavitha and G. Pradeep Kumar**

*Department of Electronics and Communication Engineering, Velammal College of Engineering and Technology, Madurai, Tamilnadu, India*

## *Abstract*

In nanoscale device design, short channel effects (SCEs) significantly deteriorate the performance of conventional MOSFETs, necessitating innovative structural and material engineering approaches. Multigate architectures, such as double-gate, triple-gate, and surrounding-gate, MOSFETs have been widely explored to enhance gate control and suppress SCEs. In parallel, gate and channel engineering strategies, including multimaterial gate structures and halo doping profiles, have further improved device electrostatics. This work presents a novel device architecture, Underlapped Triple Material Surrounding Gate (U-TMSG) MOSFET, which integrates the benefits of underlap engineering with triple-material gate technology. An analytical model of the device is developed using conformal mapping techniques, providing closed-form expressions for the surface potential and electric field distribution. The model is validated through TCAD simulations, and results demonstrate reduced leakage current, improved ON-state current, and enhanced electrostatic integrity compared to existing structures. The study highlights the U-TMSG MOSFET as a promising design-oriented solution for future nanoscale applications.

***Keywords*:** MOSFET scaling, short channel effects, multigate MOSFET, triple material gate, surrounding gate, underlap structure, TCAD simulation, nanoscale device modeling

**Corresponding author*: suveethaj@gmail.com

P. Suveetha Dhanaselvam, Naima Guenifi, Shiromani Balmukund Rahi and J. Ajayan (eds.) Field-Effect Transistor Devices: Post-CMOS Theory and Applications, (99–110) © 2026 Scrivener Publishing LLC

## 5.1 Introduction

In the nanoscale regime, using the specifications of ITRS, MOSFETs are scaled from meter scale to nano scale [1–3]. During this process of reducing the size of the transistors, there are certain negative impacts and hence corrective measures have to be taken in order to reduce the adverse effects. The term corrective measures mean modification in the device structure in order to reduce the adverse effects, so called short channel effects (SCEs). Instead of one gate, more than one gate is introduced to reduce this SCEs. Those devices are collectively remarked as multigate MOSFETs. To start with double gate [4] was introduced, followed by triple gate [5], omega gate [6] and surrounding gate [7]. Also, in addition to the introduction of more gates, the performance also can be enhanced through gate engineering and channel engineering. Gate engineering involves adding material to the poly silicon gate in order to reduce SCEs [8]. Usually in MOSFET, a single-gate material is present to control the device. But in recent years, in order to minimize the short channel effects, more than one material is introduced in the gate i.e., single material [9], dual material [10–12], triple material [13–15], quad material [16] etc. Similarly in the channel engineering, one or more layers like halo is introduced in which the doping concentration is varied. It also varies from single halo [17], double halo [18], symmetrical halo, asymmetrical halo. Apart from these measures, an underlapped structure is also a novel concept which reduces the leakage current thereby improving the ON current. This chapter embeds the underlapped concept with the triple material surrounding gate MOSFETs and discusses about the Underlapped Triple Material Surrounding Gate (U-TMSG) MOSFET.

## 5.2 Structure

The proposed device consists of three terminals namely source, drain and gate. The gate is constructed using three materials of unique work functions. The materials used are material 1 of work function $\varphi_{M1}$=4.8eV (Au), material 2 of work function $\varphi_{M2}$ =4.6eV (Mo), and gate material 3 with work function $\varphi_{M3}$ =4.4eV (Ti). The material with the high value of work function is set up at the source end and the material with lowest work function is designed at the drain end. This arrangement is required to reduce the short channel effects. The gate with a higher work function will act as a screening gate in controlling the SCEs [19]. The source and drain are extended so that it forms underlapped on both the ends. The device size is considered as 30 nm in which the underlapped regions are each 6 nm and the gate material each has 6nm.

### 5.2.1 Analytical Model

In the literature, MOSFET device is modeled using various methods and the parameters like surface potential, electric field, drain current, threshold voltage is derived. Green function, superposition [20], numerical iteration method, parabolic approximation, Fourier series [21, 22] are some of the methods used for deriving the parameters. In this paper, parabolic approximation method is used [23–31] since it is a simple method used to calculate the parameters. The proposed device U-TMSG MOSFET consists of a r in the radial direction, z in the vertical direction and an angular θ in the plane of the radial direction as shown in Figure 5.1. The quantum effects are not considered and effect of fixed charges are discarded in the model. The impurity concentration is considered to be uniform throughout the channel. The basic equations are given as

$$\frac{\partial^2}{\partial r^2}\psi_i(r,z)+\frac{1}{r}\frac{\partial}{\partial r}\psi_i(r,z)+\frac{\partial^2}{\partial z^2}\psi_i(r,z)=\frac{qN_A}{\varepsilon_{si}} \qquad i=1,2,3,4,5 \tag{5.1}$$

Using parabolic approximation method, the equations are solved and given as

$$\psi(r,z)=c_{i1}(z)+c_{i2}(z)+c_{i3}(z)r^2 \tag{5.2}$$

I represent regions 1 to 5

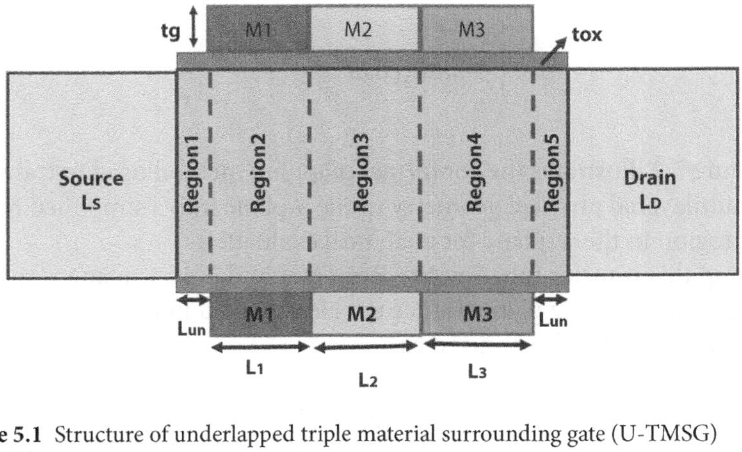


**Figure 5.1** Structure of underlapped triple material surrounding gate (U-TMSG) MOSFET.

The surface potential is represented by $\psi(r,z)$ and the voltages at the flat band are given as

$$V_{FB1} = \phi_{M1} - \phi_{si} \tag{5.3}$$

The material depends on the Region considered, accordingly the flat band voltage also varies.

$\phi_{si}$ is the silicon work function which is given by

$$\phi_{si} = \chi_{si} + \frac{E_g}{2q} + \phi_F \tag{5.4}$$

Using the boundary conditions the electric field in Regions 2, 3, 4 is given by

$$\frac{\partial \psi_2(r,z)}{\partial r} = \frac{\varepsilon_{ox}}{\varepsilon_{si}} X \frac{V_{GS-}V_{FBi} - \psi_2(r,z)}{t_{ox}} \tag{5.5}$$

The continuity equations hold good only for Region II to IV because of the variation in the thickness of the gate oxide layer. The impact of this results in decrease of electric field as we move away from the gate edges, i.e., at the source and drain end. This paper also uses conformal mapping method to transform the fringing fields from z plane to w plane and the transfer function is given as:

$$-z + j\eta(L_{un} - r) = t_{ox} sin(u + jv) \tag{5.6}$$

Where

$$\eta = \frac{t_{ox}}{L_{un}} sinh(cosh^{-1}\left(\frac{t_{ox} + t_g}{t_{ox}}\right)) \tag{5.7}$$

Figure 5.2 illustrates the conformal mapping method used to transform the multilayered physical geometry in the z-plane into a simplified rectangular region in the w-plane for analytical evaluation.

Using this transfer function the Regions I and V in w plane is mapped to region II to IV in z plane. Hence the electric field in the region near the source and drain ends are given as

$$\frac{\partial \psi_i(u,v)}{\partial u} = \frac{\varepsilon_{ox}}{\varepsilon_{si}} X \frac{V_{GS} + \phi_{si} - \psi_{s1}(v)}{m\Pi / 2} \tag{5.8}$$

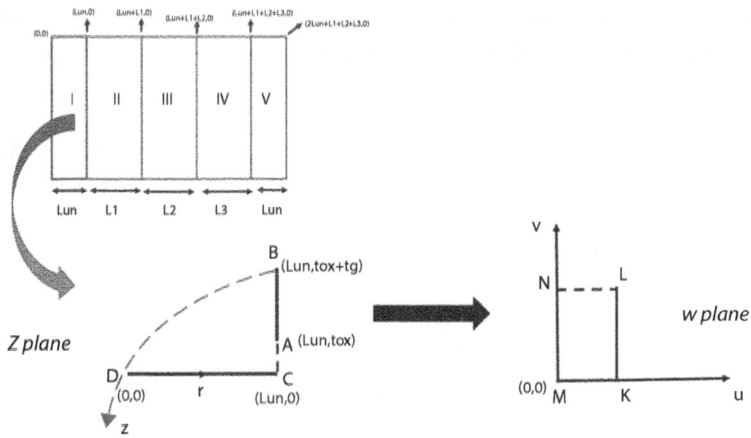


**Figure 5.2** Conformal mapping method.

where the i takes values 1 and 5

$$\frac{\partial^2}{\partial r^2}\psi_i(r,z)+\frac{1}{r}\frac{\partial}{\partial r}\psi_i(r,z)+\frac{\partial^2}{\partial z^2}\psi_i(r,z)=\frac{qN_A}{\varepsilon_{si}} \tag{5.9}$$

for regions I to 5

In the regions 1 and 5 the relationship between the r, z coordinates and u, v coordinates are given as

$$r = L_{un} - \left(\frac{t_{ox}}{\eta}\right) sinhv\ cosu \tag{5.10}$$

$$z = -t_{ox} sinu\ cosh\ cosh\ v \tag{5.11}$$

Substituting these equations in the above equation and solving we get

$$\frac{1}{cosh^2 v}\frac{d^2\psi_1(u,v)}{dv^2}-\frac{sinhv}{cosh^3 v}\frac{d\psi_1(u,v)}{dv}+\left(\frac{t_{ox}}{\eta}\right)^2 \tag{5.12}$$

$$X\frac{V_{GS}+\phi_{si}-\psi_{s1}(v)}{\frac{m\Pi}{2}}\left(\frac{1}{t_{ox}coshv}\right)=\frac{qN_A}{\varepsilon_{si}}\left(\frac{t_{ox}}{\eta}\right)^2 \tag{5.13}$$

The equation is complex and hence simplified using taylor series approximation to obtain the surface potential for each region.

$$\psi_{s1}(f) = A_1\left[1-\frac{p}{2}f^2\right]+A_2\left[f-\frac{p}{6}f^3\right]+V_{GS}+\phi_{si} \tag{5.14}$$

Where $f = \eta(L_{un} - r)/t_{ox}$

$$p = \frac{\varepsilon_{ox}}{\varepsilon_{si}}\frac{2}{t_{ox}t_{si}(\frac{m\Pi}{2})}\left(\frac{t_{ox}}{\eta}\right)^2 \tag{5.15}$$

The surface potential for region V is obtained as

$$\psi_{s5}(f) = E_1\left[1-\frac{p}{2}f^2\right]+E_2\left[f-\frac{p}{6}f^3\right]+V_{GS}+\phi_{si} \tag{5.16}$$

where f is replaced as

$$f = \eta(r - L_g - L_{un} - r)/t_{ox} \tag{5.17}$$

The surface potential of the regions 2-4 are derived as

$$\begin{aligned}\psi_{s2}(z) &= B_1 exp^{Kz} + B_2 exp^{-Kz} - \frac{M_1}{K^2}\\ \psi_{s3}(z) &= C_1 exp^{Kz} + C_2 exp^{-Kz} - \frac{M_2}{K^2}\\ \psi_{s4}(z) &= D_1 exp^{Kz} + D_2 exp^{-Kz} - \frac{M_3}{K^2}\end{aligned} \tag{5.18}$$

The electric field is obtained by differentiating the surface potential:

$$E(z) = \langle\frac{d\psi_{si}(r,z)}{dz}\rangle_{r=R} \tag{5.19}$$

$$E_n(z) = B_1 K exp^{Kz} - B_2 K exp^{-Kz} \tag{5.20}$$

where n varies from regions 2 to 5.

For the Regions 1 and 5, Electric field is obtained by differentiating (5.14) and (5.16).

The subthreshold current is calculated from the surface potential expression. In general, the current density is considered to be high towards the z direction from source to drain and hence the quasi-Fermi potential of the electron is maintained constant in the r direction. The drift ad diffusion current density is combined and given as

$$J(r,z) = -q\mu n(r,z)\frac{d\varphi_n(z)}{dz} \tag{5.21}$$

where $n(r,z)$ is the carrier concentration

$$n(r,z) = n_i e^{\frac{q\phi_j(r,z)}{K_B T}} \quad {}_{j=1\text{ to }5} \tag{5.22}$$

By integrating the current density (5.21), the sub-threshold current is calculated

$$I_{ds}(z) = \pi t_{si}\int_0^{\frac{t_{si}}{2}} J(r,z)dr \tag{5.23}$$

The drain current is obtained by substituting the current density in the above equation

$$I_{ds} = \frac{\mu\pi t_{si}\dfrac{K_B T}{q}\left(1-e^{\frac{-qV_{ds}}{K_B T}}\right)}{\int_0^L Q_j^{-1}(z)dz} \tag{5.24}$$

Substituting the limits for various Regions 1 to 5

$$I_{ds} = \frac{\mu\pi t_{si}\dfrac{K_B T}{q}\left(1-e^{\frac{-qV_{ds}}{K_B T}}\right.}{\int_0^{L_1} Q_1^{-1}(z)dz + \int_{L_1}^{L_2} Q_2^{-1}(z)dz + \int_{L_2}^{L_3} Q_3^{-1}(z)dz + \int_{L_3}^{L_4} Q_4^{-1}(z)dz + \int_{L_4}^{L_5} Q_5^{-1}(z)dz} \tag{5.25}$$

where $Q_j(z) = 2\int_0^{\frac{t_{si}}{2}} \frac{qn_i^2}{N_A} e^{\frac{q\phi_j(r,z_{j\min})}{K_BT}} dr \quad$ j=1 to 5

Numerical integration by trapezoidal rule is used considering piecewise linear curve approximation and it is calculated as

$$I_{ds} = \frac{\mu\pi t_{si}^2 n_i^2 K_B T(1 - e^{\frac{-qV_{ds}}{K_BT}})}{\left[\frac{L_1}{P_1} + \frac{L_2}{P_2} + \frac{L_3}{P_3} + \frac{L_4}{P_4} + \frac{L_5}{P_5}\right] 2mN_A} \tag{5.26}$$

The surrounding gate MOSFET is symmetrical and hence the above equations can be approximated as:

$$P_j \approx 2e^{\frac{q\phi_{j\min}}{K_BT}} \tag{5.27}$$

where the minimum surface potential is given in Equation (5.21).

The transconductance of the device is measured to analyze the efficiency of the proposed device. It is given by the equation

$$\frac{g_m}{I_{ds}} = \frac{q}{K_BT}\left(\frac{\partial\varphi_s(Z_{min})}{\partial V_{gs}}\right) \tag{5.28}$$

## 5.3 Results and Discussions

### 5.3.1 Surface Potential

The surface potential is derived using parabolic approximation method for the regions 2 to 4 and it is using conformal mapping strategy for the remaining regions. The analytically derived expressions are given from equations 5.14–5.21. It is plotted in Figure 5.3, where the Vgs is varied from 1.5 to 2.5 V. For TCAD simulation, Fermi-Dirac statistics and concentration dependent mobility models are used assuming the field mobility to be weak because of velocity saturation effect. The bandgap effects and Shockely-Read-Hall recombination are also considered but the quantum effects are comparatively negligible and ignored. In the result, it is observed that the surface potential is increasing along the channel initially

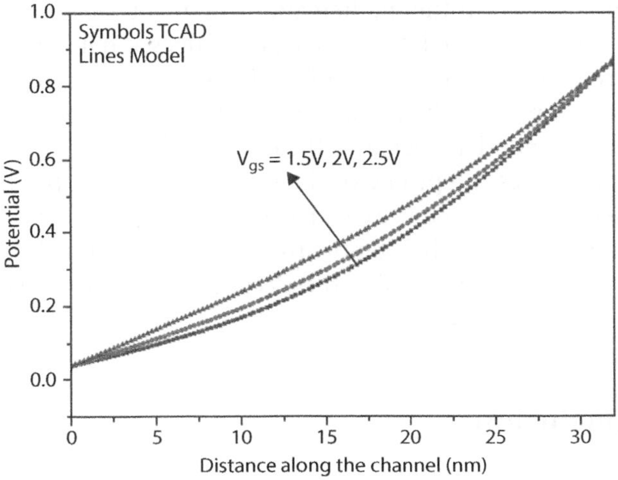


**Figure 5.3** Surface potential of U-TMSG MOSFET.

from 0.05 to 0.9 V gradually. The underlapped TMSG MOSFET shifts the surface potential upwards towards from the source to drain region. This is due to the accumulation of charges in the gate which is less initially and it is found to increase gradually towards the drain end.

### 5.3.2 Electric Field

The electric field derived using parabolic approximation method is illustrated in Figure 5.4. The surface potential is differentiated to get the electric

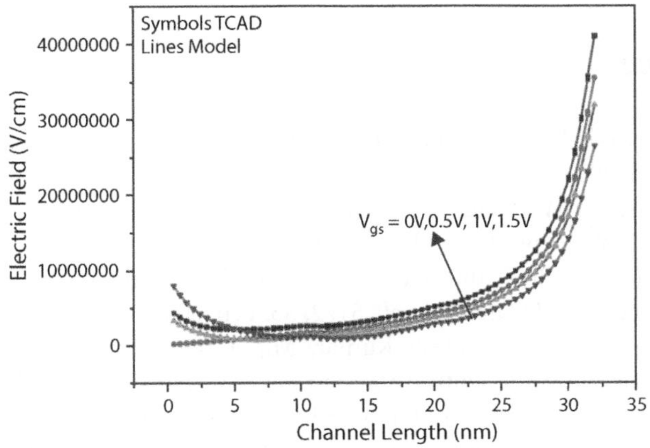


**Figure 5.4** Electric Field of U-TMSG MOSFETs for different Vgs.

field and it is shown in the mathematical modeling calculation. The plot depicts that the peak is high towards the gate edges and it seems to increase from source end to drain end. It infers that the underlap region the field tends to boost up only when it approaches the gate edges. The increasing distance of the underlap decreases the electric field. The increase in value of $V_{gs}$ also increases the profile of the surface potential. The high gate to source potential provides more electron flow through the device in the surface which provides a high potential from source to drain. This increasing profile of surface potential makes the electric field to increase proportionate with the electron flow.

## 5.4 Conclusion

This paper gives an insight in to the characteristics of a new device underlapped triple material surrounding gate (U-TMSG) MOSFET. The device is analytically modelled using equations, solved using different approaches and simulated using TCAD. The results are discussed and analyzed. This proves to be a design-oriented approach in which a novel conformal mapping method is introduced. The surface potential and electric field are derived and simulated. This makes the proposed work superior than the existing in analyzing the parameters and their associated inference.

**AI disclosure statement:** The author acknowledges the use of AI-based tools for minor language editing or grammatical correction. However, all intellectual contributions, analyses, and conclusions presented in this chapter are entirely the author's own.

## References

1. Colinge, J.P., Multiple-gate SOI MOSFETs. *Solid State Electron.*, 48, 6, 897–905, Jun. 2004.
2. Ko, Ping K. Approaches to scaling, in: *VLSI Electronics Microstructure Science*, vol. 18, pp. 1-37. Elsevier, 1989.
3. Zeitzoff, P.M., 2007 International Technology Roadmap: MOSFET scaling challenges. *Solid State Technol.*, 51, 2, 35, Feb. 2008.
4. Orouji, A.A. and Jagadesh Kumar, M., A new symmetrical double gate nanoscale MOSFET with asymmetrical side gates for electrically induced source/drain. *Microelectron. Eng.*, 83, 409–414, 2006.
5. Lemme, M.C., Mollenhauer, T., Henschel, W., Wahlbrink, T., Baus, M., Winkler, O., Granzner, R., Schwierz, F., Spangenberg, B., Kurz, H.,

Sub-threshold behavior of triple-gate MOSFETs on SOI material Sub-threshold behavior of triple-gate MOSFETs on SOI material. *Solid State Electron.*, 48, 529–534, 2004.

6. Van, N.H., Lee, J.-H., Whang, D., Kang, D.J., Low-Programmable-Voltage Nonvolatile Memory Devices Based on Omega-shaped Gate Organic Ferroelectric P(VDF-TrFE) Field Effect Transistors Using p-type Silicon Nanowire Channels. *Nano Micro Lett.*, 7, 1, 35–41, 2015.
7. Suveetha Dhanaselvam, P. and Balamurugan, N.B., Analytical approach of a nanoscale triple-material surrounding gate (TMSG) MOSFETs for reduced short-channel effects. *Microelectron. J.*, 44, 5, 400–404, May 2013.
8. Suveetha Dhanaselvam, P., Balamurugan, N.B., Ramakrishnan, V.N., A 2Dtransconductance and subthreshold behavior model for Triple Material Surrounding gate (TMSG) MOSFET. *Microelectron. J.*, 44, 1159–1164, 2013.
9. Chiang, T.K., A new two-dimensional analytical model for threshold voltage in undoped surrounding-gate MOSFETs. *IEEE Trans. Electron Devices*, 2006.
10. Jagadesh Kumar, M., Orouji, A.A., Dhakad, H., New Dual-Material SG Nanoscale MOSFET: Analytical Threshold-Voltage Model. *IEEE Trans. Electron Devices*, 53, 4, April 2006.
11. Chaudhry, A. and Kumar, M.J., Investigation of the novel attributes of a fully depleted dual- material gate SOI MOSFET. *IEEE Trans. Electron Devices*, 51, 9, 1463–1467, Sep. 2004.
12. Mohankumar, N., Syamal, B., Sarkar, C.K., Investigation of novel attributes of single halo dual-material double gate MOSFETs for analog/RF applications. *Microelectron. Reliab.*, 49, 1491–1497, Jul 2009.
13. Suveetha Dhanaselvam, P. and Balamurugan, N.B., A 2D Sub-threshold Current model for Single Halo Triple Material Surrounding Gate (SHTMSG) MOSFETs. *Microelectron. J.*, 45, 6, 574–577, 2014.
14. Suveetha Dhanaselvam, P., Balamurugan, N.B., Vivek Chakaravarthi, G.C., Ramesh, R.P., Sathish Kumar, B.R., A 2D Analytical Modeling of Single Halo Triple Material Surrounding Gate (SHTMSG) MOSFET. *J. Electr. Eng. Technol.*, 9, 4, 1355–1359, 2014.
15. Suveetha Dhanaselvam, P. and Balamurugan, N.B., Performance analysis of fully depleted triple material surrounding gate (TMSG) SOI MOSFET. *J. Comput. Electron.*, 13, 449–455, 2014.
16. Vimala, P., Suveetha Dhanaselvam, P., Nirmal, D., Arun Samuel, S., A multi-layer gate metal oxide semiconductor FET for reducing short channel effect, Indian patent No. 384801.
17. Li, Z.-C., Zhang, R.-Z., Jiang, Y.-L., Analytical modeling and Simulation of Dual-material Surrounding gate Metal-oxide-semiconductor Field effect transistors with single Halo doping. *Jpn. J. Appl. Phys.*, *48*, 2009.
18. Suveetha Dhanaselvam, P., Vimala, P., Arun Samuel, S., A 2D Analytical Modeling and Simulation of Double Halo Triple Material Surrounding Gate (DH-TMSG) MOSFET. *Silicon*, 13, 8, 2631–2637, 2021, https://doi.org/10.1007/s12633-020-00617-9.

19. Manikandan, S., Suveetha Dhanaselvam, P., Karthigai Pandian, M., A 2D Unified Subthreshold Drain Current Investigation for Junctionless Cylindrical Surrounding Gate(JCSG) Silicon Nanowire Transistor. *Silicon*, 14, 5311–5318, 2022.
20. Hsin-Kai Wang, J., Wu, S., Chiang, T.-K., Lee, M.-S., A New Two-Dimensional Analytical Threshold Voltage Model for Short-Channel Triple-Material Surrounding-Gate Metal– Oxide–Semiconductor Field-Effect Transistors. *Jpn. J. Appl. Phys.*, 51, 2012.
21. Tiwari, P.K., Dubey, S., Singh, M., Jit, S., A two-dimensional analytical model for threshold voltage of short-channel triple-material double-gate metal-oxide-semiconductor filed-effect transistors. *J. Appl. Phys.*, 108, 2010.
22. Mei, G., Hu, G., Hu, S., Gu, J., Liu, R., Tang, T., Analytical Model for Subthreshold Swing and Threshold Voltage of Surrounding Gate Metal–Oxide Semiconductor Field-Effect Transistors. *Jpn. J. Appl. Phys.*, 50, 2011.
23. Lee, H., Lee, J., Shin, H., DC and AC characteristics of sub-50- nm MOSFETs with source/drain-to-gate nonoverlapped structure. *IEEE Trans. Nanotechnol.*, 1, 4, 219–225, 2002.
24. Bansal, A. and Roy, K., Analytical subthreshold potential distribution model for gate underlap double-gate MOS transistors. *IEEE Trans. Electron Devices*, 54, 7, 1793–1798, 2007.
25. Goel, E., Pandey, A., Rahi, S.B., Samuel, A., eds. *Handbook of Advanced Semiconductor Field Effect Transistors.* John Wiley & Sons, 2025.
26. Agnihotri, H., Ranjan, A., Tiwari, P.K., Jit, S., An Analytical Drain Current Model for Short-Channel Triple-Material Double-Gate MOSFETs. *IEEE Computer Society Annual Symposium on VLSI, ISVLSI 2011*, Chennai, India, 4-6 July 2011.
27. Smaani, B., Latreche, S., Iñiguez, B., Compact drain-current model for undoped cylindrical surrounding gate metal-oxide-semiconductor field effect transistors including short channel effects. *J. Appl. Phys.*, 114, 224507, 2013.
28. Dhanaselvam, P.S., Srinivasa Rao, K., Rahi, S.B., Yadav, D.S. (Eds.), *Field Effect Transistors*, John Wiley & Sons, 2025.
29. Gautam, R., Saxena, M., Gupta, R.S., Gupta, M., Effect of localised charges on nanoscale cylindrical surrounding gate MOSFET: Analog performance and linearity analysis. *Microelectron. Reliab.*, 52, 989–994, 2012.
30. Kranti, A., Haldar, S., Gupta, R.S., Analytical model for threshold voltage and I–V characteristics of fully depleted short channel cylindrical/surrounding gate MOSFET. *Microelectron. Eng.*, 56, 241–259, 2001.
31. Li, Z.-C., Xu, J.-P., Liu, L.-L., Current modeling and simulation of dual-material surrounding-gate MOSFET with asymmetric halo. *2010 10th IEEE International Conference on Solid-State and Integrated Circuit Technology*, pp. 1904–1906, Shanghai, China, 2010, doi: 10.1109/ICSICT.2010.5667768.

# 6

# Fundamental of Carbon Nanowires and its Applications

**N. Nagarani*, P. Suveetha Dhanaselvam and B. Karthikeyan**

*Department of Electronics and Communication Engineering, Velammal College of Engineering and Technology, Madurai, Tamilnadu, India*

***Abstract***

Carbon nanowires (CNWs) are a type of nanomaterial with a one-dimensional structure, meaning they have a very small diameter typically in the nanometer range and a much longer length. These nanowires are composed entirely of carbon atoms and are related to other carbon-based nanomaterials, such as carbon nanotubes and graphene. In comparison to regular nanowires, carbon nanowires have benefits in thermal conductivity, mechanical strength, chemical stability, electrical conductivity, and nanoscale quantum effects. They are especially well-suited for cutting-edge applications where endurance and performance are crucial, such as electronics, energy storage, biomedical engineering, and materials research. This survey work provides an extensive summary of the production, properties, and uses of carbon nanowires. It examines several fabrication processes, highlighting the benefits and drawbacks of each, such as chemical vapor deposition, laser ablation, and electrochemical approaches. The numerous uses of CNWs in industries including biomedicine, energy storage, nanoelectronics, and sensor technologies are also covered in this study. Ongoing research is focused on optimizing the properties of CNWs, improving their synthesis processes, and exploring new applications, particularly in advanced electronics and nanotechnology. As a frontier material in nanotechnology and materials science, CNWs hold the potential to drive significant innovations in high-tech industries, paving the way for future advancements.

***Keywords***: Carbon nanowires, carbon nanotubes, quantum computing, consumer electronics

**Corresponding author*: rani@vcet.ac.in

P. Suveetha Dhanaselvam, Naima Guenifi, Shiromani Balmukund Rahi and J. Ajayan (eds.) Field-Effect Transistor Devices: Post-CMOS Theory and Applications, (111–128) © 2026 Scrivener Publishing LLC

## 6.1 Introduction

Carbon nanowires (CNWs) are an impressive family of nanomaterials that have recently garnered much attention due to their quasi-one-dimensional structure and seeming extraordinary properties. These super-thin wires are composed of pure carbon, usually nanometers in diameter, but can extend to much greater lengths, making them an important part of the rapidly advancing field of nanotechnology. CNWs are similar to carbon nanotubes and graphene, which are two other carbon-based nanomaterials; however, they have certain advantages associated with them, which make them different from the others. Latest research in this field includes work by Khan *et al.*, which focused on latent heat storage systems designed with extended fins and graphene nanoplatelets for enhanced thermal performance [1]. Research by Kumar *et al.* discussed a catalyst support based on carbon nanotube grown on stainless steel wire mesh for NOx reduction in diesel engines [2]. Estevez *et al.* developed hybrid fibers based on nanocarbon and magnetic microwires for microwave functionalities that can be tailored [3]. Wang *et al.* discussed the development of advanced cement-based composites for civil engineering, focusing on improved strength, durability, and smart features like self-sensing and self-healing, making them suitable for modern infrastructure needs [4]. Qu *et al.* carried out work on hybrid carbon nanoadditives to enhance the thermal conductivity of paraffin-based phase change materials [5]. Salari *et al.* discussed the integration of nanofluid-based photovoltaic thermal systems with phase change materials [6]. Raja *et al.*, at the same time, prepared aluminum and copper wires embedded in glass fiber-reinforced polymer composites, and the mechanical properties were studied [7]. In 2021, Dhand *et al.* detailed the prospects of multifunctional carbon nano-onions as potential high-performance energy storage materials in a comprehensive review [8]. In 2021, Dai *et al.* encapsulated silicon nanoparticles in a multifunctional cross-linked nanosilica/carbon hybrid matrix for high-performance anodes in lithium-ion batteries [9]. Lu *et al.* simulated and prepared interconnected carbon nanotube-nanoparticle structures; this was found to have promising versatility, as carbon nanotubes could be used innovatively to connect nanostructures [10]. Carbon nanowires are the subject of several research studies. This is the result of heavy-ion radiation experiments related to the electron field emission properties of sulphur-doped nanocomposite carbon films. González-Berríos *et al.* investigated the impact of heavy-ion radiation on the electron field emission properties of sulphur-doped nanocomposite carbon films. Their study revealed that radiation exposure significantly influences the emission behavior, affecting the material's structural and electronic properties, which is crucial for enhancing the performance of carbon-based electronic devices [11]. Ding *et al.* utilized TD-DFT, and the CASSCF method to study the electronic structures of one-dimensional carbon nanowires and rings [12]. Nadi *et al.* discussed the synthesis of nanowires by graphite by using laser ablation [13]. Guo-ping *et al.* examined the impact of various nanocarbon conductive additives on the resistance and electrochemical performance of $LiCoO_2$ composite cathodes in lithium-ion batteries. Their study highlighted how different additives influence conductivity and battery efficiency, aiding in the optimization of cathode materials for improved energy storage [14]. Janghorban, M. discussed the development of advanced cement-based composites for civil engineering, focusing on improved strength, durability, and smart features like self-sensing and self-healing, making them suitable for modern infrastructure needs. In addition, Janghorban investigated static and free vibration analyses of carbon nanowires according to the Timoshenko beam theory with differential quadrature method [15]. Chen *et al.* synthesized and analyzed electrochemical properties of LiFePO4/C composite with network connections of nanocarbon wires [16]. In this regard, Taha *et al.* studied carbon nanofibers with enhanced convective heat transfer [17]. Moreover, Dhanaselvam, P. *et al.* developed nanocarbon materials with high electrical conductivity by adding a combination of unoxidizing pyrolysis of salt-milled solid carbon sources [18]. Generally, based on the carbon nanowires across many applications including field emission properties, electronic structures, vibration analysis, electrochemical performance, enhancement in heat transfer, and interconnected structures, the overall broad field of these applications does indicate versatility and potential in a nanomaterial of this sort.

### 6.1.1 The Nanotech Revolution: Carbon Leads the Charge

Materials, being the largest class, are divided into major categories: metals, ceramics, polymers. A nanomaterial forms a very important class of materials, promising to revolutionize several key industries with their unique properties, such as enhanced strength, conductivity, and reactivity. Their range of applications goes from high-performance batteries and solar cells to advanced drug delivery systems. The ever-improving technology contributes much in the development of new innovations, and nanomaterials are not left behind in sectors such as quantum computing, biotechnology, and artificial intelligence.

### 6.1.2 Carbon Nanowires: The Next Big Thing in Nano

CNWs are one-dimensional nanomaterials composed solely of carbon atoms, with diameters in the nanometre range but much larger in length. They are known for extremely good electrical conductivity, higher mechanical strength, high thermal conductivity, and stability against most chemical environments. Due to their exceptional properties, CNWs have sparked great interest in being considered as cutting-edge materials in several different fields. The capability of efficiently transferring electricity and heat, along with their resiliency in harsh environments, makes them suitable for applications in nanoelectronics, energy storage, and bio medics. CNWs also show quantum effects at the nanoscale, thereby further exploiting the material for exciting applications in advanced technologies.

## 6.2 Properties of Carbon Nanowires

CNWs possess a set of extraordinary properties, which make it highly interesting for various applications. It is recognized to have the unique one-dimensional carbon structure that presents extraordinary advantages across the range of fields in both electronics and materials science. For example, some key properties are as follows;

### 6.2.1 Exceptional Thermal Conductivity

Conductivity is also characterized by efficient heat dissipation, a very important criterion in high-performance applications like electronics, wherein overheating results in reduced efficiency and failure of components. CNWs thereby help in the effective management and transfer of heat, making them apt for use in sophisticated systems of thermal management, such as those in microprocessors, batteries, and other electronic devices requiring enhanced cooling.

### 6.2.2 Steelier Than Steel: Exceptional Mechanical Strength

Carbon nanowires possess phenomenal mechanical strength owing to the extremely strong carbon-carbon bond that exists in their lattice. Their tensile strength is far greater and indeed much greater on a weight basis than that of steel, making them attractive for developing lightweight yet extremely strong materials. This property is quite useful in aerospace,

**Table 6.1** Comparison of mechanical properties of CNWs with other materials.

| Density(g/cm$^3$) | Tensile strength (GPa) | Material |
|---|---|---|
| ~0.7–1.7 | ~100–200 | SWNT/MWNT |
| 7.8 | 1.3 | Highly tensile steel |
| 1.75 | 3.5 | Toray carbon fibers |
| 1.44 | 3.6 | Kelvar |

automotive, and nanocomposites applications where light weight with strength are critical for performance and efficiency.

Table 6.1 highlights the superior mechanical performance of carbon nanowires (CNWs) when compared with conventional structural materials. The tensile strength of single-walled and multi-walled carbon nanotubes (SWNT/MWNT), ranging from 100 to 200 GPa, far exceeds that of highly tensile steel (1.3 GPa), Toray carbon fibers (3.5 GPa), and Kevlar (3.6 GPa). Despite their exceptional strength, CNWs maintain a low density (~0.7–1.7 g/cm$^3$), making them promising candidates for high-strength, lightweight composite applications in aerospace, defense, and advanced nanotechnology systems.

### 6.2.3 Chemical Stability of CNWs toward Fire

Carbon nanowires are stable and resistant to corrosion and degradation, even under hostile conditions. The chemical toughness in extreme conditions such as extreme temperatures, harsh chemicals, or oxidation makes CNWs useful in any application that requires such extreme conditions. Such stability means such devices as sensors, batteries, and catalysts can function for a long time in chemically challenging environments.

### 6.2.4 Electric Excellence: Conductivity at the Nanoscale

The electrical conductivity of CNWs is high, and hence appropriate for use in electronic circuits, transistors, and sensors, which carry electric current with minimum resistance. High efficiency in nanoelectronics offers the potential to carve out much smaller, faster, and more efficient devices. High electrical conductivity with nanoscale dimensions helps integrate next-generation electronics and energy storage systems, like super capacitors and batteries into CNWs.

### 6.2.5 Quantum Leap: Nanoscale Magic of CNWs

Quantum mechanical properties are manifested in CNWs at the nanoscale and describe the behavior of their material well above the macroscopic counterparts. The quantum effects of quantum tunnelling and electron confinement dominate when the size of a material drops into the nanoscale. Such properties make CNWs very fascinating for practical applications at the nanoscale level in quantum computing, nanoscale sensors, and advanced communication technologies. Due to the exclusive quantum behavior of CNWs, they are promising for breakthrough technologies requiring control over electrons in their action at the quantum level in the future.

Table 6.2 presents the detailed physical properties of carbon nanowires (CNWs), showcasing their high purity (>95%), nanoscale dimensions with an outer diameter of 40–60 nm and core diameter of 5–10 nm, and a length of 5–15 nm. Their specific surface area (40–600 $m^2/g$) and pore volume (0.8 $cm^3/g$) highlight the large reactive surface suitable for adsorption and catalytic applications. The low bulk density (~0.1 $g/cm^3$) and true density (~2.1 $g/cm^3$) emphasize their lightweight yet compact nature, while their black coloration is typical of carbon-based nanostructures. These characteristics make CNWs ideal for high-performance applications in energy storage, catalysis, and nanocomposite materials.

**Table 6.2** Accurate values of the properties of CNWs.

| Property | Value |
|---|---|
| Purity | >95% |
| Outside diameter | 40-60nm |
| Core diameter | 5-10nm |
| Length | 5-15nm |
| Specific Surface Area (SSA) | 40-600m2g-1 |
| Color | Black |
| Pore Volume | 0.8cm3g-1 |
| Bulk density | ~0.1g1cm-3 |
| True density | ~2.1gcm-3 |

## 6.3 Fabrication Techniques

### 6.3.1 Chemical Vapor Deposition

Chemical vapor deposition (CVD) is one of the most popular synthesis methods for CNWs as well as other nanomaterials. In the CVD process, a substrate is coated with a solid material through a chemical reaction involving gaseous precursors. The first step is Cleaning and pre-treatment of the substrate, silicon or glass or metal, to reach a proper surface condition for CNW deposition. This may include ultrasonic cleaning or plasma treatment to remove contaminants. The CVD system has a reaction chamber were, in the gaseous state, precursors can be admitted. Carbon precursors include those hydrocarbons containing methane, $CH_4$, acetylene, $C_2H_2$, or ethylene, $C_2H_4$. In addition, a carrier gas, such as argon or hydrogen, can be admitted to carry the precursors into the reaction chamber. The substrate is heated at very high temperature, depending upon the process being conducted, typically in the range of 600 to 1200°C. At the high temperature the gaseous precursors undergo thermal decomposition producing reactive carbon species. Decomposition of precursors releases carbon atoms along with other by-products. Carbon atoms get deposited onto the substrate. Those atoms on the substrate then start to nucleate and grow into nanowires. Catalysts: Sometimes used is the catalyst with metal nanoparticles, such as nickel, cobalt, or iron, that act as nucleation sites in initiating CNW growth. The deposition process is continued with carbon atoms aligning and bonding together to form carbon nanowires. Growth rate and morphology can be controlled by adjusting the parameters, which include temperature, pressure, precursor concentration, and reaction time. Once the required growth of CNWs has been reached, the substrate is cooled down and any remaining un reacted precursors or by products are evacuated from the system. Additional processing may involve purification or functionalization of the nanowires to alter the chemical properties and surface chemistry.

### 6.3.2 Laser Precision: Ablation and CNW Formation

Laser ablation shown in Figure 6.1 is a sophisticated synthesis technique that has now become one of the primary routes for precision in CNW formation. In this process, material from a target substrate is vaporized by focused laser energy to facilitate nanostructure formation.

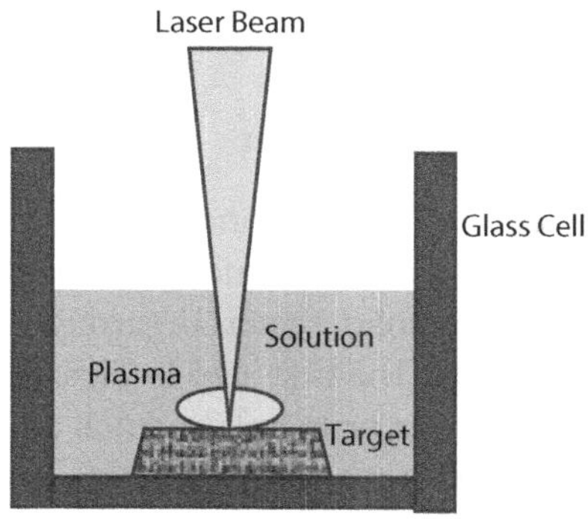


**Figure 6.1** Laser ablation method for formation of carbon nanowires.

The process of synthesis of CNWs consists of several important steps, beginning with the choice of an appropriate carbon-based precursor, for instance, graphite or carbon powder, as the properties of the precursor material significantly affect the properties of the resulting CNWs. Subsequently, a pulsed laser system, typically a Nd:YAG laser, is employed, with the wavelength and pulse duration being judiciously selected based on the absorption properties of the target material in question. The laser beam is focused onto the target, effectively concentrating energy at the interaction point to produce high-temperature plasma. The localized irradiation induces heating, enabling the target material to vaporize and create a plume consisting of carbon atoms and other byproducts. The ablation process takes place within a time interval of nanoseconds to microseconds, allowing precise material removal. Upon cooling and recombination of the vaporized carbon atoms, they self-assemble into nanowires, depending on parameters such as the environment (e.g., an inert gas atmosphere) and substrate temperature, which are optimized to favor the growth of CNWs. The resulting carbon nanowires, deposited on the substrate or collected from the ambient air, are usually subjected to a purification process or further treatment to improve their properties.

### 6.3.3 Electrochemical Approaches to CNWs

Electrochemical approaches have become prominent with their ability to synthesize CNWs with high property control. Electrochemical approaches utilize electrochemical reactions to facilitate deposition and growth of CNWs, offering an important route for the synthesis of nanomaterials. These approaches are generally carried out in an electrochemical cell composed of an anode, a cathode, and an electrolyte solution, where choice of the electrolyte is an important aspect in the determination of growth conditions and quality of CNWs. Carbon is generally used as the electrode material, and the cathode is used as the working electrode where the reduction reaction occurs, ultimately resulting in carbon deposition. A predetermined voltage is applied between the electrodes, initiating electrochemical reactions involving electron and ion transfer in the electrolyte. The cathodic reduction process is utilized to reduce carbon species in the electrolyte, resulting in carbon deposition, hydrogen evolution, and CNW formation and growth. Factors such as current density—which affects deposition rates and CNW morphology—and the electrolyte composition, including ions and additives, can have a great impact on the growth rate and characteristics of resulting CNWs. After synthesis, CNWs can exhibit an enormous variety of structures and properties depending on synthesis conditions such as voltage, current density, temperature, and electrolyte composition. These properties can be engineered for better electrical, thermal, and mechanical properties. Additional purification, functionalization, or processing steps can be utilized to further enhance the CNWs for target applications.

## 6.4 Applications of Carbon Nanowires

### 6.4.1 Smart Circuits: CNWs in Nanoelectronics

Carbon nanowires, which have higher conductivity than copper and smaller diameters with comparable strength, are emerging as novel materials capable of revolutionizing electronic devices. The discovery of carbon nanowires leads to the production of faster, smaller and more energy-efficient devices for nanotechnology applications.

CNWs have emerged as a ground-breaking alternative to traditional silicon in transistor technology. Due to their one-dimensional structure, CNWs exhibit ballistic electron transport, enabling ultra-low energy loss and extremely high switching speeds. Field-effect transistors (FETs) utilizing CNWs have demonstrated lower power consumption and enhanced performance characteristics compared to conventional silicon-based transistors, positioning them as a promising component for future high-speed computers and mobile devices. Beyond transistors, CNWs are highly suited for flexible electronics, thanks to their combination of rigidity and bendability. Unlike silicon, which is too brittle for flexible applications, CNWs can be integrated into bendable and wearable electronic devices, paving the way for innovations such as smart clothing, foldable smartphones, and flexible display technologies. Additionally, CNWs are being explored in non-volatile memory devices, particularly in resistive random-access memory (ReRAM), where their robustness and superior endurance enhance speed, power efficiency, and storage density. These advancements could lead to memory solutions that significantly outperform existing technologies in both consumer electronics and data storage systems.

In the field of energy storage, CNWs have proven to be transformative, particularly in lithium-ion batteries. When used as anode materials, CNWs offer superior electrical conductivity and the capacity to accommodate large amounts of lithium ions during charging and discharging cycles. This results in batteries with higher capacities, longer lifespans, and improved cycle stability compared to traditional graphite anodes. CNW-based anodes also support faster charging rates, making them ideal for applications in electric vehicles and portable electronic devices. Researchers have further enhanced battery performance by combining CNWs with materials like silicon or graphene, achieving significantly increased energy densities. In parallel, CNWs are revolutionizing supercapacitor technology, devices that store and release energy at rapid rates. Thanks to their high surface area and excellent conductivity, CNWs enable supercapacitors to achieve higher power densities, faster charge/discharge rates, and extended lifespans. Some CNW-based supercapacitors even approach the energy densities of conventional batteries while retaining the ability to deliver quick bursts of power, making them suitable for renewable energy systems, electric vehicles, and fast-charging consumer electronics.

The integration of CNWs across diverse technological domains highlights their versatility and transformative potential. In electronics, their use in transistors and memory devices promises faster, more energy-efficient components that can drive the next generation of computing and mobile technology as portrayed in Figure 6.2. In the realm of energy storage, CNWs are helping to bridge the gap between traditional batteries and supercapacitors, offering solutions that combine high energy density with rapid charge and discharge capabilities. This is particularly crucial in industries such as electric vehicles and renewable energy, where energy efficiency and power management are key. Moreover, the adaptability of CNWs in flexible electronics opens new possibilities for smart textiles, foldable gadgets, and wearable health monitoring devices. As research continues to refine

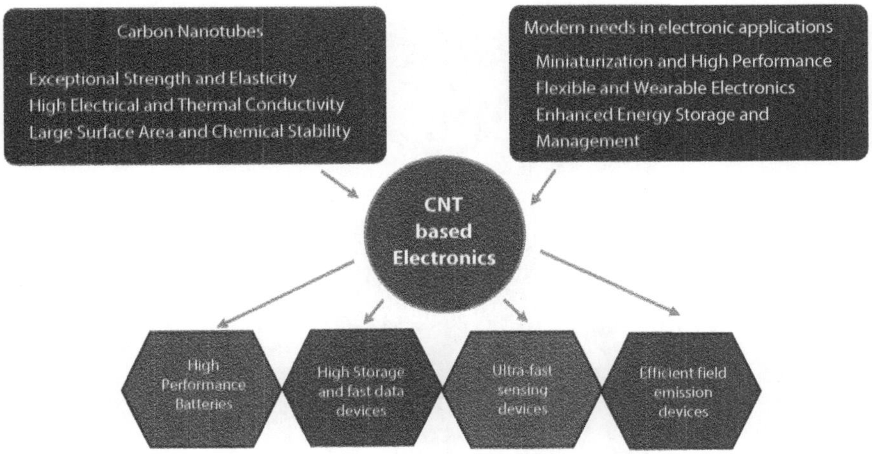


**Figure 6.2** Block representation of CNT based electronics.

CNW synthesis and integration techniques, their applications are expected to expand even further, offering new possibilities for nanotechnology, sustainable energy, and next-generation computing.

## 6.5 Ongoing Research and Future Perspectives

CNTs have been both synthesized in laboratories and discovered in natural and historical contexts as shown in Figure 6.3. Multi-walled carbon nanotubes were first observed by Iijima in 1991, followed by the identification of single-walled carbon nanotubes in 1993 by Iijima and Donald Bethune. Prior to these discoveries, structures resembling CNTs, such as carbon fibrils in 1984 and graphene whiskers in 1957, were reported. Remarkably, CNTs have also been found in ancient artefacts and natural formations. For instance, analyses of 17th century Damascus steel blades revealed the presence of carbon nanotubes and cementite nanowires, contributing to their renowned strength and flexibility. Natural occurrences of CNTs have been identified in environments like oil fields and igneous rocks, indicating that these nanostructures can form under certain geological conditions. These findings highlight that CNTs are not solely modern synthetic materials but have existed naturally and have been utilized, albeit unknowingly, in historical metallurgy. CNWs really continue to show their incredible features, and scientists immediately begin to look into new and developing

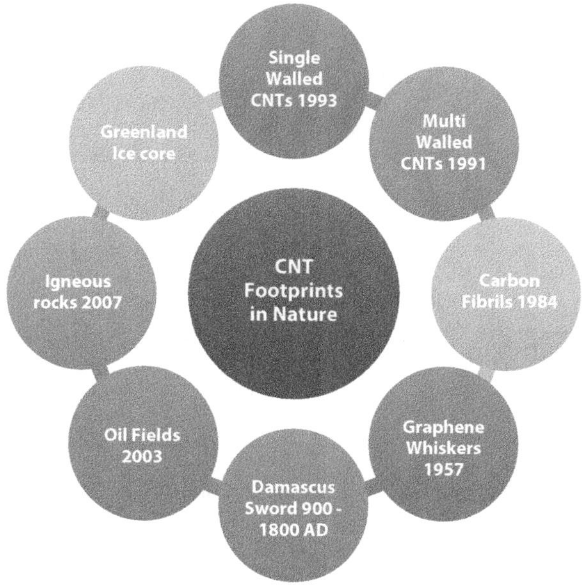


**Figure 6.3** Transformation of CNTs in various fields in nature.

applications for cutting-edge technologies. From quantum computing to space exploration, researchers are now unlocking possibilities that push the frontiers of modern science and engineering [19].

### 6.5.1 Quantum Computing, Aerospace Applications, and Photonics

CNWs hold tremendous potential in quantum electronics and quantum computing due to their nanometre-scale dimensions and high conductivity, which allow them to function as quantum wires—precursors to the fabrication of qubits, the building blocks of a quantum computer. Quantum wires can maintain quantum entanglement and quantum coherence, both essential for quantum computing. CNWs also hold promising applications in superconducting circuits functioning at very low temperatures, vital for efficient quantum systems. In space exploration and aerospace, CNWs are studied for their outstanding mechanical strength, light weight, and resistance to extreme space conditions such as radiation, vacuum levels, and extremes of temperature fluctuation. All these properties render CNWs ideal for the design of ultra-strong yet light-weight composites for spacecraft and satellites. Additionally, their high thermal conductivity is valuable in heat dissipation in space systems, overcoming the major problem of thermal management in long-duration deep space missions.

### 6.5.2 Next-Generation Photonic Devices and Wearable Technology

In photonics, CNWs have the potential to revolutionize light-based technologies through integration into optoelectronic devices such as LEDs, photodetectors, and solar cells. Their exceptional properties render them key materials for the fabrication of plasmonic devices, which manipulate light at the nanoscale for telecommunication and data storage purposes. Photonic circuits, which convey information *via* light much faster than conventional electronic circuits, can significantly benefit from the high electron mobility of CNWs. Additionally, the conductivity, resistance, and flexibility of CNWs make them ideal for wearable technology and Internet of Things (IoT) devices. In wearable electronics and smart fabrics, CNW-based sensors can be integrated to monitor real-time physiological parameters, such as heart rate, body temperature, and blood pressure. Moreover, CNWs can be utilized to power self-powered wearable devices, transducing mechanical energy from human movements into electricity, thereby enhancing the functionality and sustainability of future IoT devices.

### 6.5.3 Neuromorphic Computing and Green Energy Solutions

In neuromorphic computing, seeking to emulate the brain's structure and functionalities, CNWs are recognized as promising materials for artificial synapse development. Low-power, energy-efficient networks that can process and store information in parallel, similar to biological neurons, such systems need. The high conductivity and scalability of CNWs make them ideal for the development of such networks, which could result in computing systems far faster and more energy-efficient than existing technologies. In energy harvesting and green technology, CNWs provide new options by converting ambient energy into useful electrical power, which is vital in powering remote or off-grid devices. For instance, CNWs are used in thermoelectric generators to harvest waste heat into electricity and improve industrial processes and renewable energy systems. They are also used in piezoelectric and triboelectric generators that capture mechanical energy from vibrations or motions and provide a green method of powering small electronic devices and facilitating the growth of green energy solutions.

## 6.6 Case Study: Hardware Accelerators Based on CNWS

Hardware accelerators are specialized processing hardware programmed to perform specific functions quicker and more efficiently than general-purpose processors like CPUs. The use of CNTs in hardware accelerators significantly enhances performance, energy efficiency, and scalability over traditional silicon-based technology. Carbon nanotube field-effect transistors (CNTFETs), built from CNTs—cylindrical nanostructures made of carbon atoms in a hexagonal lattice—possess electrical, thermal, and mechanical properties superior to those of silicon, making them extremely well-suited for modern electronic devices. CNTs possess high carrier mobility, allowing increased switching speeds and reduced power consumption. Their ability to maintain ballistic transport allows electrons to travel short distances with little scattering, thus reducing resistive losses and increasing overall efficiency. The intrinsic scalability of CNTs allows for miniaturization to atomic scales, resulting in transistors that are denser, faster, and more energy efficient than their silicon-based counterparts. In addition, CNTFETs are operated at reduced voltages, translating to very low power consumption—an essential property for energy-efficient hardware accelerators.

### 6.6.1 Carbon Nanotube Transistors and Hardware Accelerators

Hardware accelerators are designed to offload some of the computational work from the CPU to special-purpose units, thereby improving performance for tasks with a parallel nature or needing rapid execution. The nature of CNT transistors makes them ideal for such accelerators, which can execute faster and use lower power. This feature is extremely useful in applications like artificial intelligence (AI), machine learning (ML), cryptography, and data center operations. In AI and ML applications, accelerators based on CNTs can perform matrix multiplication—fundamental operations in neural networks—better than in traditional architecture. Graphics Processing Units (GPUs) also benefit from CNTs, as their high-speed switching capability facilitates the inherently parallel nature of graphic rendering operations. In cryptographic calculations, CNT accelerators execute complex encryption and decryption operations at higher speeds while keeping power consumption to the minimum at the same time. In data centers, where energy efficiency is a growing demanding factor, CNT-based accelerators enhance computing efficiency while saving energy by a significant margin, making them well suited for modern cloud computing architectures.

### 6.6.2 CNT-Based Hardware Accelerator Operation Mechanism

The functional operation of CNT based hardware accelerators revolves around the integration of CNTFETs into various processing units and system architectures as illustrated in Figure 6.4. This is done by beginning with the fabrication of CNT transistors, where carbon nanotubes are grown and aligned using techniques like CVD. These CNTs replace normal silicon channels in the transistor structure and form the basis of logic circuits like AND, OR gates, and NOT gates. CNTFETs offer faster switching speeds and lower leakage currents compared to silicon-based circuits, thus enhancing overall system performance. In hardware accelerators like Field Programmable Gate Arrays (FPGAs) and graphics processing units (GPUs), CNT transistors enable parallel processing. The accelerator breaks down complex tasks into thousands of subtasks that are processed in parallel on thousands or even millions of CNTFETs, accelerating computation time by a large margin. The high-performance computing ability of CNTFETs, powered by their high electron mobility, accelerates critical operations in artificial intelligence (AI) and machine learning (ML) applications like the training of neural networks. In addition, CNT-based accelerators are highly energy-efficient, operating at lower voltages with little heat generation. This efficiency is especially beneficial for battery-powered devices and large-scale data centers, where power consumption and heat management are critical factors. The combination of speed, scalability, and energy efficiency makes CNT-based hardware accelerators a game-changing technology for the future of high-performance computing.

### 6.6.3 Advantages of CNT-Based Hardware Accelerators

- CNT transistors switch significantly faster than silicon-based transistors, enabling hardware acceleration that is significantly faster than existing technologies.

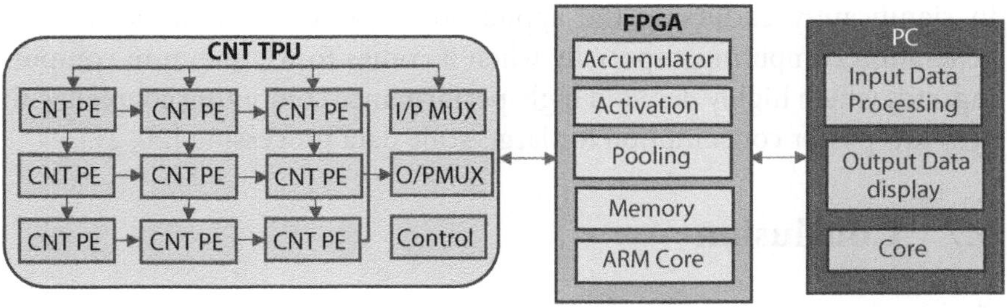


**Figure 6.4** Hardware accelerators using CNT transistors.

- CNTFETs tend to work at much lower voltages and consume much less power, making it suitable for portable devices as well as energy-conscious data centers.
- Due to the nanoscale nature, CNTs can fit more closely into silicon than transistors do, and thus enable more powerful accelerators without increasing the size of the chip.
- CNTs conduct heat better and hence decrease the thermal load on the system, allowing accelerators to carry out high-speed computations without overheating.
- Leakage currents in CNTs are much smaller compared to the existing technology based on silicon, which is a vital attribute toward the minimizing of power consumption and maximizing the reliability of the electronic systems.

### 6.6.4 Challenges in CNT Hardware Accelerator

Although promising so many good things, several issues still exist in the commercialization of CNT-based hardware accelerators:

- Uniform and defect-free CNTs in large quantities is another manufacturing challenge.
- The CNT-based components require adapting current silicon-based infrastructures, which would involve heavy investment in new technologies and production lines.
- The cost of CNT transistors has thus been a little higher than that of the traditional silicon transistors, even though it may come down with advancements in fabrication methods.

### 6.6.5 Future Prospect

Improvements in computational performance as well as energy efficiency form the basis of the future of hardware accelerators as provided by CNT-based transistors. Such progress in research and development areas helps in significantly endorsing the application of CNT transistors in next-generation computing, especially when it comes to AI, quantum computing, etc., which highly demand high-performance computing coupled with very low power consumption for large-scale data processing [20, 21].

## 6.7 Conclusion

Carbon nanowires are among the most promising nanomaterials due to their unique one-dimensional structure and excellent mechanical, electrical, and thermal properties. Such a perfect nanostructure is suitable for a wide range of applications like electronics, energy storage, biomedical technologies, and sensor systems for which it could revolutionize industries through faster, more efficient, and stronger devices. However, it has to overcome scaling up of its manufacturing capacity, cost, and integration with already existing technologies for CNTs to be widely adopted. With future research and improvement in techniques of fabrication, CNTs will have a central role to play in the near future in areas demanding both high performance and energy efficiency. Their versatile properties make them efficient contenders to drive innovation in nanotechnology and materials science.

**AI disclosure statement:** The author acknowledges the use of AI-based tools for minor language editing or grammatical correction. However, all intellectual contributions, analyses, and conclusions presented in this chapter are entirely the author's own.

## References

1. Khan, Z. and Khan, Z.A., Role of Extended Fins and Graphene Nanoplatelets in Coupled Thermal Enhancement of Latent Heat Storage System. *Energy Convers. Manage.*, 2020.
2. Kumar, A., Avasthi, P., Tripathi, G., Balakrishnan, V., Dhar, A., Carbon Nano Tube Forest on Stainless Steel Mesh as Catalyst Support for Development of Low-cost SCR for Diesel Engine NO Reduction, 2020.
3. Estevez, D. and Qin, F.X., Nano-carbon/magnetic Microwire Hybrid Fibers for Tunable Microwave Functionalities, 2020.
4. Wang, D., Zhang, W., Han, B., New Generation of Cement-based Composites for Civil Engineering, 2020.
5. Qu, Y., Wang, S., Zhou, D., Tian, Y., Experimental Study on Thermal Conductivity of Paraffin-based Shape-stabilized Phase Change Material with Hybrid Carbon Nano-additives. *Renewable Energy*, 2020.
6. Salari, A., Kazemian, A., Ma, T., Hakkaki-Fard, A., Peng, J., Nanofluid Based Photovoltaic Thermal Systems Integrated with Phase Change Materials: Numerical Simulation and Thermodynamic Analysis. *Energy Convers. Manage.*, 2020.
7. Raja, R., Jannet, S., George, L., Rao, V.M., Sebastian, N., Thomas, S.P., Synthesis of Al and Cu Wires Embedded Glass Fiber Reinforced Polymer (AWGFRP and CWGFRP) Composites and Their Mechanical Characterization. *Mater. Today Proc.*, 2020.
8. Dhand, V., Yadav, M., Hoon Kim, S., Yop Rhee, K., A comprehensive review on the prospects of multi-functional carbon nano onions as an effective, high-performance energy storage material. *Carbon*, 175, 534–5755, 2021.

9. Dai, X., Liu, H., Liu, X., Liu, Z., Liu, Y., Cao, Y., Tao, J., Shan, Z., Silicon nanoparticles encapsulated in multifunctional crosslinked nano-silica/carbon hybrid matrix as a high-performance anode for Li-ion batteries. *Chem. Eng. J.*, 418, 1294685, 2021.
10. Lu, X., Liu, J., Ding, Y., Yang, L., Yang, Z., Wang, Y., Simulation and fabrication of carbon nanotube–nanoparticle interconnected structures. *Mech. Sci.*, 12, 1, 451–459, 2021.
11. González-Berrıíos, A., Huang, D., Medina-Emmanuelli, N.M., Kristian, K.E., Ortiz, O.O., González, J.A., De Jesús, J., Vargas, I.M., Weiner, B.R., Morell, G., Effects of heavy-ion radiation on the electron field emission properties of sulfur-doped nanocomposite carbon films. *Diamond Relat. Mater.*, 13, 2, 221–225, 2004.
12. Ding, H. and Maier, J.P., Electronic structures of one-dimension carbon nano wires and rings. *J. Phys. Conf. Ser.*, 61, 1, 252, 2007. IOP Publishing.
13. El Nadi, L., Omar, M.M., Moniem, H.A., Formation of Nano Wires by Laser Ablation of Graphite, in: *AIP Conference Proceedings*, vol. 888, no. 1, pp. 173–176, AIP Publishing, 2007. https://doi.org/10.1063/1.2711107.
14. Guoping, W., Qingtang, Z., Zuolong, Y., MeiZheng, Q., The effect of different kinds of nano-carbon conductive additives in lithium ion batteries on the resistance and electrochemical behavior of the LiCoO2 composite cathodes. *Solid State Ionics*, 179, 7-8, 263–268, 2008.
15. Janghorban, M., Static and free vibration analysis of carbon nano wires based on Timoshenko beam theory using differential quadrature method. *Lat. Am. J. Solids Struct.*, 8, 463–472, 2011.
16. Chen, H., Chen, Y., Gong, W., Xiang, K., Sun, B., Liu, J., Preparation and electrochemical performance of LiFePO4/C composite with network connections of nano-carbon wires. *Mater. Lett.*, 65, 3, 559–561, 2011.
17. Taha, T.J., Thakur, D.B., Van der Meer, T.H., Towards convective heat transfer enhancement: surface modification, characterization and measurement techniques. *J. Phys. Conf. Ser.*, 395, 1, 012113, IOP Publishing, 2012.
18. Dhanaselvam, P.S., Karthikeyan, B., Nagarajan, S., Padmanaban, B., Nanowire Transistors, in: *Field Effect Transistors*, pp. 81–92, 2025.
19. Khe, N.C., Van Giat, L., Ngoc Tuan Anh, M., Thi Thu Thao, B., Quang Thuan, B., Duc Truong, H., Van Nam, B., Dinh, H., Graphene Alloy with Nano Carbon: Novel Nano Material for Biosensor?, in: *4th International Conference on Biomedical Engineering in Vietnam*, pp. 298–300, Springer Berlin Heidelberg, Berlin, Heidelberg, 2013.
20. Dhanaselvam, P.S., Karthikeyan, B., Vanitha, P., Principle and application of C-based nanomaterials: graphite, graphene, and carbon nanotube (CNT), in: *Handbook of Emerging Materials for Semiconductor Industry*, pp. 49–64, Springer Nature Singapore, Singapore, 2024.
21. Dhanaselvam, P.S., Karthikeyan, B., Anand, P., Future Prospective Beyond CMOS Technology Design, in: *Field Effect Transistors*, pp. 73–80, 2025.

# 7

# Junctionless Transistor

**S. Manikandan*, M. Karthigai Pandian and A. Srinivasan**

*NPR College of Engineering and Technology, Natham, Dindigul, Tamil Nadu, India*

***Abstract***

Junctionless transistors represent a significant breakthrough in semiconductor technology, offering a simplified yet highly effective alternative to traditional transistors. Unlike conventional Metal-Oxide-Semiconductor Field-Effect Transistors (MOSFETs), which rely on p-n junctions to control the flow of current, Junctionless transistors operate using a uniformly doped channel, either n-type or p-type. This architecture eliminates the need for junction formation, thus simplifying the fabrication process and reducing manufacturing costs, especially at nanoscale dimensions. By controlling current purely through the electric field applied to the gate, Junctionless transistors provide enhanced electrostatic control and help mitigate common short channel effects (SCEs) such as Drain-Induced Barrier Lowering (DIBL), threshold voltage roll-off, and increased leakage current. This makes them highly suitable for applications where low power consumption, high efficiency, and high performance are crucial. As the semiconductor industry continues to push the limits of scaling, particularly with the shift towards sub-nanometer technologies, the need for innovative transistor architectures becomes even more pressing. The traditional scaling of MOSFETs has reached its physical limits, requiring new device designs like gate-all-around (GAA) structures. In this context, cylindrical surrounding gate (CSG) transistors are gaining attention due to their superior electrostatic control over the channel, significantly reducing SCEs. Combining Junctionless architecture with GAA CSG structures results in devices that are highly efficient and scalable, paving the way for next-generation low-power, high-speed semiconductor devices. This book chapter focuses on the design, modeling, and analysis of Junctionless Cylindrical Surrounding Gate (CSG) Silicon Nanowire Transistors. Various gate materials, including single, double, and triple materials, are explored to improve the device's performance. The study begins by analyzing Junctionless Single Material CSG (JLSMCSG) silicon nanowire transistors, where

**Corresponding author*: manikandans@nprcolleges.org

P. Suveetha Dhanaselvam, Naima Guenifi, Shiromani Balmukund Rahi and J. Ajayan (eds.) Field-Effect Transistor Devices: Post-CMOS Theory and Applications, (129–152) © 2026 Scrivener Publishing LLC

an analytical model for drain current, transconductance, and transcapacitance is developed using the Pao-Sah gradual channel approximation. By solving the 2D Poisson's equation, the model evaluates the potential distribution in the channel, and the results are validated through MATLAB simulations and the Sentaurus TCAD numerical simulator. Subsequently, the work extends the analysis to Junctionless Double Material CSG (JLDMCSG) and Junctionless Triple Material CSG (JLTMCSG) transistors. Introducing multiple gate materials in these designs enhances electrostatic field control and reduces SCEs such as DIBL and tunneling effects. The study demonstrates that the use of double and triple material gates leads to a significant reduction in threshold voltage and improved transconductance, with the JLTMCSG design showing the most substantial improvements. The analytical modeling for these devices examines the channel potential and investigates the impact of design parameters such as oxide thickness, gate length, and doping concentration on device performance. A key focus of the chapter is the subthreshold behavior of these junctionless transistors, which is crucial for low-power applications. The ON/OFF switching characteristics, an essential feature for digital circuit design, are analyzed, demonstrating that the JLTMCSG device offers enhanced switching capabilities. The comparative analysis between single, double, and triple material gate designs reveals that the introduction of additional gate materials leads to proportional reductions in SCEs and improved overall device efficiency. In conclusion, this chapter highlights the potential of Junctionless transistors, particularly in combination with CSG structures, as promising candidates for future semiconductor devices. The findings establish that these devices meet the growing demands of the industry for low-power, high-speed, and cost-effective solutions in mobile, computing, and digital applications.

***Keywords:*** Cylindrical surrounding gate, gate all around, short channel effects

## 7.1 Introduction

Recent advances in semiconductor technology provide us with the resources to explore alternative methods for fabricating transistors with the goal of further reducing their sizes to increase transistor density and enhance performance. Conventional transistors use semiconductor junctions; they are formed by doping atoms on the silicon substrate that makes p-type and n-type regions. Decreasing the size of such transistors means that the junctions will get closer, which becomes very challenging when the size is reduced to the lower end of the nanometer scale due to the requirement of extremely high gradients in doping concentration. One of the most promising solutions to overcome this issue is realizing junctionless transistor. Recent advances in semiconductor technology provide us with the resources to explore alternative methods for fabricating transistors with the goal of further reducing their sizes to increase transistor density and enhance performance. Conventional transistors use semiconductor junctions; they are formed by doping atoms on the silicon substrate that makes p-type and n-type regions. Decreasing the size of such transistors means that the junctions will get closer, which becomes very challenging when the size is reduced to the lower end of the nanometer scale due to the requirement of extremely high gradients in doping concentration. One of the most promising solutions to overcome this issue is realizing junctionless transistor [18–22].

Transistors are the fundamental building blocks of modern electronic devices and all existing transistors contain semiconductor junctions. The most common type of junction is the p–n junction, which is formed by the contact between a p-type piece of silicon doped with impurities to create an excess of holes and an n-type piece of silicon, doped to create an excess of electrons. Other junctions include the hetero junction, which is simply a p–n junction containing two different semiconductors, and the Schottky junction between metal and semiconductor. The bipolar junction transistor contains two p–n junctions, and so does the MOSFET (metaloxide–semiconductor field-effect transistor).The JFET (junction field-effect transistor) has only one p–n junction and the MESFET (metal–semiconductor field-effect transistor) contains a Schottky junction. However, the formation of ultrasharp and shallow source/drain junctions to suppress short-channel effects is of great concern and still imposes constraints regarding doping techniques and thermal budget. Recently, the junctionless (JL) transistor has been shown to avoid these problems [18–22].

### 7.1.1 Key Characteristics of Junctionless FETs [1–3]

a) **No Junctions:** Unlike conventional FETs, which have p-n junctions (between the source, drain, and channel regions), JLFETs consist of a uniformly doped semiconductor material for the channel, source, and drain regions. This simplifies the fabrication process.
b) **Doping and Operation:** The channel is highly doped with the same type of carriers (either all electrons in n-type or all holes in p-type), which makes it conductive. However, by applying a gate voltage, you can control the current flow by depleting or accumulating carriers in the channel.
c) **Reduced Short-Channel Effects:** Because JLFETs are usually designed with a cylindrical or nanowire structure,

they show better control over the channel, reducing short-channel effects (like drain-induced barrier lowering) compared to conventional FETs. This makes them suitable for scaling down to nanometer dimensions.

d) **Subthreshold Behavior:** JLFETs exhibit unique subthreshold behavior, which can be optimized by adjusting the gate structure, gate materials, and doping levels. This behavior is particularly interesting in low-power applications, where control of the off-state current is critical.

### 7.1.2 Fundamental Principles of Junctionless FETs

In JLFETs, the gate voltage controls the current by depleting or accumulating charge carriers in the channel rather than by switching the channel between n-type and p-type regions. This results in a fundamentally different operating principle from traditional MOSFETs, where the source and drain junctions form p-n junctions with the channel.

A. **Uniform Doping:** In JLFETs, the entire channel region is uniformly doped, which simplifies fabrication and avoids the complex processes involved in forming junctions.
B. **Electrostatic Control:** Since the gate directly controls the carrier density in the channel, JLFETs can achieve high levels of electrostatic control, which helps to suppress short-channel effects and maintain performance at reduced dimensions.

This design approach offers **two main benefits**: improved scalability to smaller geometries and reduced leakage currents, which contribute to better power efficiency.

## 7.2 Junctionless FET Structures

### 7.2.1 Planar Junctionless FET

The planar junctionless FET is one of the most straightforward structures, with a flat channel beneath a planar gate. This structure closely resembles traditional MOSFET designs and offers compatibility with existing fabrication processes, making it an appealing option for rapid integration into current CMOS technology.

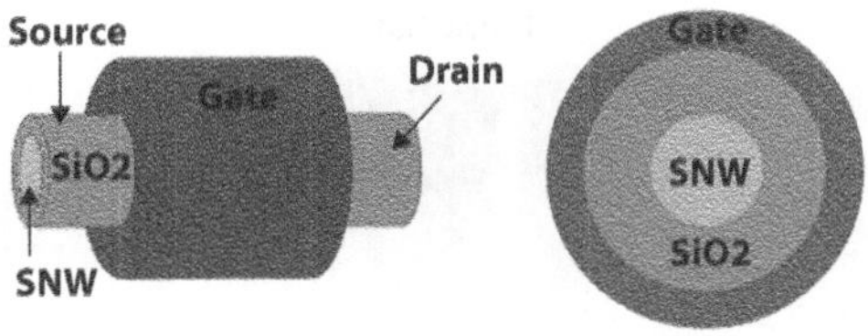


**Figure 7.1** SG silicon nanowire transistors.

### 7.2.2 Nanowire Junctionless FET

The nanowire junctionless FET structure uses a narrow, wire-like channel, which provides exceptional gate control due to the high surface-area-to-volume ratio. This structure is particularly useful in ultra-scaled nodes, where maintaining control over the channel is challenging.

There are several kinds of nanowire transistors depend mainly upon the gate material and its properties. The surrounding gate transistor shown in Figure 7.1 is also having several dimensions. They are vertical surrounding gate, CSG, dual material surrounding gates etc. The surrounding gate transistors having various dimensions and doping profiles and they differ with the parameters associated with the channel junction [4]. The threshold voltage is one of the major components which are the activation part of any transistor. So care must be taken while modeling this threshold voltage and several second order effects are affecting this threshold voltage.

There are several kinds of threshold voltage models have been proposed by various authors in the past years and they are all analyzed to get a clear idea about the threshold voltage modeling of silicon nanowire transistors.

### 7.2.3 Double-Gate Junctionless FET

In Double-gate junctionless FETs, two gates control the channel from opposite sides. This configuration improves gate control over the channel, reducing leakage currents and enhancing scalability. The double-gate structure is a precursor to more advanced configurations, such as gate-all-around structures [5]. The double gate MOSFETs are fabricated with two gates, one on the top and other on the bottom as shown in Figure 7.2. Figure 7.3 shows the cross sectional view of the Double Gate MOSFET. The symmetrical double gate transistors use the similar type of gate material to fabricate both top and bottom gate. But the asymmetrical double gate transistors are fabricated with different materials. In this model the

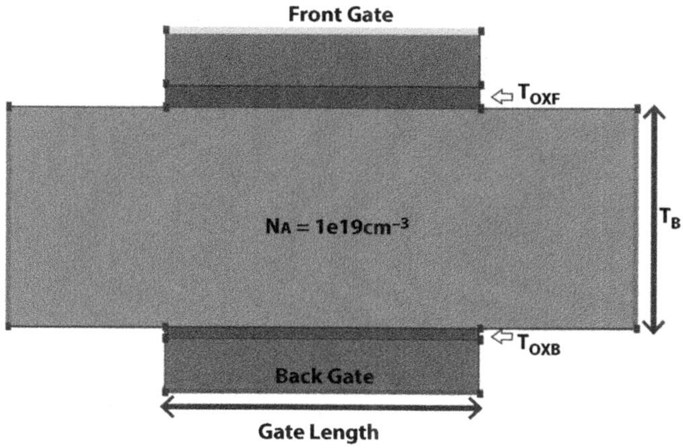


**Figure 7.2** Cross-sectional view of double gate transistor.

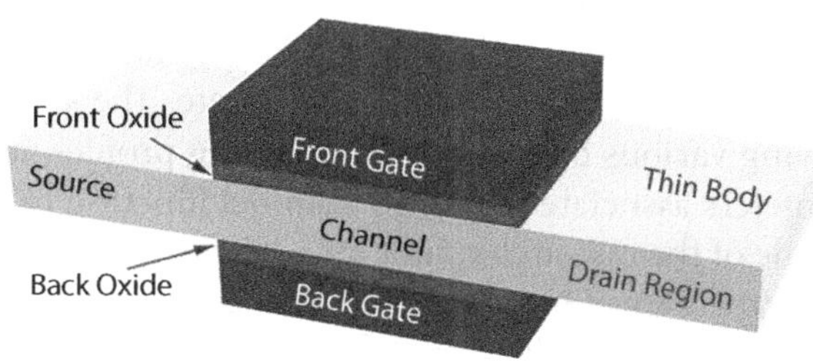


**Figure 7.3** Double-gate transistor.

current flow and the channel width is controlled by both gates; therefore, it offers better control over the channel and elevates the trapped charge effects.

### 7.2.4 Surrounding Gate (Cylindrical) JLFET

The surrounding gate junctionless FET, also known as a cylindrical or gate-all-around structure, wraps the gate entirely around the channel. This arrangement provides optimal control over the channel, allowing for extremely fine adjustments to the carrier concentration and minimizing leakage current. The Cylindrical Surrounding Gate (CSG) transistor is the most promising devices to scale down below100 nm [6]. Figure 7.4 shows the surrounding gate MOSFET where the channel is completely surrounded by the gate. The gate has an increased electrostatic control over the channel and suppresses the SCEs.

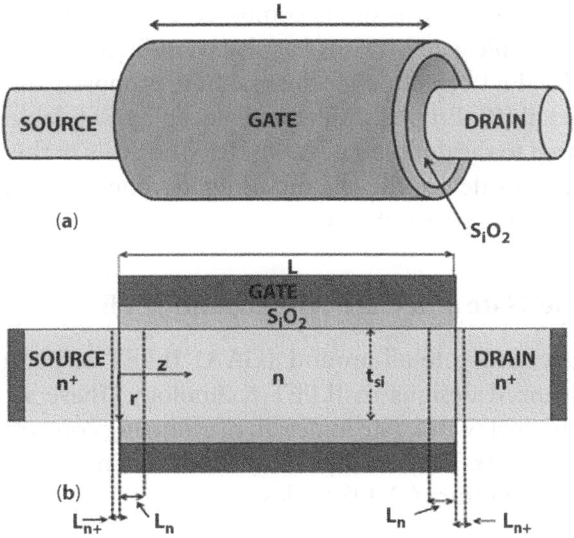


**Figure 7.4** Surrounding gate MOSFET.

### 7.2.5 Double-Material Gate JLFET

The double-material gate junctionless FET employs two different materials for the gate, allowing for better control over the electric field along the channel. This structure provides enhanced control over carrier distribution, which improves device performance by lowering short-channel effects and enhancing drive current. The merits of double metal and

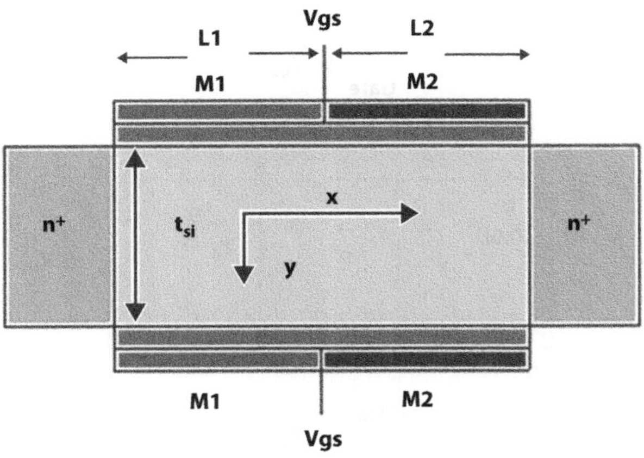


**Figure 7.5** Schematic of dual material double gate (DMDG) MOSFET.

double material gate transistors is combined and proposed a new device dual material double gate MOSFET as shown in Figure 7.5.

DMDG MOSFET structure for the first time proposed and discussed its features in detail [7]. Proposed device achieved considerable reduction in SCE's compared to the double gate MOSFET. The SCEs in this structure are suppressed. In this device the channel is lightly doped or even undopped which offers good carrier transport.

### 7.2.6 Triple-Gate and Gate-All-Around JLFET

The triple-gate and gate-all-around (GAA) JLFET structures represent the most advanced designs in JLFET technology. These structures fully enclose the channel with a gate material, maximizing control and reducing short-channel effects. They are especially effective at nanoscale dimensions and are integral to future CMOS nodes.

## 7.3 Modeling of Junctionless Fet

The two-dimensional JL single material CSGSNW Transistor shown in Figure 7.6 [1–8]. For analytical purposes, the regions of source and drain are omitted and their contacts are placed near the gate. The transistor channel is made up of silicon nanowire of diameter 10nm and the channel is junction free. The channel is lightly doped which has the doping concentration of $N_D$ $1\times10^{19}$ $cm^{-3}$. The current conduction takes place in the center of the channel which is kept along the "z" direction. The radius of the device is denoted as 'r' perpendicular to the direction of current flow

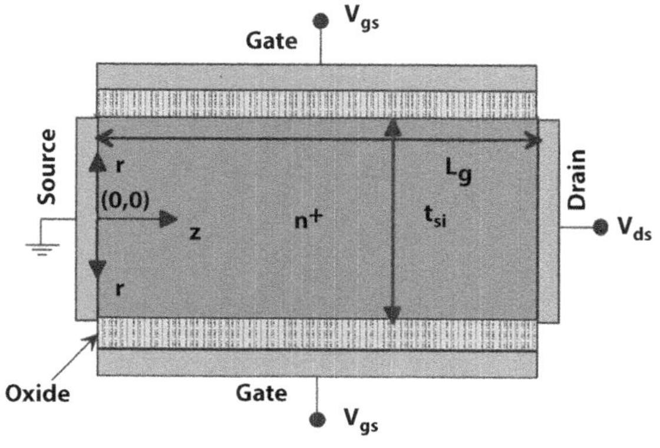


**Figure 7.6** 2-D view of JL single material CSGSNW transistor.

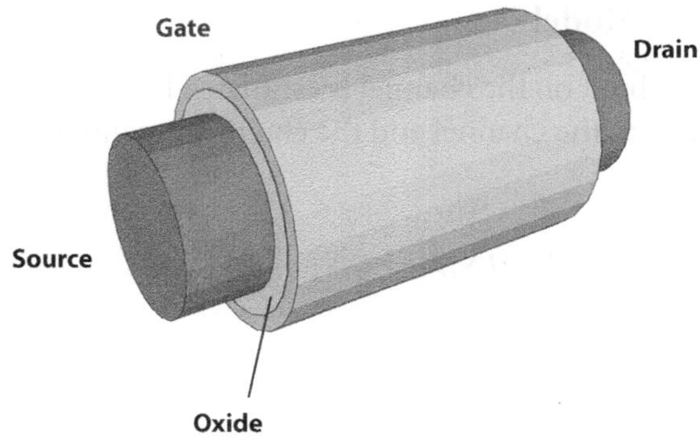


**Figure 7.7** 3-D view of JL single material CSGSNW transistor.

in the channel. The junctionless single material CSGSNW transistor suppresses the subthreshold leakage current, offers better mobility and variability from the integration of undoped (or) lightly doped silicon in the channel and the threshold voltage of the device preferably adjustable.

The 3-D view of the JL Single Material CSGSNW transistor is shown in Figure 7.7. The centre green colored cylindrical region of transistor channel is built using silicon nanowire of diameter d=10 nm. The silicon nanowire is surrounded by insulation layer of thickness $t_{ox}$=2 nm and gate metal layer is deposited over the insulation layer. One end of the channel is considered as source and another end as drain. Region from the source to drain has the same doping profile; therefore, the source, drain and the channel are equal in nature under ideal condition.

### 7.3.1 Analytical Modeling

In a junctionless nanowire transistors, the gate bias is applied to deplete the carriers at the center of the device [9]. Considering the mobile charges (electrons for NMOS) and the Pao-Sah gradual channel approximation, the potential distribution across the channel is approximated using Poisson solution [17].

$$\frac{d^2\varphi(r,z)}{dz^2} = -\frac{qN_D}{\varepsilon_{si}}\left(1-e^{\frac{(\varphi-V)}{v_T}}\right) \tag{7.1}$$

where $\varphi(r,z)$ is the channel potential along the 'z' direction $\varepsilon_{si}$ is the permittivity of the silicon channel, and V is the electron quasi Fermi potential.

### 7.3.2 Charge Model

The total charge built on the channel is estimated by integrating the charge density available in the channel and the channel potential as follows.

$$Q = qN_D t_{si} - qN_D \int_{-t_{si}}^{t_{si}} e^{\frac{(\varphi - V)}{v_T}} dz$$

$$Q = qN_D t_{si} \left(1 - e^{\frac{(\varphi_c - V)}{v_T}}\right)$$

$$Q = qN_D t_{si} \left(1 - e^{\frac{(\varphi_c - V)}{v_T}} \sqrt{\frac{\pi v_T}{\Delta\varphi}} \times Erf\left(\sqrt{\frac{\Delta\varphi}{v_T}}\right)\right) \tag{7.2}$$

$$Erf\left(\sqrt{\frac{\Delta\varphi}{v_T}}\right) \cong 1$$

The charge Equation 7.4 is not an analytical expression at this point; but, if $\Delta\varphi / v_T$ is equal to or larger than one, the error function becomes unity. Therefore a new equation relating to $\varphi_s$ (surface Potential) and $\varphi_c$ (center Potential) can be derived as

$$Q = qN_D t_{si} \left(1 - e^{\frac{(\varphi_c - V)}{v_T}} \sqrt{\frac{\pi v_T}{\Delta\varphi}}\right)$$

The charge dispersed along the channel can also found in terms of mobile charge $(Q_{mb})$ as

$$Q = \frac{1}{2}[Q_{mb} + qN_D t_{si}]$$

$$Q = \frac{4\varepsilon_{si}}{t_{si}} \Delta\varphi$$

Where $\Delta\varphi = \frac{t_{si}}{8\varepsilon_{si}}[Q_{mb} + qN_D t_{si}]$

$$2\left(\frac{4\varepsilon_{si}}{t_{si}}\Delta\varphi\right) = qN_D t_{si}\left(1 - e^{\frac{(\varphi_c - V)}{v_T}}\sqrt{\frac{\pi v_T}{\Delta\varphi}}\right) \tag{7.3}$$

The sheet charge of mobile carriers can be obtained by subtracting the dopant charge term $(qN_D t_{si})$ from Equation 7.10 and simplified as

$$Q = -qN_D t_{si} e^{\frac{(\varphi_c - V)}{v_T}}\sqrt{\frac{\pi v_T}{\Delta\varphi}} \tag{7.4}$$

The relationship between surface potential $\varphi_s$, center potential $\varphi_c$ and the gate bias $V_{gs}$ can be obtained by substituting boundary conditions.

$$Q = \frac{4\varepsilon_{si}}{t_{si}}\Delta\varphi = -\frac{\varepsilon_{ox}}{t_{ox}}\left(V_{gs} - V_{th} - V\right) \tag{7.5}$$

The closed form mobile charge expression within the channel can be obtained by equating Equations 7.4 and 7.5

$$-\frac{2\varepsilon_{ox}}{t_{ox}}\left(V_{gs} - V_{th} - V\right) = Q_{mb} - \frac{2\varepsilon_{ox} v_T}{\beta t_{ox}}\ln\left[-Q_{mb}\sqrt{\frac{Q_{mb} + qN_D t_{si}}{2\varepsilon_{si}\pi v_T q^2 N_D^2 t_{si}}}\right] \tag{7.6}$$

Where $\beta = 1 + \frac{\varepsilon_{ox} t_{si}}{4 t_{ox}\varepsilon_{si}}$

In the subthreshold region of operation when the gate bias is less than the threshold voltage, the mobile charge $Q_{mb} \ll qN_D t_{si}$ hence the Equation 7.6 can be simplified as

$$-\frac{2\varepsilon_{ox}}{t_{ox}}\left(V_{gs} - V_{th} - V\right) = Q_{mb} - \frac{2\varepsilon_{ox} v_T}{\beta t_{ox}}\ln\left[\frac{-Q_{mb}}{\sqrt{2\varepsilon_{si}\pi v_T q N_D}}\right] \tag{7.7}$$

The charge distribution on the channel is not uniform from the source to drain, it is differs based on the source and drain potential. The Drain region Charge $Q_D$ can be obtained by substituting $V_{ds}$ instead of 'V' in Equation 7.7

$$-\frac{2\varepsilon_{ox}}{t_{ox}}\left(V_{gs}-V_{th}-V_{ds}\right)=Q_{mb}-\frac{2\varepsilon_{ox}v_T}{\beta t_{ox}}\ln\left[\frac{-Q_{mb}}{\sqrt{2\varepsilon_{si}\pi v_T q N_D}}\right] \tag{7.8}$$

The source of the transistor is connected to ground with respect to positive gate bias. Therefore the Source region Charge $Q_S$ can be obtained by substituting '0' instead of 'V' in Equation 7.8

$$-\frac{2\varepsilon_{ox}}{t_{ox}}\left(V_{gs}-V_{th}\right)=Q_{mb}-\frac{2\varepsilon_{ox}v_T}{\beta t_{ox}}\ln\left[\frac{-Q_{mb}}{\sqrt{2\varepsilon_{si}\pi v_T q N_D}}\right]$$

### 7.3.3 Drain Current Model

When the gate bias reaches the threshold voltage, the channel gradually build on the center of the nanowire by depleting the holes and the electrons transport takes place from the source terminal to the drain terminal. Therefore the transistor turns 'ON' and starts conducting. The mathematical model for drain current is found by resolving the current continuity equation

$$I_{DS}=-\mu\frac{2\pi R}{L_{eff}}\int_0^{V_{ds}}Q_{mb}dv$$

$$I_{DS}=-\mu\frac{2\pi R}{L_{eff}}\int_{Q_S}^{Q_D}Q_{mb}\frac{dv}{dQ_{mb}}dQ_{mb}$$

In the above expression the '$L_{eff}$' is the effective gate length of the device. The potential difference is obtained from the charge equation

$$V=\frac{\beta t_{ox}}{2\varepsilon_{ox}}Q_{mb}-v_T\ln\left[\frac{-Q_{mb}}{\sqrt{2\varepsilon_{si}\pi v_T q N_D}}\right]+\left(V_{gs}-V_{th}\right) \tag{7.9}$$

Integrating voltage *w.r.t* mobile charge '$Q_{mb}$'

$$\frac{dV}{dQ_{mb}}=\frac{\beta t_{ox}}{2\varepsilon_{ox}}-v_T\left[\frac{1}{Q_{mb}}\right] \tag{7.10}$$

Substitute Equation 7.10 in Equation 7.9 and solve for drain current $I_{ds}$

$$I_{DS} = -\mu \frac{2\pi R}{L_{eff}} \left[ \frac{\beta t_{ox}}{4\varepsilon_{ox}} Q_{mb}^2 - v_T Q_{mb} \right]_{Q_S}^{Q_D} \tag{7.11}$$

### 7.3.4 Linear Region Drain Current

In linear operating condition the source region charge becomes very low usually negligible. Therefore the first term in Equation (7.11) on the right side can be simplified as

$$-\frac{2\varepsilon_{ox}}{t_{ox}} \left( V_{gs} - V_{th} - V \right) = Q_{mb} = Q_D \tag{7.12}$$

In the same way the drain charge in the Equation 7.11 dominates the source charge, hence $Q_S$ can be expressed as

$$I_{DS} = -\mu \frac{2\pi R}{L_{eff}} \left[ \frac{\beta t_{ox}}{4\varepsilon_{ox}} Q_{mb}^2 \right]_0^{Q_D}$$

$$I_{DS} \cong \mu \frac{2\pi R \varepsilon_{ox}}{\beta L t_{ox}} \left( V_{gs} - V_{th} - \frac{V_{ds}}{2} \right) V_{ds}$$

### 7.3.5 Drain Current in Saturation Region

The transistor experiences a pinch off at the centre of the channel near the drain region. In this condition [10], both drain and source charge is dominant, the second and first terms of the right side of charge equation are dominant, respectively. Henceforth the charge equation simplified as follows

$$Q_S => Q_{mb} = \sqrt{2\varepsilon_{si} \pi v_T q N_D} \times e^{\frac{\left(V_{gs} - V_{th} - V_{ds}\right)}{v_T}}$$

$$Q_D = Q_{mb} = -\frac{2\varepsilon_{ox}}{\beta t_{ox}} \left( V_{gs} - V_{th} \right)$$

Substitute the source charge and drain charge in Equation (7.11) and solve for saturation drain current.

$$I_{DS} = -\mu \frac{2\pi R\varepsilon_{ox}}{\beta L t_{ox}} \left[ \left(V_{gs} - V_{th}\right)^2 - \frac{\beta t_{ox}}{\varepsilon_{ox}} v_T \kappa \right]$$

$$\kappa = \sqrt{2\varepsilon_{si}\pi v_T q N_D} \times e^{\frac{\left(V_{gs} - V_{th} - V_{ds}\right)}{v_T}}$$

The subthreshold current also called as leakage current occurs when the gate bias is below the threshold voltage. In this condition the source charge dominating the drain charge and it is negligible; therefore the charge equation can be simplified as

$$Q_S => Q_{mb} = -\sqrt{2\varepsilon_{si}\pi v_T q N_D} \times e^{\frac{\left(V_{gs} - V_{th}\right)}{v_T}}$$

Substitute source charge in drain current Equation (7.11) and solve for drain current.

$$I_{DS} \cong \mu \frac{2\pi R}{L} \left[ v_T \sqrt{2\varepsilon_{si}\pi v_T q N_D} \times e^{\frac{\left(V_{gs} - V_{th}\right)}{v_T}} \psi \right]$$

$$\psi = \left(1 - e^{-\frac{\left(V_{ds}\right)}{v_T}}\right)$$

The above subthreshold current equation is effectively used to analyze the SCE behavior of the device under low bias condition.

The transconductance is the quantity to define the efficiency of the transistor. A good transistor should have high transport efficiency to build digital circuits [11]. Generally, it is defined as the ratio between output current to input voltage. Differentiating drain current with respect to gate bias voltage gives transconductance.

$$g_m = \frac{\partial I_{Ds}}{\partial V_{gs}}$$

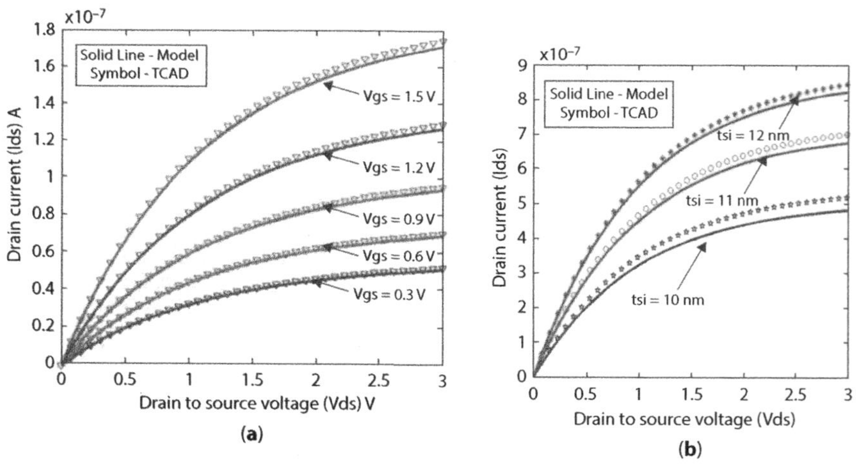

Substitute Equation (3.3.1) in above transconductance equation

$$g_m \cong \mu \frac{2\pi R}{L}\left[ v_{gs}\sqrt{2\varepsilon_{si}\pi v_T q N_D} \times e^{\frac{(V_{gs}-V_{th})}{v_T}} \psi \times \frac{\left(V_{gs}-V_{th}\right)}{v_T} \right]$$

The small signal behavior of the transistor is studied using the transconductance equation.

The gate capacitance is the combination of gate to source and gate to capacitance. The transcapacitance $C_{sg}$ & $C_{dg}$ can be obtained by substituting source charge and drain charge in drain equation. The transcapacitance can be calculated in the general case as

$$C_{ij} = \frac{\partial Q_{m,i}}{\partial V_j}$$

$$C_{sg} = \sqrt{2\varepsilon_{si}\pi v_T q N_D}\,\frac{\left(V_{gs}-V_{th}-V_{ds}\right)}{v_T^{\,2}} \times e^{\frac{(V_{gs}-V_{th}-V_{ds})}{v_T}}$$

$$C_{dg} = -\frac{2\varepsilon_{ox}}{\beta t_{ox}}$$

In particular Gate Capacitance

$$C_{gg} = C_{sg} + C_{dg}$$

## 7.3.6 Characteristics of Junctionless FET

The proposed transistor is designed and simulated using sentaurus TCAD numerical simulator and the analytical model results are validated by comparing with MATLAB results [12]. The JL single material CSGSNW transistor is designed with the specification mentioned in Table 7.1.

The drain characteristics of a JL single material CSGSNW transistor for different Gate bias is shown in Figure 7.8(a).

**Table 7.1** Parameters of JL single material CSG SNW transistor.

| Symbol | Parameter | Values |
|---|---|---|
| R | Nanowire Radius Thickness | 5 nm |
| Tsi | Silicon Thickness | 10nm |
| M | Gate Work Function | 4.8 Ev |
| L | Channel Length | 30 nm |
| Tox | Oxide Thickness | 2 nm |
| ND | Doping Concentration | 1×1019 cm-3 |
| Q | Charge | 1.6×10−19c |
| K | Boltzmann Constant | 1.38×10-23 |
| T | Temperature | 300K |

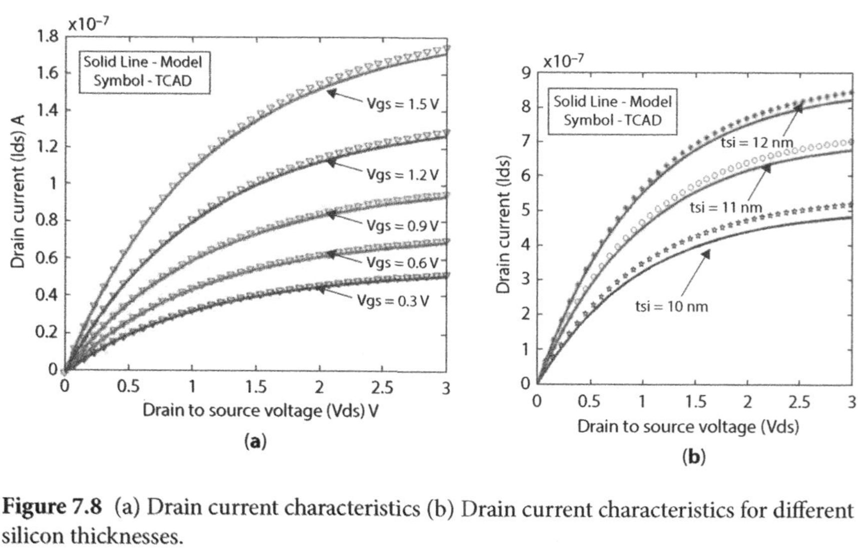


**Figure 7.8** (a) Drain current characteristics (b) Drain current characteristics for different silicon thicknesses.

The gate bias is retained constant for different values and the drain bias is steadily increased to observe the current flow in the channel. The result shows that the device operates linearly and a stable current drifts in the channel under the linear region. The drain current turns out to be constant above the threshold voltage and a device enters the saturation region. A maximum drain current of 0.18 microamperes is achieved for the value of $V_{gs}$=1.5V. The derived analytical model is validated using sentaurus TCAD software.

Figure 7.8(b) shows the drain current characteristics of a transistor with different silicon thickness (Tsi). It shows that the drain current increases proportionally with silicon thickness. A maximum drain current of 0.5 μA is achieved for the value of $t_{si}$=10nm. The device performance is better for the channel having silicon thickness of 10nm and above

Figure 7.9(a) shows the drain current versus the drain to source voltage for three different doping concentrations. A continuing rise in the drain current for increasing drain to source voltages is observed for aggregate intrinsic silicon concentrations. The result proves that the current density increases with accumulative doping concentration and operates linearly with steady current flow in the channel. Silicon nanowires are good immune to trapped charges [13], therefore it offers higher mobility with reduced velocity saturation in the channel.

Figure 7.9(b) shows the relationship between the drain current and gate to source voltages of a JLCSGSNW transistor. The transfer characteristics of the transistor are plotted as a function of different voltages. The result shows that the drain current increases nonlinearly with gate bias voltage for different $V_{ds}$ of 0.5V, 1V and 1.5V. An extreme drain current of 0.8 microamperes is witnessed for the value of $V_{ds}$=1.5V. A small amount of

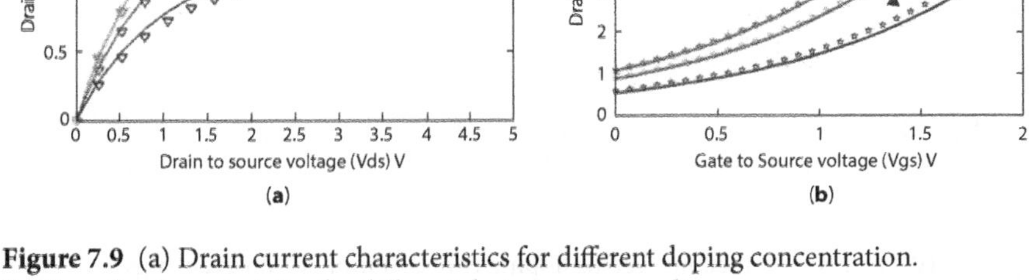


**Figure 7.9** (a) Drain current characteristics for different doping concentration. (b) Transfer characteristics for different drain to source voltage.

leakage current in the range of 0.05 micro amps flows in the channel even though there is no drain bias which is support to invert the channel quickly and increases the switching speed of the transistor. From this figure, it can be understood that the analytical model have a clear agreement with the TCAD simulation results.

Figure 7.10(a) shows the relationship between transconductance and channel doping concentration for three different drain biases. Results shows that the higher body doping concentration enhances the transconductance($g_m$). The most significant technological parameter that is connected to the interface orientation is the body doping concentration. While the doping concentration increases, the transconductance $g_m$ increases at a specific channel length. The theoretical results are plotted using MATLAB and validated with sentaurus TCAD simulation. These results will guide the designer to select the precise doping concentration for the silicon nanowire to realize the transistor with less leakage current.

Figure 7.10(b) shows the relationship between transconductance and effective channel lengths for three different drain biases. When we increase the drain bias voltage, transconductance increases with a reduction in the channel length. The reduction in the effective channel length happens due to drain induced barrier lowering which is called as SCEs. Therefore the transistor is appropriate for analog circuit applications where a higher transconductance to drain current ratio is desired.

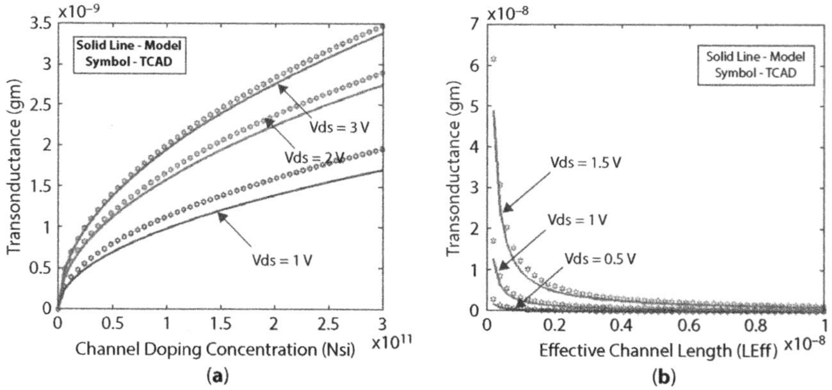


**Figure 7.10** (a) Tranconductance versus doping concentration for different Vgs (b) Tranconductance (Siemens) vs effective channel length (nm).

Figure 7.11(a) shows the relationship between the transcapacitance and oxide thickness for three different silicon thicknesses. The transcapacitance is a beneficial parameter to understand the charge density in the channel and the state of channel inversion. From the figure, it is evident that the gate capacitance (combination of gate to source capacitance and gate to drain capacitance) increases when we increasing the oxide thickness. Compared to the junction-based device the JL single material CSGSNW transistor offers higher transcapacticance which will reduce the hot carrier effects created by high electric field.

Figure 7.11(b) shows the relationship between the transcapacitance and channel silicon thickness for three different oxide thicknesses. The result shows that the transcapacitance increases when the oxide thickness increases, which tends to increase the overall gate capacitance. In our proposed transistor minimum 2nm oxide thickness is required to restrict the charges penetration in to the oxide region. If the oxide thickness is less than 1nm then the SiO2 layer loses its control and allows the current flow from the channel to the gate.

In recent trends, JL single material CSGSNW transistors are largely used to fabricate biosensor applications with low power and high sensitivity. A good agreement between numerical calculations and TCAD simulation results is achieved.

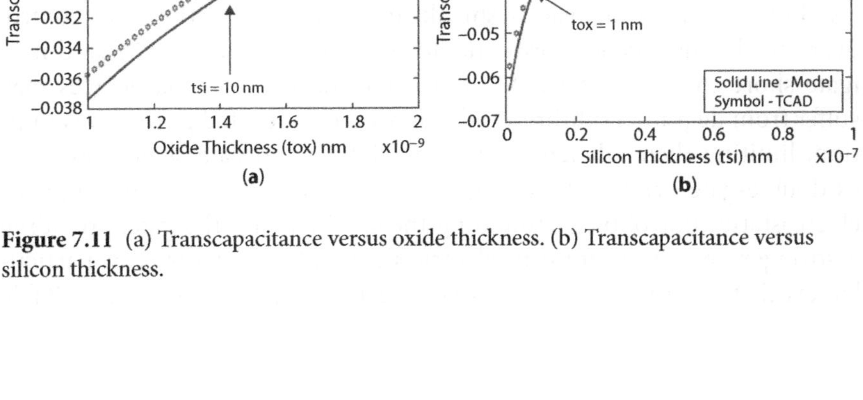


**Figure 7.11** (a) Transcapacitance versus oxide thickness. (b) Transcapacitance versus silicon thickness.

## 7.4 Challenges and Design Considerations in JLFET Fabrication

Junctionless field-effect transistors (JLFETs) offer many advantages [14–16], including simplified manufacturing and strong scalability potential. However, fabricating JLFETs comes with its own set of unique challenges and design considerations. This section explores these issues in detail, emphasizing the aspects that require precise control to achieve optimal performance at the nanoscale.

### 7.4.1 Doping Control and Uniformity

Junctionless field-effect transistors (JLFETs) depend on a uniformly doped channel to maintain junctionless operation; however, any variation in doping concentration can cause inconsistencies in device performance, particularly affecting threshold voltage and on-current. As device dimensions continue to shrink, achieving and maintaining uniform doping across the channel, source, and drain regions becomes increasingly challenging. High-precision doping techniques are essential to ensure the required uniformity, with methods like advanced ion implantation, molecular beam epitaxy (MBE), and chemical vapor deposition (CVD) offering potential solutions. While these approaches can significantly improve doping uniformity, they also add complexity and cost to the fabrication process, making efficient doping control a key consideration in JLFET design and scalability.

### 7.4.2 Gate Control and Electrostatic Integrity

As JLFET device dimensions shrink, maintaining effective gate control over the channel becomes essential to suppress leakage currents and mitigate short-channel effects. Without sufficient gate control, devices may suffer from degraded subthreshold behavior and elevated power consumption, limiting their efficiency and scalability. Multi-gate structures, such as double-gate, triple-gate, and gate-all-around (GAA) designs, enhance electrostatic control by surrounding the channel with the gate, providing a stronger and more uniform electric field. GAA structures, in particular, excel at controlling short-channel effects, making them highly suitable for ultra-scaled devices. However, while these multi-gate configurations improve performance, they also add significant complexity to the fabrication process, requiring precise alignment and advanced manufacturing techniques.

### 7.4.3 Material Selection for Gate and Channel

The choice of materials in JLFETs has a profound effect on device performance, reliability, and compatibility with conventional CMOS technology. High-mobility channel materials, like silicon-germanium (SiGe) and gallium arsenide (GaAs), can enhance performance, yet they present integration challenges within established silicon-based processes. Material selection for JLFETs therefore involves balancing mobility with ease of fabrication and thermal stability. While high-mobility materials offer performance benefits, they often necessitate advanced and costly fabrication methods. To address these challenges, researchers are also investigating emerging 2D materials, such as graphene and molybdenum disulfide ($MoS_2$), which promise further improvements in mobility and scalability, potentially supporting more compact and efficient JLFET designs.

### 7.4.4 Oxide Quality and Thickness Control

In JLFETs, ultra-thin oxide layers are required to maintain effective gate control, yet as oxide thickness decreases, the risks of oxide breakdown and tunneling leakage increase, which can compromise device reliability and longevity. To address this, high-κ dielectrics like hafnium oxide ($HfO_2$) are frequently used, as they allow for thicker insulating layers without sacrificing gate control. This approach helps mitigate leakage and improves durability. Precise thickness control, typically achieved through atomic layer deposition (ALD), is also essential to maintain consistent oxide properties across devices, ensuring reliable performance even at nanoscale dimensions.

### 7.4.5 Process Integration with CMOS Technology

To be commercially viable, JLFETs need to integrate smoothly with existing CMOS technology, yet this poses challenges, as JLFET fabrication often requires different techniques than those used for conventional MOSFETs.

Achieving compatibility without disrupting established CMOS process flows can be complex, especially when adjustments to temperature-sensitive steps and other process elements are needed. One solution is a hybrid integration approach, where JLFETs and conventional MOSFETs coexist on the same chip, allowing for incremental adoption of JLFETs. By carefully aligning fabrication steps with CMOS processes, such hybrid designs can support a gradual transition while leveraging the unique advantages of both transistor types.

**AI disclosure statement:** The author acknowledges the use of AI-based tools for minor language editing or grammatical correction. However, all intellectual contributions, analyses, and conclusions presented in this chapter are entirely the author's own.

## References

1. Nowbahari, A., Roy, A., Marchetti, L., Junctionless transistors: State-of-the-art. *Electronics*, 9, 7, 1174, 2020.
2. Colinge, J.P., Lee, C.W., Afzalian, A., Akhavan, N.D., Yann, R., Ferain, I., Razavi, P., Neill, B.O., Blake, A., White, M., Kelleher, A.M., McCarthy, B., Murphy, R., Nanowire transistors without junctions. *Nat. Nanotechnol.*, ED-5, 225–229, 2010.
3. Karthigai Pandian, M., Balamurugan, N.B., Manikandan, S., Analytical modeling of junction less surrounding gate silicon Nanowire Transistors. *J. Nanoelectron. Optoelectron.*, 9, 1–6, 2014.
4. Dehdashti Akhavan, N., Ferain, I., Razavi, P., Yu, R., Colinge, J.-P., Improvement of carrier ballisticity In junctionless nanowire transistors. *Appl. Phys. Letter*, 98, 10, 103510, 2011.
5. Te-Kuang, C., A Compact Model for Threshold Voltage of Surrounding-Gate MOSFETs with Localized Interface Trapped Charges. *IEEE Trans. Electron Devices*, 58, 2, 567–571, 2011.
6. Auth, C.P. and Plummer, J.D., A Simple Model for Threshold Voltage of Surrounding-Gate MOSFETs. *IEEE Trans. Electron Devices*, 45, 11, 2381–2383, 1998.
7. Reddy, G.V. and Jagadesh Kumar, M., A new dual-material double-gate (DMDG) nanoscale SOI MOSFET-two-dimensional analytical modeling and simulation. *IEEE Trans. Nanotechnol.*, 4, 260–268, 2005.
8. Manikandan, S., Suveetha Dhanaselvam, P., Karthigai Pandian, M., A Quasi 2-D Electrostatic Potential and Threshold Voltage Model for Junctionless Triple Material Cylindrical Surrounding Gate Si Nanowire Transistor. *J. Nanoelectron. Optoelectron.*, 16, 318–323, 2021.

9. Kaur, H., Kabra, S., Haldar, S., Gupta, R.S., An Analytical Drain Current Model for Graded Channel Cylindrical/Surrounding Gate MOSFET. *Microelectron. J.*, 38, 3, 352–359, 2007.
10. Colinge, J.P., Lee, C.W., Afzalian, A., Dehdashti, N., Yan, R., Ferain, I., *et al.*, SOI gated resistor: CMOS without junctions. *IEEE Int. SOI Conf.*, 2, 1–2, 2009.
11. Lee, C.-W., Ferain, I., Afzalian, A., Yan, R., Akhavan, N.D., Razavi, P., *et al.*, Performance estimation of junctionless multigate transistors. *Solid State Electron.*, 54, 2, 97–103, 2010.
12. Suveetha Dhanaselvam, P. and Balamurugan, N.B., Analytical approach of a nanoscale triple- material surrounding gate (TMSG) MOSFETs for reduced short-channel effects. *Microelectron. J.*, 44, 5, 400–404, 2013.
13. Nowak, E.J., Aller, I., Ludwig, T., Kim, K., Joshi, R., Chuang, C.-T., *et al.*, Turning silicon on its edge. *IEEE Circuits Device Mag.*, 20, 1, 20–31, 2004.
14. Duarte, J.P., Choi, S.J., Moon, D.I., Choi, Y.K., Simple analytical bulk current model for long-channel double-gate junctionless transistors. *IEEE Electron Device Lett.*, 32, 6, 704–706, 2011.
15. Dhanaselvam, P.S., Srinivasa Rao, K., Rahi, S.B., Yadav, D.S. (Eds.), *Field Effect Transistors*, John Wiley & Sons, 2025.
16. Manikandan, S., Suveetha Dhanaselvam, P., Karthigai Pandian, M., A 2D unified subthreshold drain current investigation for junctionless cylindrical surrounding gate (JCSG) silicon nanowire transistor. *Silicon*, 14, 10, 5311–5318, 2022.
17. Arun Samuel, T.S., Karthigai Pandian, M., Shenbagavalli, A., Arumugam, A., Analytical Modeling and Simulation of Advanced Silicon Nanowire Transistors. *Journal of Experimental and Theoretical Nanotechnology Specialized Researches (JETNSR)*, 2, 3, 2018.
18. Song, Y.S., Laxman, R.T., Tayal, S., Rahi, S.B., Samuel, T.S.A., eds. Handbook of Emerging Materials for Semiconductor Industry. Springer, 2024.
19. Rahi, S.B., Pranav, A., Shoubhik, G., Heterogate junctionless tunnel field-effect transistor: future of low-power devices. *J. Comput. Electron.*, 16, 1, 30–38, 2017.
20. Singh, P., Singh, N., Miao, J., Park, W.T., Kwong, D.L., Gate-all-around junctionless nanowire MOSFET with improved low-frequency noise behavior. *IEEE Electron Device lett.*, 32, 12, 1752–1754, 2011.
21. Rahi, S.B. and Bahniman Ghosh, B., High-k double gate junctionless tunnel FET with a tunable bandgap. *RSC Advances*, 5, 67, 54544–54550, 2015.
22. Asthana, P.K., Ghosh, B., Rahi, S.B.M., Goswami, Y., Optimal design for a high performance H-JLTFET using HfO2 as a gate dielectric for ultra low power applications. *RSC Advances*, 4, 43, 22803–22807, 2014.

# 8

# Field-Effect Transistor-Based Biosensors in Medical Field: Principles and Material Innovations

**Shivangi Srivastava and Sajal Agarwal***

*Department of Electrical and Electronics Engineering, Rajiv Gandhi Institute of Petroleum Technology, Jais, India*

***Abstract***

In current toxic and polluted scenario, biosensing in various fields, particularly in medicine, is playing a crucial role as numerous diseases affecting human body. Real-time monitoring and detection of these diseases may save many human lives as early detection of these diseases is essential for diagnosis. Human body expel out many volatile organic compounds (VOCs) through stools, urine, saliva, breath, etc. These VOCs provide important insights of an individual's metabolism and by analyzing their level and concentrations, one can identify the diseases and abnormalities associated with that specific human body. Unique VOC signatures show information about various health conditions such as diabetic, respiratory system diseases, kidney functionality and even cancer. To detect composition and concentration of these VOCs, portable and handy biosensors are required as they are convenient to use and can integrate with other systems. Although various invasive and non-invasive techniques exist, such as needle-based electrochemical sensors, biopsies, implantable biosensors, chromatography and mass spectrometry (GC-MS). But FET-based sensors are appropriate for the application of biosensing due to its high efficiency, cost effectiveness, wide detection range and compact size. These biosensors are basically an analytical tool that can transform a biochemical signal into a measurable electrical quantity such as change in current, transconductance, etc. This chapter provides a detailed and analytical explanation of the principles underlying bio-field effect transistors (bio-FETs) and their sensing mechanisms are provided along with the VOCs associated with specific disease and their origin. This chapter also covers different VOCs, such as ammonia

**Corresponding author*: sagarwal@rgipt.ac.in

P. Suveetha Dhanaselvam, Naima Guenifi, Shiromani Balmukund Rahi and J. Ajayan (eds.) Field-Effect Transistor Devices: Post-CMOS Theory and Applications, (153–174) © 2026 Scrivener Publishing LLC

($NH_3$), nitrogen oxide ($NO_2$), acetone ($CH_3COCH_3$), hydrogen sulfide ($H_2S$), which are commonly associated with specific diseases and health conditions. In addition, it explores the variety of nanomaterials employed as sensing layers in bio-FETs, such as graphene, transition metal dichalcogenides (TMDCs), polymers, and metal oxides. Each of these materials offers unique properties, such as high surface area, tunable electronic characteristics, and strong interaction with target analytes, which enhance their sensitivity and selectivity for detecting specific VOCs. Their integration into bio-FET platforms enables the development of highly sensitive, real-time detection systems for medical diagnostics and environmental monitoring.

***Keywords:*** Bio-FET, biosensing, volatile organic compounds, real-time disease detection, portable biosensors, nanomaterials, medical diagnostics, environmental monitoring

## 8.1 Introduction

Detecting diseases through many volatile organic compounds (VOCs) or biomarkers is one of the most prominent methods, as different diseases are associated with distinct VOCs found in human stool, saliva, urine, and breath, among other sources [1]. Basically, these VOCs are the byproducts of different metabolic functions which indicate biological and biochemical reactions that happen inside our body and excel out from human body through different pathways as previously discussed [2]. This will further be discussed in the third section.

Sensing different VOCs from distinct human residue is known as biosensing. This approach involves detecting specific VOCs from the collected sample which is the indicative measure of individual health and condition of body. Traditional biosensing techniques are often invasive, require large setups, and are time-consuming, leading to discomfort for patients [3]. Complex sample preparation, high-end sophisticated specific laboratory equipment and need of technical staff are some limitations which are not possible for practical use. As demand of portable, more compact, invasive biosensors increases, FET-based sensors are becoming quite popular [4]. These sensors translate the biochemical signal into measurable electrical signals. In a FET-based sensor, sensing material is employed on a channel that interacts with the target gas, leading to changes in the transistor's parameters [5].

Early and accurate detection is critical for diagnosing various diseases, especially in cases of cancer or organ damage, enables clinical diagnosis and treatment, which can significantly increase survival rates of patients and even save many lives [6]. Sensitive detection of volatile organic compounds (VOCs) is a crucial aspect of clinical diagnosis. Since VOCs serve as biomarkers that represent metabolic changes and disease progression, their precise measurement is essential for early diagnosis and treatment. To achieve this, highly specific and sensitive sensors are required, which can be realized using advanced nanomaterials. High surface area, tunable structures, and enhanced reactivity are some properties of nanomaterials by which performance of sensors significantly improves [7]. As every disease produces its own specific VOCs, for detecting these VOCs, different nanomaterials are used, depending on VOCs. Studies say, among three-dimensional (3-D), two-dimensional (2-D) and one-dimensional (1-D), 2-D nanomaterials show superior response as a biosensor is concerned [6]. Some of 2-D nanomaterials are graphene, reduced graphene oxide, transition metal dichalcogenides (TMDCs), different metal oxide, and conductive polymers as shown in Figure 8.1.

Biosensors have some prerequires such as high sensitivity, high selectivity, high stability, low limit of detection (LOD), low response time and reusability [8] as shown in Figure 8.2, as accurate and specific detection of VOCs are required. Sensitivity and selectivity of sensors refers to the ability to detect specific and smallest change in analyte while LOD represents the lowest concentration that can be detected by sensor, which is essential for disease detection.

This chapter illustrates the basic principle of biosensors specifically bio-FET as well as different VOCs responsible for various diseases and their sources in the human body. Additionally, chapter covers different nanomaterials that had been involved in VOCs sensing along with its advantages

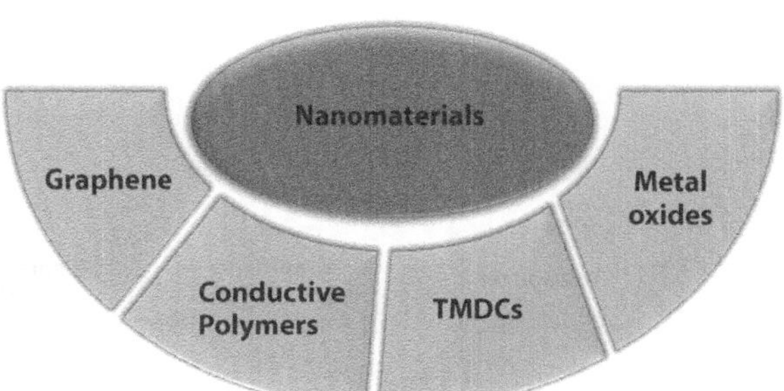


**Figure 8.1** Some nanomaterial used in VOCs sensing (discussed in this chapter).

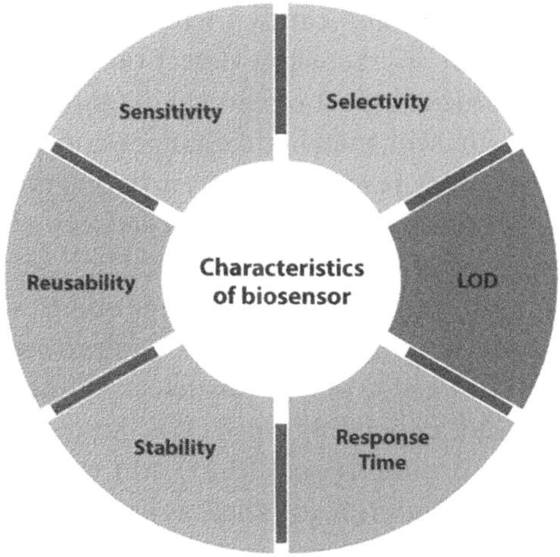


**Figure 8.2** Different characteristics of biosensor.

and limitations. Finally, it elaborates further improvement that has been made or will be made. Chapter brief about ongoing research, future aspect for these sensors.

## 8.2 Working Principle of Bio-FET

This section illustrates the sensing process of bio-FET. As discussed earlier, it translates the biological and chemical signal into measurable electrical signal. Generally, Biosensor comprises three components: bio-recognition elements to recognize the VOCs, a transducer to convert biological signal to measurable signal and then an amplifier with processing unit to

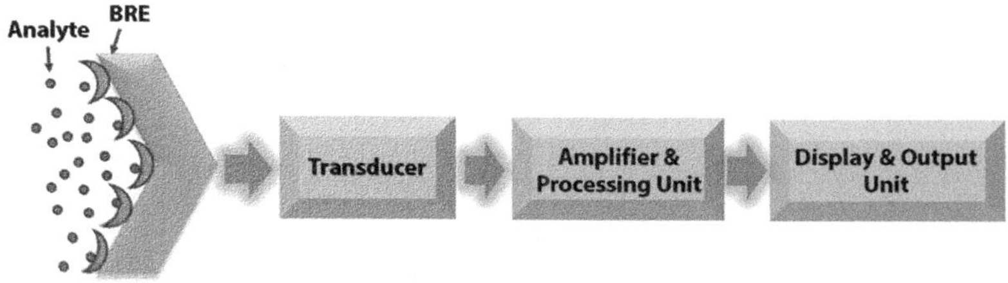


**Figure 8.3** Different components of biosensor.

further amplify the signal [9, 10] as shown in Figure 8.3. Basically, biosensor is the integration of sensor surface, transducer and a processing unit within the single equipment and based on the transducer it is classified as optical, electrochemical, resistive, and electronic biosensor [11]. In bio-FET, FET is working as a transducer where target gas causes the change in local electrical parameters of channel (nanomaterial) mainly charge carrier density. This change further can be measured by change in the drain current, threshold voltage, mobility, transconductance, etc [9]. By measuring this change in parameters of transistor, one can easily determine the presence of analyte gas and its concentration. The relationship between drain current and total charge transfer in the presence of gas is given by Eq. 8.1.

$$I_D = \frac{\mu W}{L}\left\{C_{ox}(V_{GS} - \frac{V_{DS}}{2}) \pm Q_m\right\}V_{DS} \tag{8.1}$$

In Equation 8.1, μ has denoted for charge carrier mobility, W and L are MOSFET width and channel length, $C_{OX}$ is oxide capacitance. Whereas $Q_m$ in Equation 8.1 is the total charge transfer between analytes gas and nano material (sensing layer), is an additional term which is not present is conventional MOSFET expression, showing the adsorption of target gas on the channel [12]. The sensing mechanism of bio-FET is efficiently illustrated by Figure 8.4.

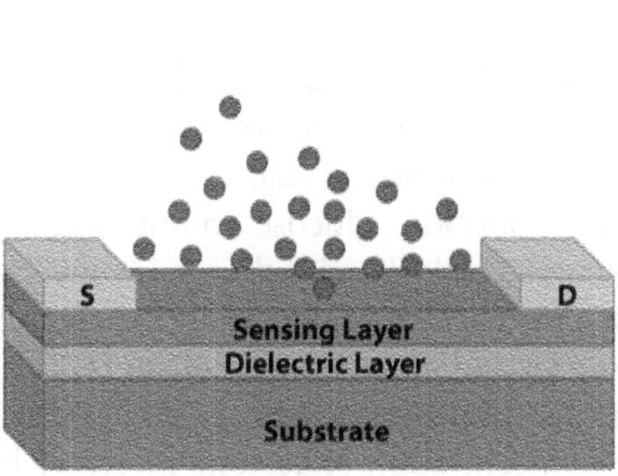


**Figure 8.4** Sensing through bio-FET.

## 8.3 Distinct VOCs Related to Diseases and Its Human Body Origins

VOCs is the type of biomarker which is expel out from human body from different routes and as per world health organization (WHO) "VOCs are characterized as organic chemical substances that exhibit a high vapor pressure at ambient temperatures. This characteristic results in a significant number of molecules entering the air from solid or liquid forms of the compound, leading to health risks and environmental concerns." Abnormal levels of these VOCs are helpful in measuring different diseases, for example higher amount of ammonia ($NH_3$) is responsible for kidney related issues [13]. These VOCs can be endogenous or exogenous depending on the source. Exogenous volatile organic compounds (VOCs) are mainly acquired from outside sources, including industrial emissions, environmental pollutants, everyday household items, and tobacco smoke. These substances can enter the body through various routes, including inhalation, ingestion, or dermal absorption. After their entry, VOCs undergo metabolic processes, resulting in the production of metabolites, which are subsequently eliminated from the body through exhalation, urine, or perspiration [14–16]. Conversely, endogenous VOCs are generated within the body as byproducts of cellular and tissue metabolic processes [16]. These internal VOCs travel through the bloodstream and may interact with various organs and tissues. This section briefly explains different VOCs and associated diseases as represented by Figure 8.5.

a. **Acetone ($CH_3COCH_3$):** Diabetes is becoming the most common disease in world. As per the record of WHO, about 422 million people worldwide have diabetes, and 1.5 million deaths are directly attributed to diabetes each year. Hence it is mandatory to monitor sugar level, especially for sugar patients. But periodic glucose monitoring with conventional glucometers is a painful and inconvenient process. Acetone is a primarily biomarker for diabetes, produced due to increased fat and glucose metabolism [17]. Normal range of acetone for healthy persons is $\geq 0.9$ parts per million (ppm). As concentration of acetone increases by 0.9 ppm it is an alarming situation. Condition becomes worst, when it becomes serval tens of ppm. Although acetone formation is a normal process in liver, where fatty acids are broken down into acetyl-CoA, which is then converted into acetoacetate.

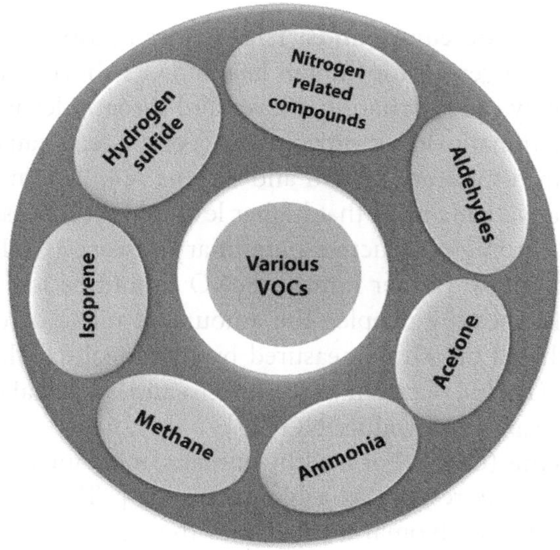


**Figure 8.5** Various VOCs associated with different diseases.

But in diabetic patients (both type1 and type 2), the body produces energy by breaking down fat, leading to excess acetone production, as glucose is not fully utilized [18].

b. **Ammonia ($NH_3$):** Excessive ammonia is toxic to the human body, and if not effectively eliminated, it can lead to various types of damage [19]. Generally, high levels of ammonia indicate kidney related diseases but it continuously high level may damage kidney, liver, and other organ [20]. Hence early detection of excessive ammonia is mandatory to avoid further consequences. In 2012, R. Adrover reported that ammonia level should be around 151.4 parts per billion (ppb) for healthy humans as compared to the patients suffering from chronic kidney diseases (CKD) [21]. In another study, Peter *et al.* illustrated that the excessive amount of ammonia is not only responsible for CKD but also for peptic ulcers and COVID-19 and carbon dioxide and acetone are some interferences while detecting ammonia [22]. Therefore, selective detection of ammonia is essential for human lives. Ammonia is produced by liver as a byproduct of amio acid, produced by digestion of protein. In normal case, ammonia is converted into urea which is filtered from kidney and came out with urine [23].

c. **Nitric oxide compounds:** High concentration of $NO_2$ is much dangerous for human health as its short-term exposure may cause irritation of eyes and throat. Elevated levels of fractional exhaled nitric oxide (FeNO) lead cause irritation in respiratory system and asthma [24, 25]. In 2018, Y. Guo *et al.* concluded that higher level of FeNO is associated with chronic obstructive pulmonary disease (COPD) and asthma [26]. Higher limit of FeNO should be less than 25 ppb for healthy people. The amount of nitric oxide (NO) in exhaled breath is measured by fractional exhaled nitric oxide (FeNO). Certain immune cells and epithelial cells lining the airways produce NO [27].
d. **Methane ($CH_4$):** For healthy humans, methane is not present in their residue as a biomarker [28]. The presence of methane as a biomarker indicates the excessive amount of methane in body. Methane is produced by a methanogen in human gut during the conversion of hydrogen and carbon dioxide. Higher level of methane associated with bowel disease, obesity, gaseous disorder and many infections in intestine [29, 30]. Expect that methane is a prominent biomarker for heart disease and breast cancer [31, 32]. Threshold upper limit of methane may vary depending on age, diet and other factors. But for infected adults it should not exceed 10 ppm.
e. **Hydrogen sulfide ($H_2S$):** $H_2S$ is another biomarker which smells like a rotten egg. $H_2S$ can be detected in human breath, serum, urine stool and sputum supernatants [33]. Production of $H_2S$ is through breaking down the surfer containing amino acid and digestion of protein in the presence of gut bacteria in human intestine. Normal range of $H_2S$ is around 0.1 to 0.5 ppm and its elevated level may indicate the traces of asthma [33], halitosis [34], lung inflammation, and fibrosis [33].
f. **Isoprene:** Isoprene is a prominent biomarker, its lower and higher level associated with different health issues. Chronic heart failure [35], cystic fibrosis [36], lung cancer [37], and muscle dystrophy [38] are some diseases caused by lower concentration of isoprene whereas higher order of isoprene is responsible for various metabolic disorders such as diabatic [39], CKD [40], chronic liver disease [41], and gastric cancer [42]. Isoprene is produced by the human body

through various metabolic processes. Cholesterol synthesis, biochemical reactions during lipid peroxidation, and metabolic activities in muscle tissues are some of the major sources of isoprene [43]. For healthy adults, concentration of isoprene should be 100-500 ppb, it may influence different factors.

g. **Aldehydes:** It is an exogenous as well as endogenous VOCs, generally not present in human body at normal condition [44]. Inhalation from polluted air, absorbed in skin with cosmetic or industrial residue are main external sources of exposure of aldehydes. Package and contaminated food and drinks are also responsible for elevated levels of aldehydes. It causes various deadly diseases such as oxidative stress, liver cancer and lung cancer.

## 8.4 Nanomaterials Used for Sensing VOCs

As discussed previous sections, nanomaterials used as sensing layer in different biosensors including bio-FET. This section explores some of the popular nanomaterials and its studies on gas sensing.

a. **Graphene & its derivatives:** Graphene is a carbon atom-based structure arranged in honeycomb lattice with $sp^2$ hybridization [45]. Its unique physical properties ensure good response towards the adsorption of specific VOCs. Larger surface to volume ratio, high conductivity, low production cost, high stability and high porosity are some characteristics which are favorable for analyte binding [46]. As compared to the contaminated graphene pure graphene has higher carrier mobility (four times) and reduced doped carrier density (two third times). Moreover, for some VOCs exhibit better response for doped or contaminated graphene [47].

   Graphene based biosensors are quite popular, as they are used to detect various diseases including diabetes, lung cancer. asthma, kidney diseases and many more [48]. In 2009, Lu and Dan found that the detection of $NH_3$ and $NO_2$ at room temperature is possible [47] through reduced graphene oxide (rGO). In another study, Madhav *et al.* monitored a shift in Dirac peak of graphene for various concentration of $NH_3$ [49]. This experiment was performed for different gate voltage and exposure time of $NH_3$ at dry air environment. Shift in Dirac peak increases linearly with temperature while response of FET is also increasing linearly as both temperature as well as concentration of $NH_3$ increases. Total 0.039 ± 0.001 electrons get transferred (per $NH_3$ molecule) on graphene surface, as $NH_3$ has a reducing nature. According to the studies, most of the sensing of $NH_3$ is executed in vacuum, which is not possible in real time applications, as atmospheric air interrupting the sensing process [47, 50, 51]. Recovery is also an important concern for commercial sensors, as sensors are not made for single use. It is believed that infrared (IR) radiation exposure increases the free electron density on the graphene surface, facilitating oxygen adsorption and inherently improving the device's recovery [52, 53]. In this case, in the presence of IR light, $NH_3$ is converted into nitrogen oxides by reacting with oxygen. Recently in 2023, Liu *et al.* proposed a $NH_3$ sensor where carbon nanotubes are used as a sensing layer [54]. This sensor has an ability to amplify weak signals which enhances sensitivity of sensor. Additionally, sensor exhibited improved selectivity with comparatively lower LOD (100 ppb).

As discussed above. it is clear that graphene is highly popular for detecting $NH_3$ and $NO_2$. Except these two gases, graphene and its derivative are also sensitive towards other VOCs such as acetone [55], ethanol [56, 57], methanol [56], alcohol [58], hydrogen [59], formaldehyde [5], etc.

Despite various favorable properties for VOCs sensing, graphene-based sensors have some restrictions. Absence of bandgap, toxicity, low selectivity, and poor performance in oxidative environments are some barriers which limit the real time performance of these sensors. This can be improved by using hybrid structures such as nanocomposite or heterojunction with other nanomaterials [60], decoration of nanoparticles [61] etc. Decoration of gold nanoparticles on graphene sheet enhances the adsorption of $NH_3$ and $NO_2$, is one of the examples of it [62, 63].

b. **TMDCs:** Different transition metal dichalcogenides (TMDCs) demonstrate exceptional potential for VOC sensing due to their unique and advantageous properties. Their tunable bandgap allows precise control over electronic characteristics, while their high on/off ratio, excellent subthreshold swing (SS) and environmental stability ensure reliable performance under various conditions. TMDC is a layered structure where a transition metal layer is placed between two chalcogen layers. TMDCs based sensors are basically based on two methods namely: charge modulated method and dielectric modulated method. Charge modulated method concern about the charge transfer mechanism takes place between analyte gas and sensor surface, while in dielectric modulated method, capacitance of sensor changes due to interaction of analyte and sensing layer [64, 65]. Both methods involve oxidation or reduction of charge depending on types of analyte gas. Molybdenum disulfide ($MoS_2$), Molybdenum diselenide ($MoSe_2$), tungsten disulfide ($WS_2$), tungsten diselenide ($WSe_2$), niobium disulfide ($NbS_2$), molybdenum ditelluride ($MoTe_2$) and rhenium disulfide ($ReS_2$) are examples of TMDCs employed for sensing volatile organic compounds (VOCs). $MoS_2$ is most utilized TMDCs in area of gas sensing due to its high mobility and other characteristics. Li *et al.* introduced a FET based nitric oxide ($NO$) sensor with single as well as multilayer $MoS_2$ at room temperature. 4-layer $MoS_2$ exhibited better sensing response with LOD of 0.8 ppm [66]. Another study showed that vertical $MoS_2$ nanoflakes based FET structure used for sensitive and fast sensing of ethanol and methanol vapors. Device reported better sensing response at 50°C with sensing response of 9.7 for 1 ppm ethanol and 4.6 for 1 ppm methanol [66]. It is observed from studies that mechanical exfoliated $MoS_2$ showed excellent sensitivity and recovery for $NO_2$ and $NH_3$ gas particles [67]. Sometimes aluminum oxide ($Al_2O_3$) on the top of $MoS_2$ as a passivation layer improved sensing response of sensor. This is because of the shift in threshold voltage of device [68]. Another experiment conducted by SooáKang concluded that tri-layer of CVD grown $MoS_2$ has an excellent ability to detect 1000 ppm of oxygen ($O_2$) and $CO_2$. Response/recovery time of sensor, for $CO_2$ were 37/53 seconds whereas for $O_2$ it were 18/47 seconds. Additionally, device showed LOD of 100 ppm and sensitivity of 3.95% for $CO_2$ [69].

$MoSe_2$ is another popular TMDC used for gas sensing. Baek *et al.* proposed an extremely sensitive $NO_2$ sensor with sensitivity ~1,907 using CVD grown $MoSe_2$ films.

Additionally, its high response/recovery time (20/30 minutes) offering real-time response and fast switching for 300 ppm $NO_2$ [70]. The $NO_2$ sensing mechanism based on changes in gap states, where adsorption of $NO_2$ at varying concentrations (0–300 ppm) increases the hole current, contributing to additional band states. It is believed that $WS_2$ [71, 72] and $MoTe_2$ [73] have demonstrated strong specificity and stability in detecting ethanol [71], methanol [72], and ketone [71, 72] compounds. In the presence of light, TMDC-based gas sensors show an improved sensing response. In 2021, a $ReS_2$ based exceptional selective and sensitive VOC sensor was introduced. Device consisted of 6 nm thick N-type $ReS_2$ layer which were effectively sense acetone, methanol, ethanol and isopropanol. Sensor was exposed to different concentrations of VOCs ranging from 1 to 100 ppm under the light illumination [73].

In summary, different TMDCs show good sensing response for various VOCs as discussed earlier. Despite various properties and its popularity in VOCs sensing, its lower electrical conductivity, slow recovery, less long-term stability and limited scalability are some challenges encounters with these materials. These challenges can be addressed by employing hybrid structures or heterojunctions that combine TMDCs with metal oxides or graphene.

c. **Metal oxides:** Semiconductor metal oxides (SMOs) are crystalline structure where metal and oxygen atoms are ionic bonded. Its large surface area, fast response, compact size and cost-efficient fabrication make it suitable candidate for VOCs sensing applications. Ethanol, toluene, acetone, methane, formaldehyde, isopropanol, $NO_2$, $H_2S$ are some gases sensed by SMO. SMO can be N-type or P-type, depending on its majority charge carriers. Basically, adsorption of target VOCs causes change in the resistance of SMO. For P-type SMO, interaction of reducing gas causes an increase in resistance of material and interaction of oxidating gas decrease material resistance depending on the concentration of analyte VOC. However, for N-type SMO, its reverse is true.

Tin oxide ($SnO$) tin (IV) oxide ($SnO_2$), zinc oxide ($ZnO$), indium oxide ($In_2O_3$), tungsten trioxide ($WO_3$) and titanium dioxide ($TiO_2$) are some prominent used SMO for VOCs sensing. In 2019, Jeong *et al.* [74] proposed a $NO_2$ gas sensor using P-type tin oxide ($SnO_X$) in thin film transistor (TFT). Device exhibited selective detection of $NO_2$ among $NH_3, CO_2, CO$ and $H_2S$. Maximum response time of device for 10 ppm $NO_2$ was recorded as 19.4 at room temperature. Another study proposed by Surya showed superior sensing response of $SnO_2$ for $H_2S$ as compared to $ZnO$. This $P3HT$ based organic FET exhibited selective response of gas at room temperature among methanol, ethanol, $NH_3$ and acetic acid [75]. In this series, Hunag introduced a floating gate structure for $CO$ detection using $SnO_2$/CNT. Sensor showed good sensing response (201.25%) with LOD of 1 ppb and response time of sensor is 300 seconds at 150$^0$C. Additionally, among sulfur dioxide ($SO_2$), $NH_3, NO_2, H_2S,$ and $CH_4$, response of $CO$ is much higher ensuring the selectively of sensor [76].

Another metal oxide, $ZnO$ with its good sensing response, selectivity and stability is also known for different VOCs sensing. $ZnO$ is N-type SMO, generally not used at room temperature due to its low response and recovery time. According to Hang, selective detection of $NO_2$ is executed through a floating gate device by utilizing 10 nm $ZnO$ as a sensing layer at 180°C. Device is tested for total of seven gases namely $H_2S, NO_2, SO_2, NH_3, CH_4$, Propane ($C_3H_8$). Among them $NO_2$ showed highest sensitivity of 176% with response and recovery time of 90 seconds and 580 seconds respectively [77]. In 2021, Singh *et al.* proposed highly sensitive hydrogen sensor with extended gate structure using $ZnO$. This junctionless device worked on the principle of change in work function. A total improvement of 2.59% in the 'off' current and 24.85% in the 'on' current was recorded, indicating the enhanced sensitivity of the device [78]. According to Jung, selective and sensitive (increased by $\sim 3.7$ times) detection of $SO_2$ is made by 18 nm thin $WO_3$ layer [79].

Some latest studies related to SMO based gas sensing are illustrated by this section. It can be further extended. Apart from various properties of SMO, requirement of high operating temperature, limited lifetime of device, less immune to humidity and other atmospheric conditions are some restrictions for practical use of SMO based VOCs sensors. These can be improved by using hybrid structures. Usage of conductive polymers in hybrid structure with different SMO is one of the good alternatives and can inherently increase the performance of these sensors. It is further discussed in the next subsection.

d. **Conducting Polymers and its hybrid structure with different nanomaterials:** Conducting polymers (CP) are organic materials also referred as conductive plastics. They are popular for their unusual characteristics which are generally not present in plastic. As its name suggests, conductive polymers are conducting in nature. Some popular CPs are polyaniline (PANI), Polypyrrole (PPy), Polythiophene, Poly(3,4-ethylenedioxythiophene) (PEDOT). As discussed earlier, higher operating temperature is the biggest limitation of semiconductor metal oxides in VOCs detection. CP is an excellent alternative, due to its ability to detect at room temperature. Additionally, CP are flexible in nature which can be easily synthesized and can sense wide range of organic vapors. Interaction of analyte gas and CP causes either P-type or n-type doing in polymer chain, depending on redox reaction or protonation. CP based gas sensors can be used for sensing $NH_3, NO_2, H_2S,$ alcohol and amines etc. In 2015, Ryu *et al.* [80] demonstrated diketopyrrolopyrrole (DPP)- based high performance flexible $NH_3$ sensor. This bottom-gate/top contact device is showing an increment in resistance (~100 times), as concentration of gas increases. Recorded change in resistance is from $10^7$ to $10^9$ at $V_{DS}$ = -5V and $V_{GS}$ = -5V. In 2011, Chen used single polyaniline nanofibers for sensing different concentration of $NH_3$. This bio-FET exhibited high sensitivity (7%) against 1 ppm of target VOC [81]. Hybrid structure of different CPs with different nanomaterial are also popular. In this series, a nanocomposite of PANI/$TiO_2$ exhibited improved sensing response (better stability and sensitivity). This $NH_3$ and $CO$ sensor showed better humidity resistance as compared to pure structure [82]. In 2024, MS Mehrolia proposed ag doped PBTTT-C14 as a sensing layer for $H_2S$. Device exhibited 80% sensitivity for 5 ppm $H_2S$ along with LOD of 151.17 ppb which is excellent. Additionally, its response time and recovery time are 5 second and 26 second respectively [83].

This chapter covers various studies of FET based VOCs sensing, although there are some advanced materials as well as some latest studies (not FET-based structure), which cannot be covered by this chapter. Borophene [84], antimony [85], manganese [86], and their hybrid structure with different nanomaterials or nanoparticles are some of the examples exhibiting potential advancement in VOCs sensing.

## 8.5 Conclusion

FET-based VOCs sensing is an emerging and next-generation technology that enables fast and sensitive detection of early detection of various diseases. This chapter highlights various aspects of bio-F*ET al*ong with studies involving different nanomaterials. Through a detailed analysis, key factors like sensitivity, selectivity, and limit of detection (LOD) have been identified as crucial parameters of sensor efficiency. This review explores various biomarkers and their biological mechanisms, highlighting that VOC analysis offers significant promise for the early detection of diseases, enabling proactive medical intervention. It is evident that surface modification improves the gas sensing performance of biosesnor. The use of heterojunctions and hybrid structures further boosts sensor efficiency, offering potential for improved detection capabilities.

**AI disclosure statement:** The author acknowledges the use of AI-based tools for minor language editing or grammatical correction. However, all intellectual contributions, analyses, and conclusions presented in this chapter are entirely the author's own.

## References


1. Velusamy, P., Su, C.H., Ramasamy, P., Arun, V., Rajnish, N., Raman, P., Gopinath, S.C., Volatile organic compounds as potential biomarkers for non-invasive disease detection by nanosensors: A comprehensive review. *Crit. Rev. Anal. Chem.*, *53*, 8, 1828–1839, 2023.
2. Chen, H., Zheng, Y., Wang, M., Wu, Y., Yao, M., Gene-Regulated Release of Distinctive Volatile Organic Compounds from Stressed Living Cells. *Environ. Sci. Technol.*, *56*, 13, 9546–9555, 2022.
3. Li, P., Lee, G.H., Kim, S.Y., Kwon, S.Y., Kim, H.R., Park, S., From diagnosis to treatment: recent advances in patient-friendly biosensors and implantable devices. *ACS Nano*, *15*, 2, 1960–2004, 2021.

4. Nguyen, T.T.H., Nguyen, C.M., Huynh, M.A., Vu, H.H., Nguyen, T.K., Nguyen, N.T., Field effect transistor based wearable biosensors for healthcare monitoring. *J. Nanobiotechnol.*, *21*, 1, 411, 2023.
5. Zhang, P., Xiao, Y., Zhang, J., Liu, B., Ma, X., Wang, Y., Highly sensitive gas sensing platforms based on field effect Transistor-A review. *Anal. Chim. Acta*, *1172*, 338575, 2021.
6. Srivastava, S. and Agrawal, S., Potential of nanomaterials in FET-based breath sensing: A comprehensive review & future direction. *Micro Nanostruct.*, 207830, 2024.
7. Noah, N.M., Design and synthesis of nanostructured materials for sensor applications. *J. Nanomater.*, *2020*, 1, 8855321, 2020.
8. Manimekala, T., Sivasubramanian, R., Dharmalingam, G., Nanomaterial-based biosensors using field-effect transistors: a review. *J. Electron. Mater.*, *51*, 5, 1950–1973, 2022.
9. Elli, G., Hamed, S., Petrelli, M., Ibba, P., Ciocca, M., Lugli, P., Petti, L., Field-effect transistor-based biosensors for environmental and agricultural monitoring. *Sensors*, *22*, 11, 4178, 2022.
10. Gupta, S. and Kakkar, V., Development of environmental biosensors for detection, monitoring, and assessment, in: *Nanomaterials and Environmental Biotechnology*, pp. 107–125, 2020.
11. Chepyala, R., Applications and success of MIPs in optical-based nanosensors, in: *Nanofabrication for Smart Nanosensor Applications*, pp. 89–121, Elsevier, 2020.
12. Tang, X., Debliquy, M., Lahem, D., Yan, Y., Raskin, J.P., A review on functionalized graphene sensors for detection of ammonia. *Sensors*, *21*, 4, 1443, 2021.
13. Chan, M.J., Li, Y.J., Wu, C.C., Lee, Y.C., Zan, H.W., Meng, H.F., Breath ammonia is a useful biomarker predicting kidney function in chronic kidney disease patients. *Biomedicines*, *8*, 11, 468, 2020.
14. Papaefstathiou, E., Stylianou, M., Andreou, C., Agapiou, A., Breath analysis of smokers, non-smokers, and e-cigarette users. *J. Chromatogr. B*, *1160*, 122349, 2020.
15. Miekisch, W., Sukul, P., Schubert, J.K., Diagnostic potential of breath analysis–focus on the dynamics of volatile organic compounds. *TrAC, Trends Anal. Chem.*, 117977, 2024.
16. Cen, Z., Huang, Y., Li, S., Dong, S., Wang, W., Li, X., Advancing Breathomics through Accurate Discrimination of Endogenous from Exogenous Volatiles in Breath. *Environ. Sci. Technol.*, 2024.
17. Obeidat, Y., The most common methods for breath acetone concentration detection: A review. *IEEE Sens. J.*, *21*, 13, 14540–14558, 2021.
18. Méndez, J.D., Acetone, the third ketone body. I. Presence in the human body in health and disease, 2021.
19. Zeng, X., Liu, R., Li, Y., Li, J., Zhao, Q., Li, X., Bao, J., Excessive ammonia inhalation causes liver damage and dysfunction by altering gene networks

associated with oxidative stress and immune function. *Ecotoxicol. Environ. Saf.*, *217*, 112203, 2021.
20. Dasarathy, S., Mookerjee, R.P., Rackayova, V., Rangroo Thrane, V., Vairappan, B., Ott, P., Rose, C.F., Ammonia toxicity: from head to toe? *Metab. Brain Dis.*, *32*, 529–538, 2017.
21. Adrover, R., Cocozzella, D., Ridruejo, E., Garcia, A., Rome, J., Podesta, J.J., Breath-ammonia testing of healthy subjects and patients with cirrhosis. *Dig. Dis. Sci.*, *57*, 189–195, 2012.
22. Ricci, P.P. and Gregory, O.J., Sensors for the detection of ammonia as a potential biomarker for health screening. *Sci. Rep.*, *11*, 1, 7185, 2021.
23. Hibbard, T. and Killard, A.J., Breath ammonia analysis: Clinical application and measurement. *Crit. Rev. Anal. Chem.*, *41*, 1, 21–35, 2011.
24. Ricciardolo, F.L.M., Sorbello, V., Fontana, R.B., Schiavetti, I., Ciprandi, G., Exhaled nitric oxide in relation to asthma control: a real-life survey. *Allergol. Immunopathol.*, *44*, 3, 197–205, 2016.
25. Idavain, J., Julge, K., Rebane, T., Lang, A., Orru, H., Respiratory symptoms, asthma and levels of fractional exhaled nitric oxide in schoolchildren in the industrial areas of Estonia. *Sci. Total Environ.*, *650*, 65–72, 2019.
26. Guo, Y., Hong, C., Liu, Y., Chen, H., Huang, X., Hong, M., Diagnostic value of fractional exhaled nitric oxide for asthma-chronic obstructive pulmonary disease overlap syndrome. *Medicine*, *97*, 23, e10857, 2018.
27. Feng-Jia, C., Xin-Yan, H., Geng-Peng, L., Yang-Li, L., Can-Mao, X., Validity of fractional exhaled nitric oxide and small airway function indices in diagnosis of cough-variant asthma. *J. Asthma*, *55*, 7, 750–755, 2018.
28. Pathak, A.K., Swargiary, K., Kongsawang, N., Jitpratak, P., Ajchareeyasoontorn, N., Udomkittivorakul, J., Viphavakit, C., Recent advances in sensing materials targeting clinical volatile organic compound (VOC) biomarkers: a review. *Biosensors*, *13*, 1, 114, 2023.
29. Polag, D. and Keppler, F., Global methane emissions from the human body: Past, present and future. *Atmos. Environ.*, *214*, 116823, 2019.
30. de Lacy Costello, B.P.J., Ledochowski, M., Ratcliffe, N.M., The importance of methane breath testing: a review. *J. Breath Res.*, *7*, 2, 024001, 2013.
31. Szulejko, J.E., McCulloch, M., Jackson, J., McKee, D.L., Walker, J.C., Solouki, T., Evidence for cancer biomarkers in exhaled breath. *IEEE Sens. J.*, *10*, 1, 185–210, 2009.
32. Kar, D., Mandal, S.K.D., Bhattacharya, T.K., Volatile Organic Gaseous Compound Biomarkers for Cardiovascular Diseases: A Review, in: *2024 International Conference on Innovations and Challenges in Emerging Technologies (ICICET)*, 2024, June, IEEE, pp. 1–6.
33. Chung, K.F., Hydrogen sulfide as a potential biomarker of asthma. *Expert Rev. Respir. Med.*, *8*, 1, 5–13, 2014.
34. Choi, K.I., Kim, H.J., Kang, Y.C., Lee, J.H., Ultraselective and ultrasensitive detection of H2S in highly humid atmosphere using CuO-loaded SnO2

hollow spheres for real-time diagnosis of halitosis. *Sens. Actuators, B, 194*, 371–376, 2014.
35. Lawrence T, M., Robin, P., Bernard, S., Breath isoprene in patients with heart failure. *Eur. J. Heart Fail., 3*, 4, 423–427, 2001.
36. McGrath, L.T., Patrick, R., Mallon, P., Dowey, L., Silke, B., Norwood, W., Elborn, S., Breath isoprene during acute respiratory exacerbation in cystic fibrosis. *Eur. Respir. J., 16*, 6, 1065–1069, 2000.
37. Poli, D., Carbognani, P., Corradi, M., Goldoni, M., Acampa, O., Balbi, B., Mutti, A., Exhaled volatile organic compounds in patients with non-small cell lung cancer: cross sectional and nested short-term follow-up study. *Respir. Res., 6*, 1–10, 2005.
38. King, J., Mochalski, P., Unterkofler, K., Teschl, G., Klieber, M., Stein, M., Baumann, M., Breath isoprene: muscle dystrophy patients support the concept of a pool of isoprene in the periphery of the human body. *Biochem. Biophys. Res. Commun., 423*, 3, 526–530, 2012.
39. Neupane, S., Peverall, R., Richmond, G., Blaikie, T.P., Taylor, D., Hancock, G., Evans, M.L., Exhaled breath isoprene rises during hypoglycemia in type 1 diabetes. *Diabetes Care, 39*, 7, e97–e98, 2016.
40. Lirk, P., Bodrogi, F., Raifer, H., Greiner, K., Ulmer, H., Rieder, J., Elective haemodialysis increases exhaled isoprene. *Nephrol. Dial. Transplant., 18*, 5, 937–941, 2003.
41. Sehnert, S.S., Jiang, L., Burdick, J.F., Risby, T.H., Breath biomarkers for detection of human liver diseases: preliminary study. *Biomarkers, 7*, 2, 174–187, 2002.
42. Xu, Z.Q., Broza, Y.Y., Ionsecu, R., Tisch, U., Ding, L., Liu, H., Haick, H., A nanomaterial-based breath test for distinguishing gastric cancer from benign gastric conditions. *Br. J. Cancer, 108*, 4, 941–950, 2013.
43. Mochalski, P., King, J., Mayhew, C.A., Unterkofler, K., A review on isoprene in human breath. *J. Breath Res., 17*, 3, 037101, 2023.
44. Hubbard, H.F., Sobus, J.R., Pleil, J.D., Madden, M.C., Tabucchi, S., Application of novel method to measure endogenous VOCs in exhaled breath condensate before and after exposure to diesel exhaust. *J. Chromatogr. B, 877*, 29, 3652–3658, 2009.
45. Allen, M.J., Tung, V.C., Kaner, R.B., Honeycomb carbon: a review of graphene. *Chem. Rev., 110*, 1, 132–145, 2010.
46. Wang, T., Huang, D., Yang, Z., Xu, S., He, G., Li, X., Zhang, L., A review on graphene-based gas/vapor sensors with unique properties and potential applications. *Nano Micro Lett., 8*, 95–119, 2016.
47. Dan, Y., Lu, Y., Kybert, N.J., Luo, Z., Johnson, A.C., Intrinsic response of graphene vapor sensors. *Nano Lett., 9*, 4, 1472–1475, 2009.
48. Tripathi, K.M., Kim, T., Losic, D., Tung, T.T., Recent advances in engineered graphene and composites for detection of volatile organic compounds (VOCs) and non-invasive diseases diagnosis. *Carbon, 110*, 97–129, 2016.

49. Gautam, M. and Jayatissa, A.H., Graphene based field effect transistor for the detection of ammonia. *J. Appl. Phys.*, *112*, 6, 2012.
50. Gautam, M. and Jayatissa, A.H., Gas sensing properties of graphene synthesized by chemical vapor deposition. *Mater. Sci. Eng., C*, *31*, 7, 1405–1411, 2011.
51. Xu, H., Xiang, J.X., Lu, Y.F., Zhang, M.K., Li, J.J., Gao, B.B., Gu, Z.Z., Multifunctional wearable sensing devices based on functionalized graphene films for simultaneous monitoring of physiological signals and volatile organic compound biomarkers. *ACS Appl. Mater. Interfaces*, *10*, 14, 11785–11793, 2018.
52. Sun, P., Zhu, M., Wang, K., Zhong, M., Wei, J., Wu, D., Zhu, H., Photoinduced molecular desorption from graphene films. *Appl. Phys. Lett.*, *101*, 5, 2012.
53. Trung, T.Q., Tien, N.T., Kim, D., Jung, J.H., Yoon, O.J., Lee, N.E., High thermal responsiveness of a reduced graphene oxide field-effect transistor. *Adv. Mater.*, *24*, 38, 5254–5260, 2012.
54. Liu, J., Hu, J., Zhang, Y., A carbon-based FET-type sensor with floating sensing gate for ppb-level ammonia detection at room temperature. *IEEE Sens. J.*, 2023.
55. Hostert, L., Orlandi, M.O., Bueno, P.R., Low-limit acetone detection system combining quantum conductance and capacitance signal analyses derived from oxidized single-layer graphene. *Sens. Actuators, B*, *397*, 134651, 2023.
56. Hayasaka, T., Lin, A., Copa, V.C., Lopez Jr., L.P., Loberternos, R.A., Ballesteros, L.I.M., Lin, L., An electronic nose using a single graphene FET and machine learning for water, methanol, and ethanol. *Microsyst. Nanoeng.*, *6*, 1, 50, 2020.
57. Nozaki, R., Ikuta, T., Ueno, K., Tsukakoshi, K., Ikebukuro, K., Maehashi, K., Ethanol Detection at the Parts per Billion Level with Single-Stranded-DNA-Modified Graphene Field-Effect Transistors. *Phys. Status Solidi B*, *257*, 2, 1900376, 2020.
58. Luo, S., Wang, R., Wang, L., Qu, H., Zheng, L., Breath alcohol sensor based on hydrogel-gated graphene field-effect transistor. *Biosens. Bioelectron.*, *210*, 114319, 2022.
59. Huang, L., Zhang, Z., Li, Z., Chen, B., Ma, X., Dong, L., Peng, L.M., Multifunctional graphene sensors for magnetic and hydrogen detection. *ACS Appl. Mater. Interfaces*, *7*, 18, 9581–9588, 2015.
60. Bae, G., Jeon, I.S., Jang, M., Song, W., Myung, S., Lim, J., An, K.S., Complementary dual-channel gas sensor devices based on a role-allocated ZnO/graphene hybrid heterostructure. *ACS Appl. Mater. Interfaces*, *11*, 18, 16830–16837, 2019.
61. Singhal, A.V., Charaya, H., Lahiri, I., Noble metal decorated graphene-based gas sensors and their fabrication: a review. *Crit. Rev. Solid State Mater. Sci.*, *42*, 6, 499–526, 2017.

62. Gautam, M. and Jayatissa, A.H., Ammonia gas sensing behavior of graphene surface decorated with gold nanoparticles. *Solid State Electron.*, *78*, 159–165, 2012.
63. Rattan, S., Kumar, S., Goswamy, J.K., Gold nanoparticle decorated graphene for efficient sensing of NO2 gas. *Sens. Int.*, *3*, 100147, 2022.
64. Chen, X., Liu, C., Mao, S., Environmental analysis with 2D transition-metal dichalcogenide-based field-effect transistors. *Nano Micro Lett.*, *12*, 1–24, 2020.
65. Kanungo, S., Introduction to dielectrically modulated biological field effect transistor, in: *2018 International Symposium on Devices, Circuits and Systems (ISDCS)*, 2018, March, IEEE, pp. 1–8.
66. Li, H., Yin, Z., He, Q., Li, H., Huang, X., Lu, G., Zhang, H., Fabrication of single-and multilayer MoS2 film-based field-effect transistors for sensing NO at room temperature. *Small*, *8*, 1, 63–67, 2012.
67. Late, D.J., Huang, Y.K., Liu, B., Acharya, J., Shirodkar, S.N., Luo, J., Sensing behavior of atomically thin-layered MoS2 transistors. *ACS Nano*, *7*, 6, 4879–4891, 2013.
68. Im, H., Al Mutairi, A., Kim, S., Sritharan, M., Kim, S., Yoon, Y., On MoS2 thin-film transistor design consideration for a NO2 gas sensor. *ACS Sens.*, *4*, 11, 2930–2936, 2019.
69. SooáKang, B. and YongáSong, J., Sensor based on chemical vapour deposition-grown molybdenum disulphide for gas sensing application. *RSC Adv.*, *6*, 79, 75839–75843, 2016.
70. Baek, J., Yin, D., Liu, N., Omkaram, I., Jung, C., Im, H., Kim, S., A highly sensitive chemical gas detecting transistor based on highly crystalline CVD-grown MoSe 2 films. *Nano Res.*, *10*, 1861–1871, 2017.
71. Huo, N., Yang, S., Wei, Z., Li, S.S., Xia, J.B., Li, J., Photoresponsive and gas sensing field-effect transistors based on multilayer WS2 nanoflakes. *Sci. Rep.*, *4*, 1, 5209, 2014.
72. Yong, A.S.E., Sofer, Z., Mayorga-Martinez, C.C., Ambrosi, A., Pumera, M., Metallic 1T-WS2 for Selective Impedimetric Vapor Sensing, 2015.
73. Wu, E., Xie, Y., Yuan, B., Hao, D., An, C., Zhang, H., Zhang, D., Specific and highly sensitive detection of ketone compounds based on p-type MoTe2 under ultraviolet illumination. *ACS Appl. Mater. Interfaces*, *10*, 41, 35664–35669, 2018.
74. Jeong, H.S., Park, M.J., Kwon, S.H., Joo, H.J., Kwon, H.I., Highly sensitive and selective room-temperature NO2 gas-sensing characteristics of SnOX-based p-type thin-film transistor. *Sens. Actuators, B*, *288*, 625–633, 2019.
75. Surya, S.G., Ashwath, B.N., Mishra, S., Karthik, A.R.B., Sastry, A.B., Prasad, B.L.V., Rao, V.R., H2S detection using low-cost SnO2 nano-particle Bi-layer OFETs. *Sens. Actuators, B*, *235*, 378–385, 2016.
76. Huang, Y., Zhou, Y., Liu, Y., Chen, J., Wei, X., Cao, J., SnO2/Carbon Nanotube Floating-Gate Field-Effect Transistor Gas Sensor for ppb-Level CO Detection. *ACS Appl. Nano Mater.*, *7*, 14, 16119–16126, 2024.

77. Hong, Y., Kim, C.H., Shin, J., Kim, K.Y., Kim, J.S., Hwang, C.S., Lee, J.H., Highly selective ZnO gas sensor based on MOSFET having a horizontal floating-gate. *Sens. Actuators, B, 232,* 653–659, 2016.
78. Singh, N.K., Kar, R., Mandal, D., Simulation and analysis of ZnO-based extended-gate gate-stack junctionless NWFET for hydrogen gas detection. *Appl. Phys. A, 127,* 4, 290, 2021.
79. Jung, G., Jeong, Y., Hong, Y., Wu, M., Hong, S., Shin, W., Lee, J.H., SO2 gas sensing characteristics of FET-and resistor-type gas sensors having WO3 as sensing material. *Solid State Electron., 165,* 107747, 2020.
80. Ryu, G.S., Park, K.H., Park, W.T., Kim, Y.H., Noh, Y.Y., High-performance diketopyrrolopyrrole-based organic field-effect transistors for flexible gas sensors. *Org. Electron., 23,* 76–81, 2015.
81. Chen, D., Lei, S., Chen, Y., A single polyaniline nanofiber field effect transistor and its gas sensing mechanisms. *Sensors, 11,* 7, 6509–6516, 2011.
82. Harb, M.E., Ebrahim, S., Soliman, M., Shabana, M., Fabrication of organic transistors using nanomaterials for sensing applications. *J. Electron. Mater., 47,* 353–358, 2018.
83. Mehrolia, M.S., Kumar, D., Verma, A., Singh, A.K., Fabrication and Characterization of Self-Assembled Low Voltage Operated OTFT for H2S Gas Sensor for Oil and Gas Industry. *IEEE Trans. Electron Devices, 71,* 1, 769–776, 2024.
84. Hou, C., Tai, G., Liu, Y., Liu, X., Borophene gas sensor. *Nano Res.,* 1–8, 2022.
85. Li, Y., Zhang, Y., Zhou, Y., Zhao, L., Yan, X., Liu, F., Sun, P., Wearable Gas Sensor Based on Reticular Antimony-Doped SnO2/PANI Nanocomposite Realizing Intelligent Detection of Ammonia within a Wide Range of Humidity. *ACS Sens., 8,* 11, 4132–4142, 2023.
86. John, R.A.B., Shruthi, J., Reddy, M.R., Kumar, A.R., Manganese doped nickel oxide as room temperature gas sensor for formaldehyde detection. *Ceram. Int., 48,* 12, 17654–17667, 2022.

9

# Wearable Device Analysis and Applications

**Sarin Vijay Mythry**

*Department of Electronics and Communication Engineering, Vignan Institute of Technology and Science, Hyderabad, India*

***Abstract***

This work presents the design of a two-stage ultra-low-power operational amplifier. To find the best design, the suggested operational amplifier is examined from a number of angles. It has a 76° phase margin, a maximum gain of roughly 75.33 dB, and a dissipation of 2.08 mW at 1.5 V supply voltage. For biological applications, where high gain at little power is needed, it is hence ideal. A circuit simulator based on SPICE has been used to construct the suggested operational amplifier. Multiplexing is done by passing one analog or digital input signal into a single line from a variety of input signals using an electrical device known as a multiplexer. The design of a low power consumption 90 nm CMOS 4:1 multiplexer using n-Pass Time Division Techniques is covered in this study. This study compares and simulates a few critical multiplexer factors, including as size, transistor count, and power dissipation. We shall identify Vdd as the constant power supply in static CMOS logic. Since energy recovery does not occur as it does in the case of static CMOS logic, the capacitor loses all the energy that is across the resistor, retaining very little of it. The design of a CMOS-based fourth-order low pass biquad filter for biomedical applications is covered in this short. Cascading the second order P-FVF and N-FVF biquad results in the construction of the fourth-order filter structure. The recommended filter functions in a region below the threshold. 114.4 Hz is the cut-off frequency of this circuit. The proposed design is shown in the Cadence environment with 90-nm CMOS technology.

***Keywords:*** Wearable medical devices, LNA, filter, multiplexer, CMOS

*Email:* sarinmythry@gmail.com

P. Suveetha Dhanaselvam, Naima Guenifi, Shiromani Balmukund Rahi and J. Ajayan (eds.) Field-Effect Transistor Devices: Post-CMOS Theory and Applications, (175–188) © 2026 Scrivener Publishing LLC

## 9.1 Introduction

All country's healthcare systems need to adjust to the challenges presented by emerging illnesses, food preferences, lifestyle modifications, environmental shifts, and other elements. The demands of the consumer healthcare sector, which are impacted by statistically significant features of human populations (such age or income), present researchers with new and exciting difficulties [1, 2]. According to a World Health Organization (WHO) estimate, cardiovascular diseases (CVDs) are the world's leading cause of mortality, with 17.90 million deaths predicted to occur each year [3]. Information on the prevalence of heart disease, stroke, and other conditions worldwide may be found in the Stroke and Heart Disease Statistics Update Fact Sheet [4]. In the globe, about 32 out of every 100 fatalities from CVD are caused by strokes and heart attacks. 1.56 billion individuals worldwide will have hypertension, sometimes referred to as high blood pressure (BP), in 2030. Survival rates and healthcare expenses can be greatly impacted by early diagnosis and treatment of medical issues.

More adaptable, dependable, and wearable electronics may now be developed because to the electrical industry's rapid downsizing advances. This might lead to a worldwide shift in health systems toward ongoing monitoring of physical health issues. Wearable and implanted therapeutic devices are two subcategories of biomedical devices. A WMD is a stand-alone, non-invasive device used for a particular medical or diagnostic function [5]. Between 2003 and 2023, more study was done on WMDs, and our review of the literature indicates that this field is expanding. Over the past ten years, wearable medical technology has seen significant advancements, particularly in remote patient monitoring. A WMD is a risk-detection tool that has to be lightweight, portable, and cozy [6]. Consumer electronics, defines, and health care are predicted to be the three main industry areas driving the wearable market. According to this poll, 54% of wearables are stylish, useful, wrist-mounted gadgets with low battery consumption.

In Figure 9.1, a crucial component of a WMD is displayed. Wearable sensors make it possible to swiftly identify aberrant signals and deliver medical assistance. The area where the biological signal is collected from the skin surface has a significant impact on the constant electrode or sensor placement [7]. The main components of an AFE are analog-to-digital converter (ADC) blocks, amplifiers, and filters. The amplifier and filter blocks detect, magnify, filter, and preprocess input voltages from electrode sensors. An ADC subsequently transforms the output voltage into a digital signal, which is subsequently sent to a transmitter [8]. The sent signal is

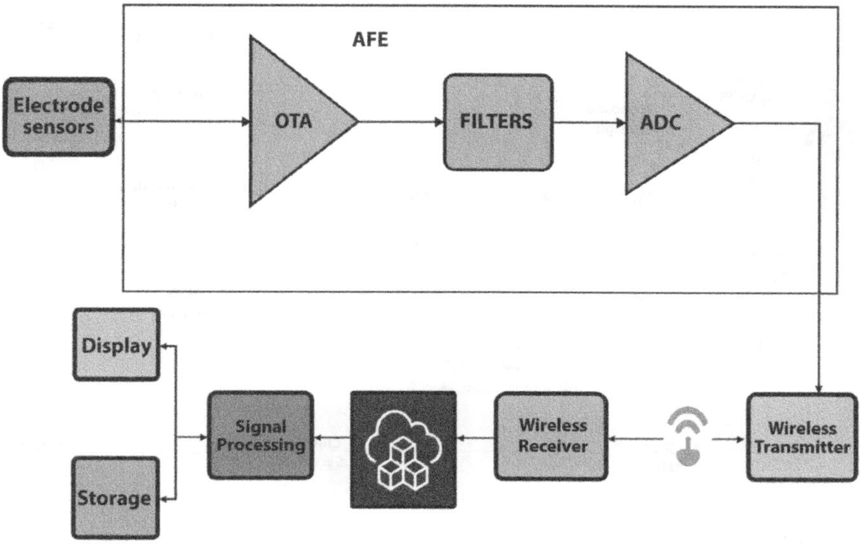


**Figure 9.1** General block diagram of WMD.

received by a smartphone or similar device for real-time viewing. After that, the information will be transferred to the cloud for further processing and reporting to clinicians in the area.

Rapid multiplexing has been used to build the proposed WMD architecture, as seen in Figure 9.2, by sharing amplifiers and filters among many electrodes. The use of EMG, ECG, pH, and temperature data can aid in the diagnosis of biophysical symptoms such as pain, exhaustion, and cardiac issues. The diagnosis of the underlying cause of diseases connected to stress or pain is made easier by the simultaneous gathering of several signals. A 4*1 MUX in the AFE collects bioinformation from temperature, pH, ECG, and EMG sensors. In WMDs, AFE is crucial for signal amplification and anomalous human identification.

A traditional AFE requires a separate amplifier, filter, and ADC converter for each electrode, which presents several design difficulties including the huge active area of resistors, complexity, scalability, and power consumption. This paper proposes the MUX-based AFE as a potential remedy for these issues. For the suggested MUX-based AFE, a low-power Schmitt trigger, a single-ended fully differential OTA, and an analog 6T n-pass time division multiplexer have all been constructed. Utilizing CMOS technology, the performance of every component has been assessed. A wearable gadget that has to be worn constantly is used to monitor individuals. This allows the user to take appropriate action by detecting deviations from the

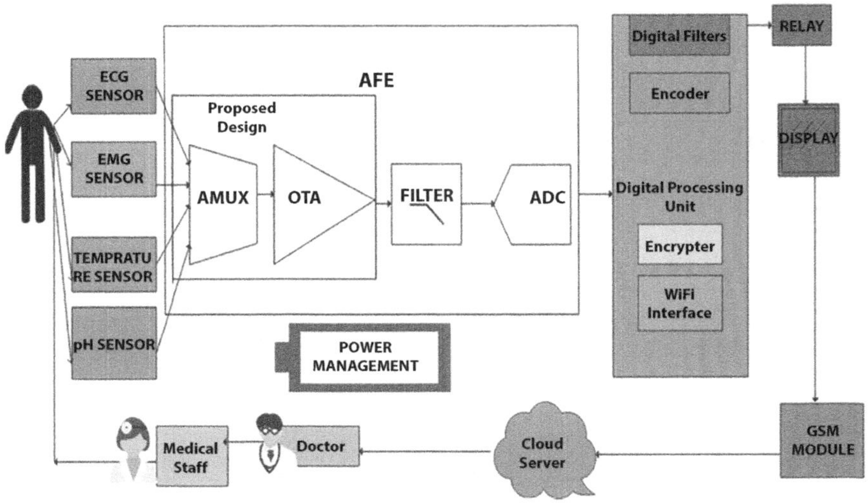


**Figure 9.2** The proposed architecture of WMD with AFE.

normal pre-set values and alerting the user. The patients using the Internet of Things (IoT) in their existing conditions are routinely observed by a crew and medical specialists. One potential solution may be a continuous monitoring system that alerts the crew to potential medical issues and protects against fatalities.

## 9.2 Methodology

Utilizing n-pass time division methods, we have developed a 4:1 Multiplexer. Figure 9.3 illustrates the suggested design for a 4 × 1 n-pass time division analog multiplexer utilizing 90 nm CMOS technology. Using synchronized switches, the 4 × 1 time division multiplexing (TDM) technique progressively delivers four different signals across a signal line. Low power pass transistor logic (LPPTL), which requires only six transistors, has been used in the multiplexer's design. Electrodes with dimensions of L = 45 nm and W = 120 nm are linked to the source and drain terminals of n-pass transistors as inputs. a reversible gate called the R gate is used to build a 4:1 multiplexer and 1:4 demultiplexer in reversible logic. The paper presents a reversible logic implementation of a 4:1 multiplexer and 1:4 demultiplexer using the R gate, a reversible gate.

The sensors supply the sample input values: the temperature is measured at 100 mV, the pH is measured at 63 mV, the EEG is measured at 1

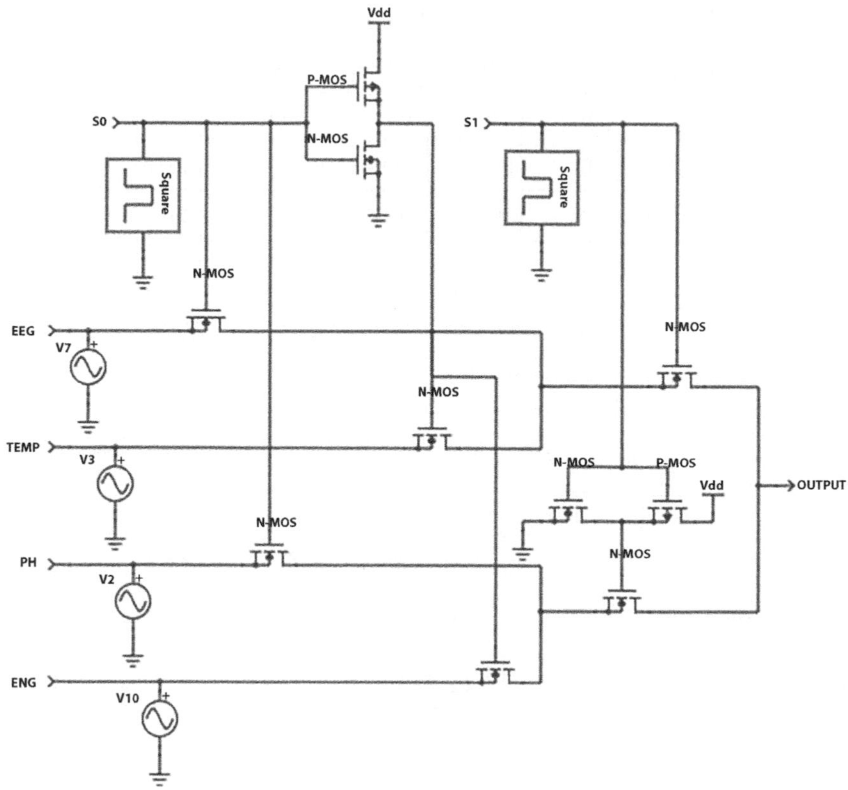


**Figure 9.3** Proposed schematic of MULTIPLEXER.

u V, and the ENG is measured at 10 mV. Table 9.1 displays the truth table for the 4:1 n-pass time division analog multiplexer that is utilized in the circuit schematic. The inputs that should be transmitted to the output are chosen by the selection lines S_0 and S_1. Equation (9.1) yields a potential for each input.

$$O/P = \overline{S_0}\,\overline{S_1}ENG + \overline{S_0}S_1\mathrm{TEMP} + S_0\overline{S_1}PH + S_0S_1EEG.. \tag{9.1}$$

Figure 9.4 depicts the layout of a two-stage Miller-compensated Op-Amp. A two-stage Miller-compensated operational amplifier (Op-Amp) is an operational amplifier that consists of two amplification stages connected in series. Typically, cascade or differential amplifiers comprise the stages. The compensating capacitor, also known as the Miller capacitor, is commonly connected to both the first stage's input and output. A two-stage

**Table 9.1** 4:1 MUX truth table.

| $S_0$ | $S_1$ | EEG | PH | TEMP | ENG | O/P(V) |
|---|---|---|---|---|---|---|
| 0 | 0 | 0 | 0 | 0 | 10 m | 10 m |
| 0 | 1 | 0 | 0 | 100m | 0 | 100 m |
| 1 | 0 | 0 | 63 m | 0 | 0 | 63 m |
| 1 | 1 | 1 u | 0 | 0 | 0 | 1 u |

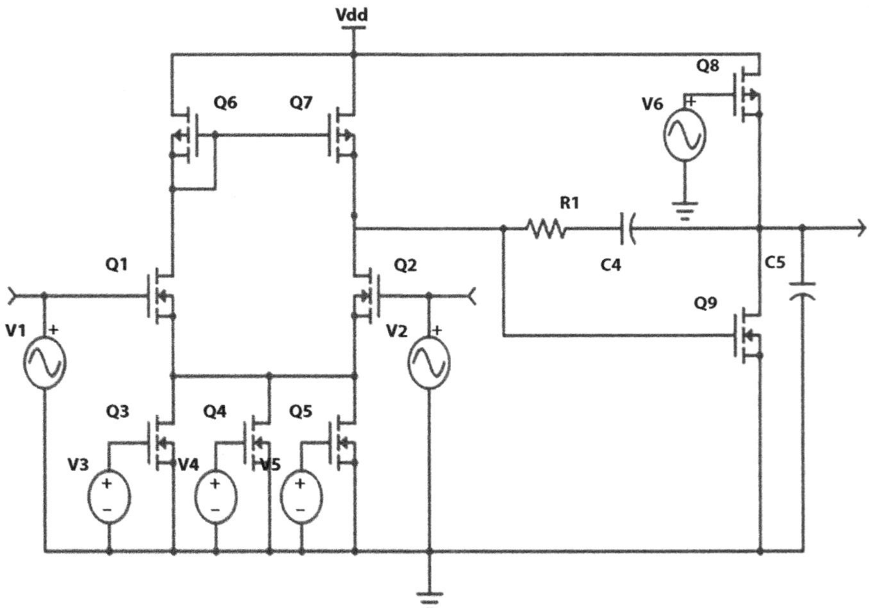


**Figure 9.4** Two-stage Miller adjusted Op-Amp schematic diagram.

Miller-compensated Op-Amp's high gain, wide bandwidth, and stability make it perfect for a range of analog signal processing applications, including audio amplification, instrumentation, and communication systems. the bulk driven operational transconductance amplifier's (OTA) low power design. It describes static, short circuit, and dynamic power dissipation and emphasizes the importance of power dissipation in CMOS devices.

The term "design goals" refers to the overarching aims that direct the process of designing any electrical component, including operational transconductance amplifiers (OTAs). These objectives capture the functionality, limitations, and desirable performance characteristics that the design seeks to accomplish. Table 9.2 enumerates these objectives.

Operational transconductance amplifiers (OTAs) are constructed using various circuit topologies and design methodologies to meet certain performance goals. The aspect ratio of each transistor is determined by going through a few phases in the implementation process. The first method computes the slew rate to determine the drain current (ID). Next, for each transistor stage, the (W/L) ratio is ascertained by,

$$I_D = \frac{W}{L} I_0 \exp \frac{V_{GS} - V_{TH}}{mV_T} (1 - exp \frac{-V_{DS}}{V_T}) \tag{9.2}$$

The thresholding voltage (VTH) is determined after accounting for the body's decreasing effects on barriers.

$$V_{TH} = V_{TH0} - \lambda_D V_{DS} - \lambda_B V_{BS} \text{ with } \lambda_D \text{ and } \lambda_B > 0 \tag{9.3}$$

The transconductance gm equation may be obtained using,

$$g_{m=} \frac{I_D}{mV_T} \tag{9.4}$$

$$\frac{W}{L} = \frac{I_D}{I_0} exp \frac{V_{TH} - V_{GS}}{mV_T} \tag{9.5}$$

Utilizing the phase compensation capacitance, which was discovered by, the equation for the necessary Unity Gain Bandwidth (UGD) was preserved.

**Table 9.2** Design specifications.

| | |
|---|---|
| Open-loop gain | ≥70 dB |
| Unity Gain Bandwidth (UGB) | ≈40 kHz |
| Maximum Power Dissipation | ≈100 nW |
| Phase Margin | ≥70° |
| Slew rate | ≈25 mV/µs |

$$C_C = \frac{g_{m1}}{2\pi.UGB} \tag{9.6}$$

where gm1 is M1's transconductance and CC is the compensating capacitor. The compensation resistance Rc increases the required open loop phase margin while stabilizing the Op-Amp.

$$R_C = \frac{1}{2g_{m7}}(1+\sqrt{1+\frac{4g_{m7}C_L}{g_{m1}C_1\tan(\phi_m)}}) \tag{9.7}$$

## 9.3 Simulation Result

Figure 9.5 and Figure 9.6 shows proposed schematic of MULTIPLEXER with OTA and FILTER and the circuit schematic respectively. Figures 9.7, 9.8, 9.9, 9.10 and 9.11 show the Filter's transient response, AC response, gain, noise, and power, in that order.

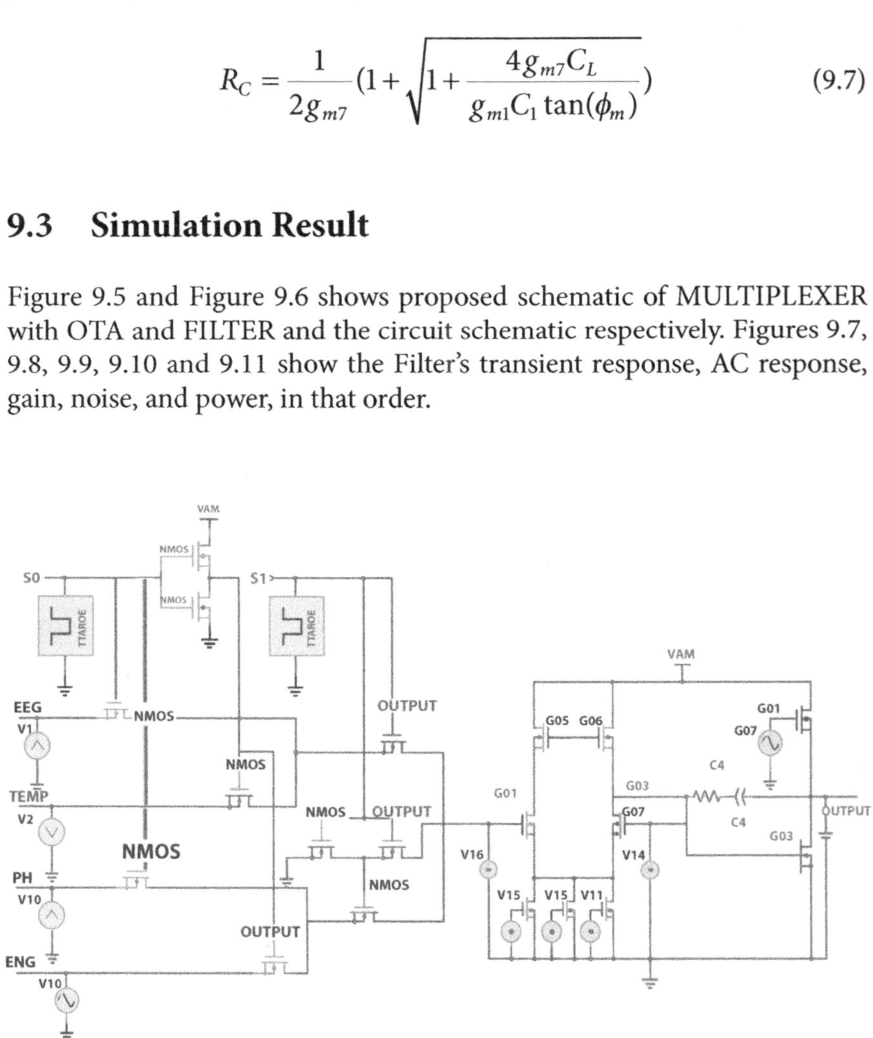


**Figure 9.5** Proposed schematic of MULTIPLEXER with OTA and FILTER.

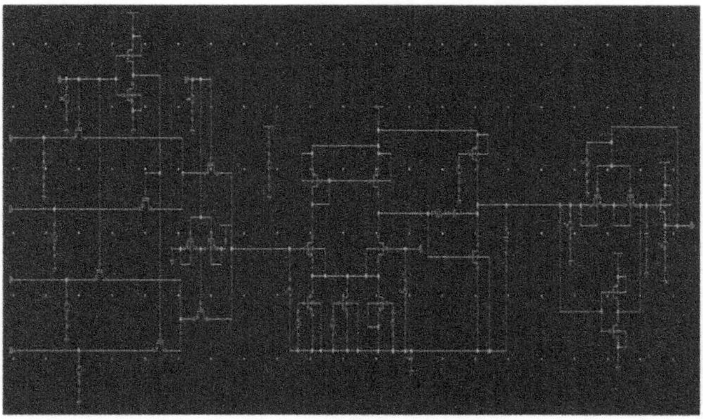

**Figure 9.6** Schematic diagram of OTA.

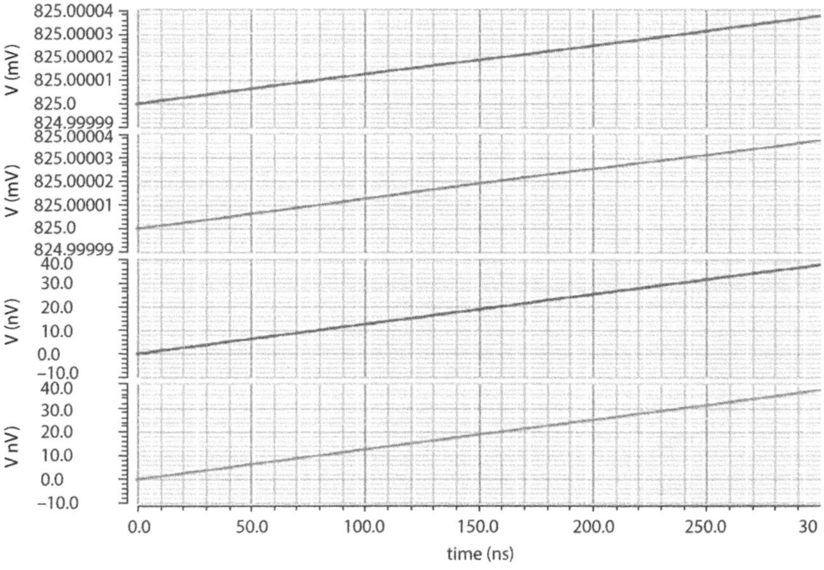


**Figure 9.7** Transient response.

The circuit's gain, noise, and power variations in response to variations in voltage (Vdc), temperature, and load capacitance are shown in detail in Tables 9.3, 9.4, 9.5.

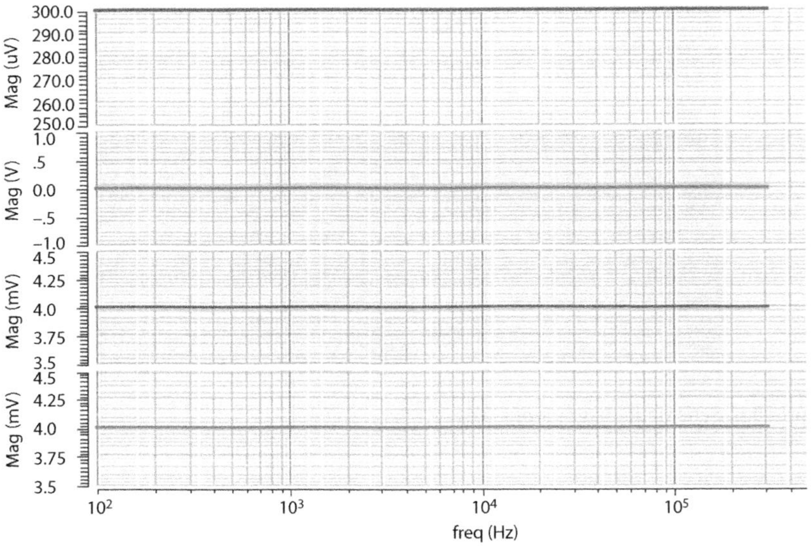


**Figure 9.8** AC response.

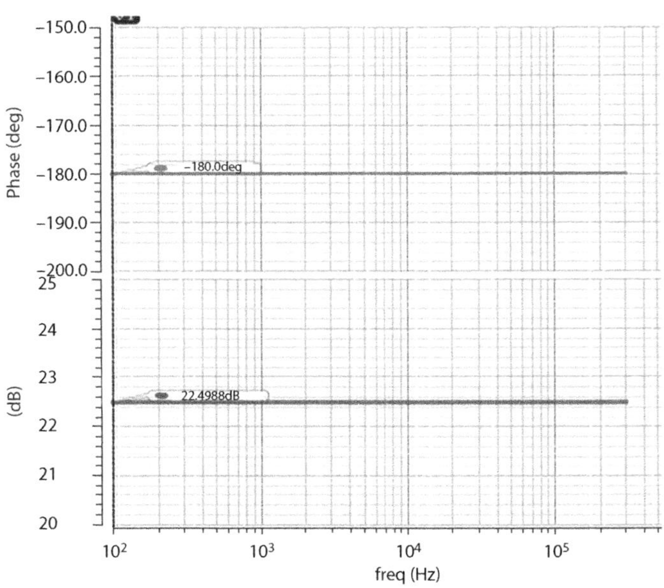


**Figure 9.9** Gain.

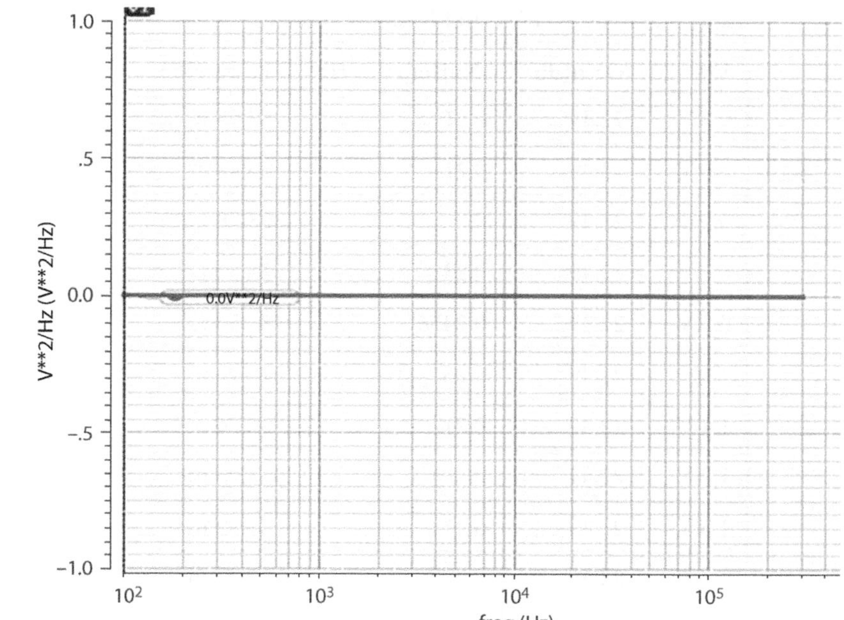


**Figure 9.10** Noise.

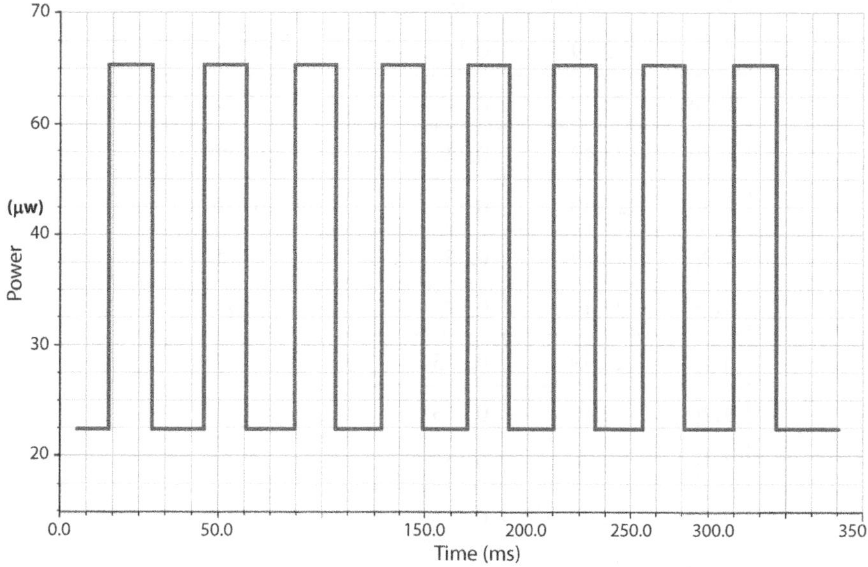


**Figure 9.11** Power.

**Table 9.3** MUX-OTA-FILTER supply voltage.

| Vdc | Gain(db) | Noise($\mu V/\sqrt{Hz}$) | Power(W) |
|---|---|---|---|
| 0.5 V | 22.49 | 0.21 | 37.95 μ |
| 0.6 V | 22.49 | 0.21 | 66.88 μ |
| 0.7 V | 22.49 | 0.2 | 282.89 μ |
| 0.8 V | 22.49 | 0.23 | 2.77 m |
| 0.9 V | 22.49 | 0.2456 | 8.75 m |
| 1 V | 22.49 | 0.25 | 16.96 m |
| 1.1 V | 22.49 | 0.21 | 26.74 m |
| 1.2 V | 22.49 | 0.2 | 38.05 m |
| 1.3 V | 22.49 | 0.21 | 50.92 m |
| 1.4 V | 22.49 | 0.22 | 65.35 m |
| 1.5 V | 22.49 | 0.221 | 81.34 m |

**Table 9.4** MUX-OTA-FILTER capacitive load (CL).

| CL | Gain(dB) | Noise ($\mu V / \sqrt{Hz}$) | Power(μW) |
|---|---|---|---|
| 0.5p | 22.49 | 0.2 | 67.88 |
| 0.6p | 22.49 | 0.21 | 67.88 |
| 0.7p | 22.49 | 0.22 | 67.88 |
| 0.8p | 22.49 | 0.23 | 67.88 |
| 0.9p | 22.49 | 0.25 | 67.88 |
| 10p | 22.49 | 0.27 | 67.88 |
| 1.1p | 22.49 | 0.255 | 67.88 |
| 1.2p | 22.49 | 0.265 | 67.88 |
| 1.3p | 22.49 | 0.222 | 67.88 |
| 1.4p | 22.49 | 0.278 | 67.88 |
| 1.5p | 22.49 | 0.256 | 67.88 |

**Table 9.5** MUX-OTA-FILTER Temp.

| Temp | Gain(dB) | Noise($\mu V/\sqrt{Hz}$) | Power(mW) |
|---|---|---|---|
| -75 | 22.49 | 0.2 | 68.10 |
| -58.3 | 22.49 | 0.22 | 70.16 |
| -41.6 | 22.49 | 0.21 | 72.27 |
| -25 | 22.49 | 0.23 | 74.43 |
| -8.33 | 22.49 | 0.2 | 76.62 |
| 8.33 | 22.49 | 0.21 | 78.84 |
| 25 | 22.49 | 0.23 | 81.07 |
| 41.67 | 22.49 | 0.2 | 83.32 |
| 58.33 | 22.49 | 0.2 | 85.58 |
| 75 | 22.49 | 0.2 | 87.84 |

## 9.4 Conclusion

To sum up, developing a Low-Noise Pre-amplifier for Wearable Biomedical Seizure Measurement System is a big project that needs to carefully consider a lot of important factors. The OTA is a crucial part of many medical devices, including as implanted medical devices, ENG systems, and ECG monitors, as biomedical applications require accurate, low-noise, and dependable biosignal amplification. The developed OTA may satisfy the stringent criteria of biological applications and advance medical technology if these factors are completely taken into account. This study effectively used n-Pass Time Division Techniques to analyse the 4:1 multiplexer utilizing 90 nm technology. Even though the research only looks at four input multiplexers for specifications, these also apply to larger input multiplexers. Consequently, a less complex and power-hungry circuit was used in the paper's conclusion. Phone networks, computer memory, and communication systems may all benefit from the application of this performance-enhancing MUX. The amount of electricity utilized is 1.5 units. In this work, we presented an SSF-based pseudo-RC filter with Sallen-Key configuration that may be used for neuronal and biological signals at low frequencies. Easily adjusted to a very low frequency, the cut-off frequency nevertheless performs comparably to other modern arts. The GC approach, which keeps the IRN and PRs' linearity intact, has solved the gain-loss issue of SSF brought on by the body effect. The results of THD and corner simulations under various bias schemes show that the suggested self-biased PRs may maintain low-impact PVT fluctuations and strong linearity. With its nW-class power consumption and very tiny chip area, the suggested LPF shows a lot of promise for wearable biomedical applications.

**AI disclosure statement:** The author acknowledges the use of AI-based tools for minor language editing or grammatical correction. However, all intellectual contributions, analyses, and conclusions presented in this chapter are entirely the author's own.

## References


1. Jusoh, W.M.E.A.W. and Ruslan, S.H., Design of Low-Power Bulk-Driven Balanced OTA in 90nm CMOS Technology, in: *Proceedings of the 978-1-5386-9175-5/18/$31.00*, IEEE.
2. Mittal, D. and Niranjan, A., Designing of Multiplexer and De-Multiplexer using different Adiabatic Logic in 90nm Technology, in: *Proceedings of the IEEE - 43488 9th ICCCNT 2018*, IISC, Bengaluru, India, July 10-12, 2018.
3. Karima, G., Néjib, H., Thouraya, E., Kamel, B., A low voltage and low power OTA using Bulk-Driven Technique and its application in Gm-c Filter, in: *Proceedings of the 2018 15th International Multi-Conference on Systems, Signals & Devices (SSD)*.
4. Carvajal, R., Ramirez-Angulo, J., Lopez-Martin, A., Torralba, A., Galan, J., Carlosena, A., Chavero, F., The flipped voltage follower: A useful cell for low-voltage low-power circuit design. *IEEE Trans. Circuits Syst. I Regul. Pap.*, 52, 7, 1276–1291, July 2005.
5. Maia, N.J., Sanchez, A.R., Moreno, R.L., Pimenta, T.C., Carvalho Ferreira, L.H., A Low Power CMOS Operational Transconductance Amplifier with Improved CMRR, in: *Proceedings of the 2020 32nd International Conference on Microelectronics (ICM)*.
6. Lee, S.Y. and Cheng, C.J., Systematic design and modeling of a OTA–C filter for portable ECG detection. *IEEE Trans. Biomed. Circuits Syst.*, 3, 1, 53–64, 2009.
7. Mann, S. and Jain, R., Design and Analysis of Reversible Multiplexer and Demultiplexer using R-Gates. *Proceeding International Conference on Recent Innovations is Signal Processing and Embedded Systems (RISE-2017)*, 27-29 October, 2017.
8. Kumar, N., Mittal, P., Mittal, M., Performance Analysis of FinFET based 2:1 Multiplexers for Low Power Application. *2020 IEEE Students' Conference on Engineering & Systems (SCES)*.

# 10

# Perovskites: Pioneering the Next Generation of Semiconductor Technologies

**P. Suveetha Dhanaselvam[1], B. Karthikeyan[1*], S. Nagarajan[2], K. Kavitha[1], P. Karthikeyan[1] and G. Pradeep Kumar[1]**

[1]*Department of Electronics and Communication Engineering, Velammal College of Engineering and Technology, Madurai, Tamilnadu, India*
[2]*Department of Electronics and Communication Engineering, Easwari Engineering College, Chennai, Tamilnadu, India*

## *Abstract*

Perovskite materials have emerged as a ground-breaking class of semiconductors, offering unprecedented opportunities to revolutionize modern technology. With their unique crystal structure, tunable electronic and optical properties, and cost-effective fabrication processes, perovskites are poised to address many of the limitations faced by traditional semiconductor materials. This chapter explores the science behind perovskites, highlighting their exceptional properties such as high charge carrier mobility and defect tolerance. It delves into their transformative applications, including high-efficiency photovoltaics, bright and flexible light-emitting diodes (LEDs), sensitive photodetectors, and their potential in spintronics and quantum computing. Despite their remarkable advantages, challenges such as environmental stability, toxicity, and scalability remain significant hurdles to widespread adoption. Recent advances in material engineering, lead-free alternatives, and hybrid designs are paving the way for more robust and sustainable perovskite-based technologies. This chapter provides a comprehensive overview of the current state of perovskite research, its challenges, and its potential to shape the future of semiconductor technologies.

***Keywords*:** Perovskites, semiconductor materials, next-generation technologies, photovoltaics, solar energy, light-emitting diodes, photodetectors

**Corresponding author*: om.bkarthikeyan@gmail.com

P. Suveetha Dhanaselvam, Naima Guenifi, Shiromani Balmukund Rahi and J. Ajayan (eds.) Field-Effect Transistor Devices: Post-CMOS Theory and Applications, (189–208) © 2026 Scrivener Publishing LLC

## 10.1 Introduction

Semiconductor technology has experienced significant advancements over the past few decades, with the continuous search for novel materials that offer superior performance, cost-effectiveness, and ease of fabrication. Among these emerging materials, perovskites have garnered immense interest owing to their exceptional electronic and optical properties. Their versatility, tunability, and compatibility with various device architectures position them at the forefront of next-generation semiconductor technologies. Perovskites are a class of materials that has a common $ABX_3$ crystal structure, where A represents a large organic or inorganic cation (e.g., methylammonium, cesium), B is a metal cation (commonly lead or tin), and X is a halide anion (such as iodine, bromine, or chlorine) [1, 2]. This unique structure imparts perovskites with remarkable electrical conductivity, tunable bandgaps, and high light absorption efficiency, designating them vastly suitable for applications in solar cells, LEDs, photodetectors, and transistors. The ability to engineer perovskite compositions enables tailored properties, leading to enhanced device performance.

The concept of perovskites dates back to 1839, when the mineral calcium titanium oxide ($CaTiO_3$) was first discovered by Gustav Rose and named after the Russian mineralogist Lev Perovski. However, the application of perovskites in semiconductors gained momentum much later. In 1978, early research into organic-inorganic halide perovskites laid the groundwork for future developments. The true breakthrough came in 2009, when Kojima *et al.* [26] proved the potential of perovskites in solar cells, realizing an efficiency of 3.8%. Over the next few years, rapid advancements pushed perovskite solar cell efficiencies beyond 20% by 2015. In the 2020s, ongoing research focused on enhancing stability, exploring tandem solar cell architectures, and advancing commercial-scale production. This rapid evolution highlights perovskites as one of the capable materials for novel semiconductor devices.

Perovskites have gathered significant consideration in the scientific and industrial communities owing to their unique properties, which often surpass those of traditional semiconductor materials like silicon, gallium arsenide, and cadmium telluride. Perovskites exhibit high absorption coefficients, enabling them to absorb light more efficiently than traditional semiconductors. This makes them ideal for applications like solar cells and photodetectors. They possess excellent charge carrier mobility, allowing for efficient transport of electrons and holes, which is critical for high-performance devices. Perovskites exhibit high absorption coefficients, enabling them to absorb light more efficiently than traditional semiconductors. This makes them ideal for applications like solar cells and photodetectors. They possess excellent charge carrier mobility, allowing for efficient transport of electrons and holes, which is critical for high-performance devices. The chemical composition of perovskites can be easily modified to tailor their bandgap, enabling customization for specific applications such as infrared photodetectors [3]. This tunability extends to their optical and electronic properties, making them adaptable to a wide range of technologies.

Unlike traditional semiconductors that require high-temperature processing and expensive fabrication techniques, perovskites can be synthesized using low-temperature solution-based methods. This reduces production costs and energy consumption, making them accessible for large-scale manufacturing. Perovskites are remarkably tolerant to defects in their crystal structure, which would typically degrade the performance of traditional semiconductors [4]. Perovskites can be deposited on flexible substrates, enabling the development of lightweight, bendable, and wearable devices [5]. While perovskites offer these compelling advantages, it is important to acknowledge their challenges, such as stability issues and environmental concerns, which are addressed later in the chapter. Nevertheless, their unique combination of properties makes them a promising candidate for the next generation of semiconductor technologies.

## 10.2 The Science of Perovskites

Perovskites with unique crystal structure and remarkable physical properties that make them greatly resourceful for a wide range of applications. This section delves into the fundamental science behind perovskites, exploring their structure, electronic and optical properties, design flexibility, and inherent stability mechanisms.

### 10.2.1 Crystal Structure and Composition

The term "perovskite" originally refers to the mineral $CaTiO_3$ which was discovered in the Ural Mountains, but it has since been expanded to describe any material with a similar crystal structure. The general formula for a perovskite is $ABX_3$ with a three-dimensional network of corner-sharing $BX_6$ octahedra, with the A-site cations occupying the interstitial spaces as shown in Figure 10.1. This arrangement creates a highly symmetric and stable lattice, which is key to the material's exceptional properties. The cubic or tetragonal symmetry of perovskites contributes to

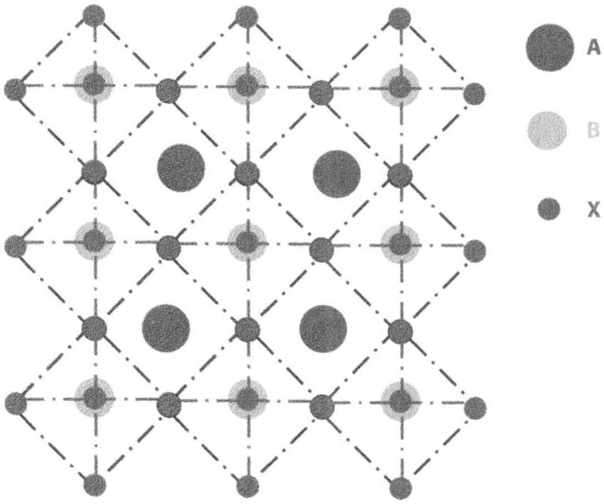


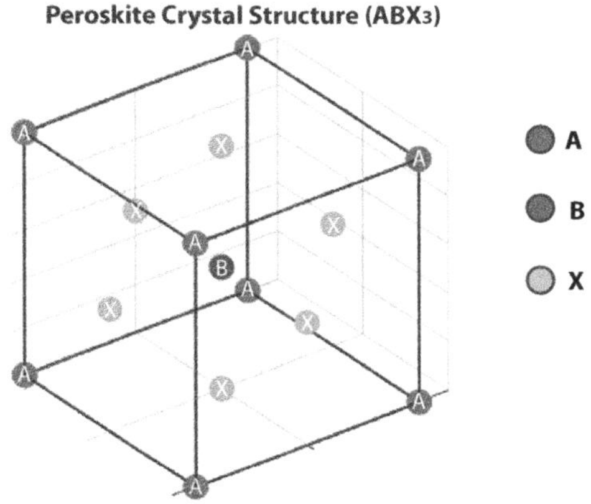


**Figure 10.1** Perovskite basic crystal structure. (a) 2D view. (b) cubic structure.

their excellent electronic and optical properties. The stability of the perovskite structure is determined by the Goldschmidt tolerance factor, which depends on the ionic radii of the A, B, and X ions [6]. A tolerance factor close to 1 indicates a stable perovskite structure. The A, B, and X sites can be substituted with a variety of ions, allowing for extensive tuning of the material's properties.

One of the most compelling aspects of perovskites is their tunability. By modifying the A, B, or X sites in the $ABX_3$ structure, researchers can tailor the material's properties to suit a wide range of applications. For instance, replacing the A-site cation with cesium instead of an organic molecule can enhance thermal stability, while substituting lead with tin at the B-site can reduce toxicity. Similarly, changing the halide anion from iodine to bromine can shift the bandgap, enabling the material to absorb or emit light at different wavelengths. This compositional flexibility extends to the creation of hybrid perovskites, which combine organic and inorganic components to achieve enhanced properties such as improved charge transport and mechanical flexibility.

### 10.2.2 Electronic and Optical Properties

The electronic and optical properties of perovskites are nothing short of extraordinary. These materials exhibit high charge carrier mobility, enabling efficient transport of electrons and holes, which is vital for applications like solar cells and transistors. Unlike traditional semiconductors, perovskites are ambipolar, meaning they can transport both electrons and holes with comparable efficiency [7]. This property makes them highly versatile for use in LEDs and field-effect transistors. Another standout feature is their defect tolerance. In most semiconductors, defects in the crystal structure can severely degrade performance, but perovskites are remarkably resilient to such imperfections. This defect tolerance, combined with their high absorption coefficients across a broad spectrum of light, makes them exceptionally efficient in photovoltaic and photodetector applications. Moreover, perovskites exhibit bright photoluminescence, with some materials achieving near-unity photoluminescence quantum yields, making them ideal for light-emitting applications such as LEDs and lasers. The bandgap of perovskites can also be finely tuned by adjusting their composition, allowing researchers to optimize them for specific wavelengths of light.

Figure 10.2 illustrates the bandgap tuning in perovskite materials as a function of halide composition. The x-axis represents different halide compositions, ranging from pure $MAPbI_3$ (methylammonium lead iodide) to pure $MAPbBr_3$ (methylammonium lead bromide), with mixed halides in between. The y-axis shows the corresponding bandgap energy in electron volts (eV). The plot demonstrates how the bandgap ascents gradually from ~1.5 eV for $MAPbI_3$ to ~2.3 eV for $MAPbBr_3$ as the bromine content increases. This tunability is a key feature of perovskites, enabling their optimization for specific applications. For example, $MAPbI_3$, with its lower bandgap, is ideal for solar cells, while $MAPbBr_3$, with its higher bandgap, is better suited for LEDs and displays. Annotations on the graph highlight these applications, making it clear how compositional changes directly impact the material's functionality [8–10].

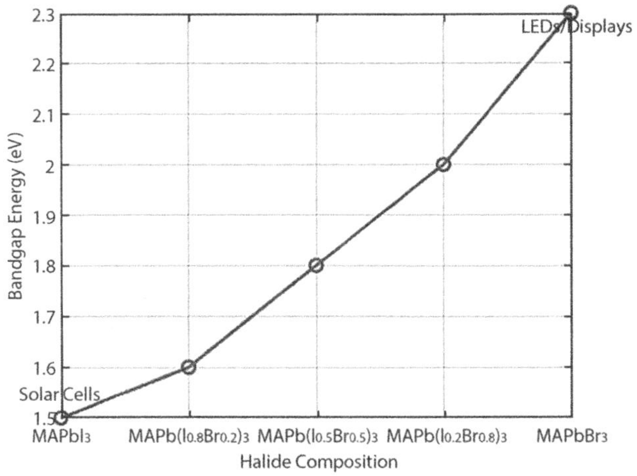


**Figure 10.2** Bandgap tuning in perovskites.

Figure 10.3 displays the absorption spectra of the same perovskite compositions, simulated using Gaussian peaks centered at the midpoint of their respective absorption ranges. The x-axis denotes the wavelength of light in nanometers (nm), and the y-axis represents the absorption intensity. Each curve corresponds to a specific halide composition, with the absorption peak shifting to shorter wavelengths as the bromine content increases. For instance, $MAPbI_3$ absorbs light across a broad range (400–800 nm), making

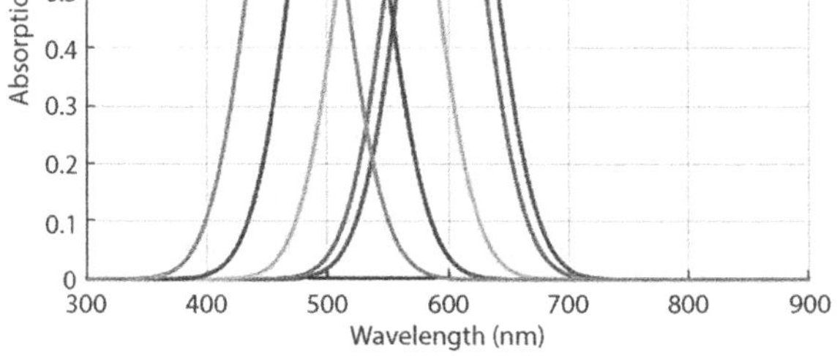


**Figure 10.3** Absorption spectra of perovskites.

it suitable for solar energy harvesting, while $MAPbBr_3$ absorbs at shorter wavelengths (400–550 nm), aligning with the visible spectrum for LED applications [11, 12]. The distinct colors of the curves and the accompanying legend help differentiate between the compositions, visually demonstrating how halide tuning affects the optical properties of perovskites. Together, Figures 10.1 and 10.2 provide a comprehensive understanding of how compositional changes in perovskites enable thorough control over their electronic and optical properties, making them preferable for a wide range of optoelectronic applications.

### 10.2.3 Stability and Defect Tolerance

Stability remains a significant concern, particularly in environmental conditions. Exposure to moisture, oxygen, or UV light can cause degradation, limiting the long-term performance of perovskite-based devices. Thermal instability is another issue, especially for hybrid organic-inorganic perovskites, which can decompose at elevated temperatures. However, the inherent defect tolerance of perovskites provides a silver lining. Defects in the crystal structure, which would typically cripple the performance of traditional semiconductors, have a minimal impact on perovskites due to the shallow nature of defect states in their bandgap. Some perovskites even exhibit self-healing properties, where defects are passivated or repaired over time.

Encapsulation is a decisive strategy for enhancing the stability and longevity of perovskite materials, mainly in practical applications where exposure to environmental factors like moisture, oxygen, and UV light can lead to degradation. This process involves coating or embedding the perovskite material within protective layers or matrices to shield it from external stressors, thereby improving its durability and performance. This coating or barrier can be made of organic, inorganic, or hybrid materials and is designed to prevent the ingress of moisture and oxygen, which are primary causes of perovskite decomposition. It also helps mitigate the effects of UV light, thermal stress, and mechanical damage, making it a vital step in the development of stable perovskite-based devices. Thin films of inorganic materials, such as aluminum oxide ($Al_2O_3$), silicon oxide ($SiO_2$), or silicon nitride ($Si_3N_4$) which are dense, impermeable, are deposited onto the perovskite surface using techniques like atomic layer deposition (ALD) or chemical vapor deposition (CVD). Polymers such as poly (methyl methacrylate) (PMMA), polyethylene terephthalate (PET), or epoxy resins are used to coat the perovskite. Polymer encapsulation is economical and flexible, rendering it right for large-area devices and flexible electronics.

Combining inorganic and organic materials can provide the benefits of both. For example, a thin inorganic layer can be deposited first to block moisture, followed by a polymer layer to add mechanical flexibility and additional protection. This approach is often used in high-performance devices where long-term stability is critical. Glass sheets are used to sandwich the perovskite device, creating a hermetic seal that is highly effective at blocking moisture and oxygen. This method is commonly used in perovskite solar cells, where long-term stability is essential. Encapsulation significantly improves the environmental stability of perovskites, enabling their use in real-world applications. Devices with proper encapsulation can maintain their performance for extended periods, even under harsh conditions.

The stability comparison plot in Figure 10.4 visually demonstrates the effectiveness of encapsulation in enhancing the durability of perovskite solar cells under various environmental stressors. The x-axis represents time (in hours), while the y-axis shows the power conversion efficiency (PCE) retention (%), which indicates how well the solar cells maintain their performance over time. The plot compares encapsulated and non-encapsulated devices under three conditions: ambient air, elevated temperature, and

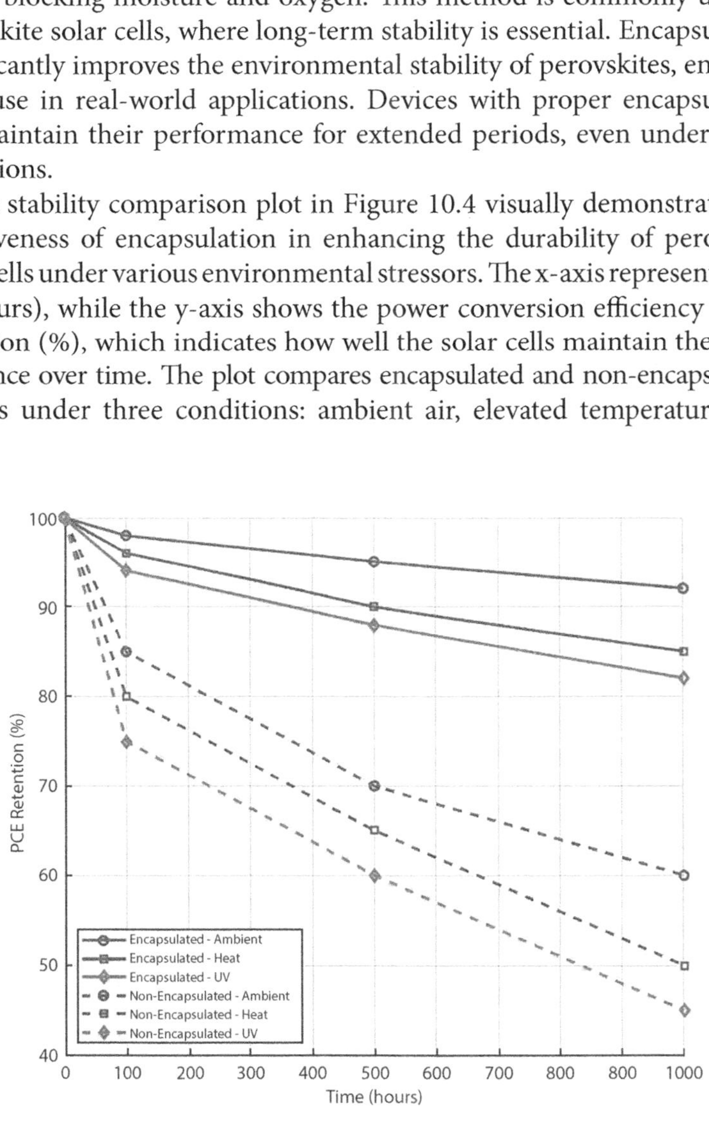


**Figure 10.4** Stability comparison of encapsulated vs. non-encapsulated perovskites.

UV exposure. Encapsulated devices are represented by solid lines, while non-encapsulated devices are shown with dashed lines, and different colors and markers distinguish between the environmental conditions. The plot reveals that encapsulated devices consistently outperform their non-encapsulated counterparts, retaining a higher percentage of their initial PCE over time. For example, under ambient air, encapsulated devices retain ~92% of their PCE after 1000 hours, compared to only ~60% for non-encapsulated devices. Similarly, under elevated temperature and UV exposure, encapsulated devices show significantly better stability, retaining ~85% and ~82% of their PCE, respectively, while non-encapsulated devices drop to ~50% and ~45%. This stark contrast highlights the critical role of encapsulation in protecting perovskite solar cells from environmental degradation, such as moisture, heat, and UV light. The plot underscores the importance of encapsulation strategies in improving the long-term stability and commercial viability of perovskite-based technologies, making them more suitable for real-world applications.

Compositional engineering has become a cornerstone in advancing the stability and performance of perovskite materials, addressing some of the key challenges that have hindered their widespread adoption. By carefully selecting and modifying the chemical composition of perovskites, researchers can significantly enhance their thermal and environmental stability. For instance, replacing organic cations like methylammonium or formamidinium with more stable inorganic cations such as cesium has proven effective in improving thermal resilience. Cesium-based perovskites, such as $CsPbI_3$, exhibit greater resistance to heat and humidity, making them more suitable for real-world applications. Similarly, mixed-cation approaches, where organic and inorganic cations are combined, offer a balance between stability and performance, allowing for tailored solutions to specific challenges. Another strategy involves substituting lead at the B-site with less toxic alternatives like tin or germanium, which not only addresses environmental concerns but also enhances stability in certain conditions. Mixed halide compositions, such as $MAPb(I_{1-X}Br_X)_3$, have also been widely explored, as they reduce halide migration and phase segregation while enabling precise tuning of the bandgap for optimal performance in various applications. These compositional adjustments, combined with the inherent flexibility of the perovskite structure, provide a powerful toolkit for designing materials that are both stable and efficient [13, 14].

In addition to compositional engineering, the use of additives has emerged as a highly effective strategy for further enhancing the durability and robustness of perovskite materials. Additives, which can be polymers, small molecules, or even inorganic compounds, are incorporated into the perovskite precursor solution or applied as interfacial layers to address specific weaknesses. One of the primary roles of additives is defect passivation. Perovskites often contain defects for instance vacancies, interstitials, and grain boundaries, which can perform as recombination centers and degrade device performance. Additives like phenethylammonium iodide or potassium iodide can fill these vacancies or bond with under-coordinated ions, effectively neutralizing defects and improving charge carrier lifetimes. Polymers such as polyethyleneimine or poly (methyl methacrylate) (PMMA) are also used to passivate grain boundaries and promote the growth of high-quality perovskite films with larger grain sizes and fewer defects. Beyond defect passivation, additives can enhance environmental stability by forming protective layers or networks within the perovskite film. For example, hydrophobic polymers like polystyrene can repel moisture, while crosslinking agents such as silane-based molecules can create a robust network that improves mechanical stability and resistance to environmental stressors. Additives can also improve thermal stability by reinforcing the crystal structure and preventing decomposition at high temperatures. For instance, graphene oxide has been used as an additive to enhance thermal conductivity and stability in perovskite solar cells. The synergistic effects of compositional engineering and additive strategies have already led to substantial improvements in the stability and implementation of perovskite-based devices.

### 10.2.4 Defect Tolerance

Defect tolerance is one of the most remarkable and defining characteristics of perovskite materials, setting them apart from traditional semiconductors like silicon or gallium arsenide. In most conventional semiconductors, defects in the crystal structure traps charge carriers and significantly degrade device performance. Perovskites exhibit an unusual resilience to such defects, maintaining high efficiency and functionality even in the presence of imperfections [15]. This defect tolerance is largely accr0065dited to the unique electronic structure of perovskites, where defect states tend to be shallow rather than deep within the bandgap [16]. Shallow defects have minimal impact on the movement of electrons and holes, allowing charge carriers to recombine radiatively and contribute to the material's optoelectronic performance.

The origins of defect tolerance in perovskites can be traced to their ionic nature and the specific arrangement of their crystal lattice. In the $ABX_3$ structure, the A-site cations, B-site metal ions, and X-site halides interact in a way that localizes defects and prevents them from spreading through the material. For example, halide vacancies, which are common in perovskites, create shallow trap states that do not significantly hinder charge transport. Also, the dynamic and soft nature of the perovskite lattice allows it to self-heal to some extent, mitigating the impact of defects over time. This self-healing behaviour is further enhanced by the material's ability to undergo ion migration, which can passivate defects or redistribute charge carriers in a way that maintains overall performance. These inherent mechanisms make perovskites highly robust, even when synthesized using low-cost, solution-based methods that typically result in higher defect densities compared to high-temperature, vacuum-based processes used for traditional semiconductors.

Despite their defect tolerance, perovskites are not entirely immune to the effects of defects, especially in extreme conditions or over long operational periods. For instance, defects can still contribute to non-radiative recombination, particularly at grain boundaries or surfaces, where defect densities are often higher. To address this, researchers have developed strategies to further enhance defect tolerance and minimize performance losses. One common approach is the use of additives, such as small molecules or polymers, which can passivate defects by filling vacancies or bonding with under-coordinated ions. For example, potassium iodide (KI) has been shown to passivate grain boundaries in perovskite solar cells, refining both efficiency and stability. Another strategy involves compositional engineering, where the A, B, or X sites in the perovskite structure are modified to reduce defect formation. For instance, mixed-cation or mixed-halide perovskites often exhibit lower defect densities and improved tolerance to environmental stressors. Additionally, advanced fabrication techniques, such as anti-solvent methods or vapor-assisted crystallization, can produce perovskite films with larger grain sizes and fewer defects, further enhancing their performance.

The defect tolerance of perovskites has profound implications for their practical applications. In solar cells, for example, it allows for high efficiencies even with relatively low material purity, reducing production costs and making perovskites a viable alternative to traditional silicon-based photovoltaics. In LEDs, defect tolerance contributes to high photoluminescence quantum yields, enabling bright and efficient light emission. Moreover, the ability to maintain performance in the presence of defects makes perovskites highly suitable for flexible and wearable electronics, where mechanical stress and environmental exposure can introduce imperfections. However, it is important to note that defect tolerance does not eliminate the need for careful material design and device engineering. While perovskites can withstand a higher density of defects compared to traditional semiconductors, minimizing defects through compositional tuning, additive engineering, and optimized fabrication processes remains essential for achieving the best possible performance and long-term stability.

## 10.3 Advantages of Perovskites over Traditional Semiconductors

Perovskites superior performance stem from their exceptional optoelectronic properties, cost-effective fabrication methods, and versatility in applications. One of the most compelling benefits of perovskites is their cost-effectiveness and scalability. Unlike traditional semiconductors, which often require expensive, high-temperature processing and vacuum-based techniques, perovskites can be synthesized using low-cost, solution-based methods at relatively low temperatures. This makes them accessible for large-scale production and reduces energy consumption during fabrication. In addition, the raw materials used in perovskites are abundant and inexpensive, further driving down costs. These factors make perovskites an attractive option for industries seeking affordable and scalable solutions for optoelectronic devices.

Another key benefit of perovskites is their high absorption coefficients and charge carrier mobility [17]. Perovskites exhibit exceptional light absorption capabilities, allowing them to capture a broad spectrum of sunlight with much thinner layers compared to traditional semiconductors like silicon. This high absorption efficiency translates to reduced material usage and lower production costs. Perovskites demonstrate excellent charge carrier mobility, enabling the rapid and efficient transport of electrons and holes. This property is crucial for applications such as solar cells, where efficient charge separation and collection are essential for high performance. The combination of strong light absorption and efficient charge transport makes perovskites highly competitive with conventional materials in terms of energy conversion efficiency.

The solution workability and low-temperature fabrication of perovskites further set them apart from traditional semiconductors. Perovskites can be deposited onto substrates using simple techniques such as spin-coating, inkjet printing, or blade coating, which are compatible with flexible and lightweight substrates. This contrasts sharply with the high-temperature, energy-intensive processes required for silicon-based devices. The ability to process perovskites at low temperatures not only lessens manufacturing costs but also facilitates the integration of perovskites with other materials and technologies, such as flexible electronics and wearable devices. This flexibility in fabrication opens up new possibilities for innovative applications that were previously unattainable with rigid, traditional semiconductors. The versatility in multifunctional applications of perovskites is a major advantage that underscores their potential to revolutionize multiple industries. Their tunable bandgap, achieved through compositional engineering, allows them to be optimized for specific wavelengths of light, making them suitable for diverse optoelectronic applications. For instance, in solar cells, perovskites have accomplished remarkable power conversion efficiencies, rivalling those of silicon-based cells. In LEDs, they offer bright and efficient light emission, while in photodetectors, they provide high sensitivity and fast response times. This multifunctionality, combined with their ease of fabrication and cost-effectiveness, positions perovskites as a transformative material for next-generation technologies.

The radar chart in Figure 10.5 provides a detailed and intuitive comparison of perovskites with traditional semiconductors like silicon and gallium arsenide, highlighting their performance across five key properties: efficiency, cost, defect tolerance, tunability, and scalability [18, 19].

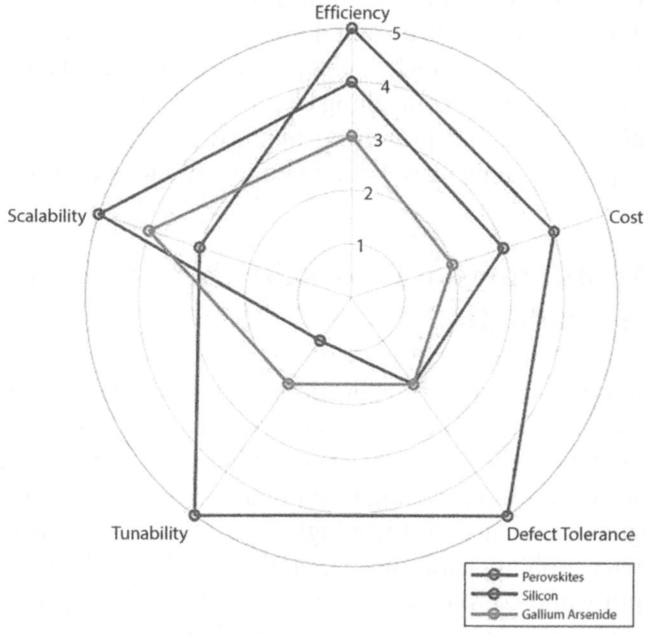


**Figure 10.5** Comparison of perovskites vs. traditional semiconductors.

Each property is represented by an axis on the chart, with ratings on a scale of 1 to 5. Perovskites, represented by a blue line, excel in multiple areas, achieving a perfect score in efficiency, defect tolerance, and tunability. Their high efficiency makes them ideal for applications like solar cells and LEDs, while their defect tolerance allows them to maintain performance even with imperfections, reducing the need for high-purity materials. In addition, perovskites' tunability enables precise control over their properties, such as bandgap, making them adaptable to a wide range of applications. They also score well in cost, thanks to their low-cost, solution-based fabrication methods, but face challenges in scalability, where they score moderately due to issues like stability and large-scale manufacturing. In contrast, silicon, represented by a red line, scores highly in scalability, benefiting from well-established manufacturing processes, but lags behind perovskites in tunability and defect tolerance. Its efficiency is slightly lower than perovskites, and its fabrication costs are higher due to energy-intensive processing. Gallium arsenide, shown in green, performs moderately in efficiency and scalability but scores poorly in cost, defect tolerance, and tunability, limiting its use to specialized applications like high-frequency electronics. The chart clearly illustrates the trade-offs between these materials, emphasizing perovskites' unique advantages in efficiency, cost-effectiveness, and flexibility, while also highlighting the areas where traditional materials like silicon still dominate, such as scalability. Overall, the radar chart underscores why perovskites are considered a transformative technology, offering a compelling combination of properties that make them highly versatile and promising for next-generation optoelectronic devices, despite the challenges that remain in scaling up their production.

## 10.4 Applications of Perovskites in Semiconductor Technologies

One of the most prominent applications is in photovoltaics, where perovskites are revolutionizing solar energy. Perovskite solar cells have achieved remarkable power conversion efficiencies, surpassing 25% in just a decade of research, rivaling traditional silicon-based solar cells. This rapid progress is because of their high absorption coefficients, excellent charge carrier mobility, and tunable bandgaps, which allow them to efficiently capture and convert sunlight into electricity. Also, perovskites can be fabricated using low-cost, solution-based methods at relatively low temperatures, significantly reducing production costs compared to silicon solar cells. Still, challenges remain in commercializing perovskite solar cells, mostly in terms of long-term stability and scalability. Issues such as environmental sensitivity to moisture, oxygen, and UV light, as well as thermal instability, need to be addressed to ensure their durability and performance in real-world conditions. Despite these challenges, the potential for economic, high-efficiency solar energy makes perovskites a promising candidate for the future of renewable energy.

Another exciting application of perovskites is in LEDs and displays. Perovskites exhibit bright and efficient light emission, with some materials achieving near-unity photoluminescence quantum yields. This makes them ideal for use in LEDs, where they can produce vibrant colors with high energy efficiency. The tunability of their bandgap allows for the creation of LEDs that release light in a wide breadth of wavelengths, from visible to near-infrared. Moreover, the solution-processability of perovskites enables their deposition on flexible substrates, opening up possibilities for flexible and wearable displays. These lightweight, bendable devices could revolutionize the display industry, enabling new form factors such as foldable screens and wearable technology. Though, challenges such as improving the operational lifetime and stability of perovskite LEDs under continuous use need to be tackled to fully realize their potential in commercial applications.

Perovskites also show great promise in photodetectors and sensors, where their high sensitivity and broadband response make them highly effective. Photodetectors based on perovskites can detect light across a wide spectrum, from ultraviolet to visible and near-infrared, making them suitable for a variety of applications, including imaging, communication, and environmental monitoring. Their fast response times and high detectivity enable the development of high-performance sensors for use in cameras, optical communication systems, and medical imaging devices. Additionally, the flexibility and low-temperature processing of perovskites consent for the creation of lightweight and conformable sensors, which can be incorporated into wearable devices or Internet of Things (IoT) applications. Despite these advantages, tasks such as improving the stability and reproducibility of perovskite photodetectors need to be addressed to ensure their reliability in practical applications.

The chart in Figure 10.6 provides a visual comparison of various emerging applications of perovskites based on three key metrics: cost, scalability, and efficiency. Each bubble represents a specific application such as solar cells, LEDs, photodetectors, spintronics, and quantum computing plotted by its cost (on the reversed X-axis, where lower is better) and scalability (Y-axis, where higher is better). The size and color of each bubble reflect the

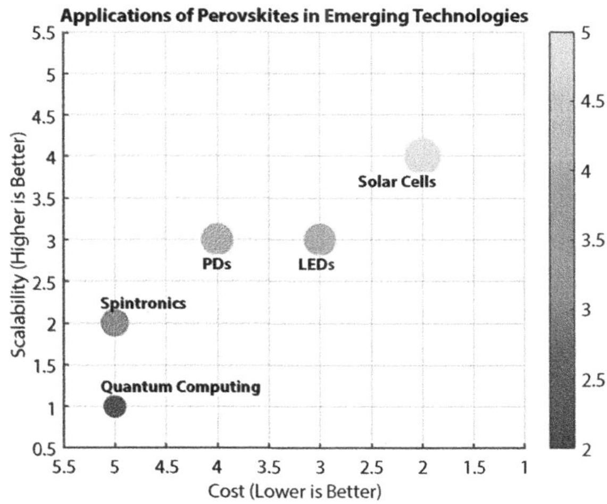


**Figure 10.6** Applications of perovskites.

efficiency of the application, with larger and more vividly colored bubbles indicating higher efficiency. This visualization highlights that perovskites show exceptional promise in applications like solar cells and LEDs, which balance high efficiency and scalability with relatively low cost. Conversely, applications such as quantum computing and spintronics, while still promising, face greater challenges in cost-effectiveness and scalability at present [20, 21].

In the realm of spintronics and quantum computing, perovskites are being explored for their unique spin properties and potential to enable next-generation computing technologies. Spintronics, which relies on the spin of electrons rather than their charge, could benefit from the spin-orbit coupling and long spin lifetimes observed in certain perovskite materials. These properties make perovskites a promising candidate for spintronic devices, such as spin valves and magnetic memory storage [22, 23]. In quantum computing, perovskites are being investigated for their potential to host quantum bits (qubits) and facilitate quantum information processing. The tunability of their electronic properties and the ability to engineer defects for specific quantum states offer exciting opportunities for advancing quantum technologies [24, 25]. Yet, significant trials remain, with enhancing the coherence times of qubits and creating scalable fabrication methods for perovskite-based quantum devices. Overall perovskites are leading the way for ground breaking developments in semiconductor technologies and as these challenges are resolved, perovskites are in position to play a key role in shaping the future of technology, enabling innovative solutions across industries and driving the next wave of technological innovation.

## 10.5 Conclusion

Perovskites have arisen as a transformative class of materials, offering unmatched opportunities to revolutionize semiconductor technologies. Their unique crystal structure, exceptional optoelectronic properties, and remarkable tunability make them highly versatile for a wide variety of applications, from high-efficiency solar cells and bright LEDs to sensitive photodetectors and next-generation quantum computing. The ability to tailor their composition and properties through compositional engineering, combined with innovative strategies like encapsulation and defect passivation, has significantly improved their stability and performance, addressing many of the challenges that once hindered their widespread adoption. However, the journey of perovskites from laboratory breakthroughs to commercial success is not without hurdles. Issues such as environmental sensitivity, thermal instability, and scalability remain critical areas of research. Despite these challenges, the rapid progress in perovskite technology evidenced by their soaring efficiencies, cost-effective fabrication methods, and adaptability to flexible and wearable devices, underscores their immense potential. With continued innovation and investment, perovskites have the potential to not only revolutionize technology but also contribute to a more sustainable and energy-efficient world. This chapter has explored the science, applications, and future prospects of perovskites, offering a glimpse into the exciting possibilities that lie ahead for this ground breaking material.

**AI disclosure statement:** The author acknowledges the use of AI-based tools for minor language editing or grammatical correction. However, all intellectual contributions, analyses, and conclusions presented in this chapter are entirely the author's own.

## References

1. Bhalla, A.S., Guo, R., Roy, R., The perovskite structure—a review of its role in ceramic science and technology. *Mater. Res. Innovations*, 4, 1, 3–26, 2000.
2. Luo, S. and A. Daoud, W., Crystal structure formation of CH3NH3PbI3-xClx perovskite. *Materials*, 9, 3, 123, 2016.

3. Tasleem, S. and Tahir, M., Recent progress in structural development and band engineering of perovskites materials for photocatalytic solar hydrogen production: A review. *Int. J. Hydrogen Energy*, 45, 38, 19078–19111, 2020.
4. Wang, F., Bai, S., Tress, W., Hagfeldt, A., Gao, F., Defects engineering for high-performance perovskite solar cells. *npj Flexible Electron.*, 2, 1, 22, 2018.
5. Tabassum, M., Zia, Q., Zhou, Y., Wang, Y., Reece, M.J., Su, L., A review of recent developments in smart textiles based on perovskite materials. *Textiles*, 2, 3, 447–463, 2022.
6. Tidrow, S.C., Mapping comparison of Goldschmidt's tolerance factor with Perovskite structural conditions. *Ferroelectrics*, 470, 1, 13–27, 2014.
7. Giorgi, G. and Yamashita, K., Organic–inorganic halide perovskites: an ambipolar class of materials with enhanced photovoltaic performances. *J. Mater. Chem. A*, 3, 17, 8981–8991, 2015.
8. Tan, M., Wang, S., Rao, F., Yang, S., Wang, F., Pressures tuning the band gap of organic–inorganic trihalide perovskites (MAPbBr 3): A first-principles study. *J. Electron. Mater.*, 47, 7204–72115, 2018.
9. Roy, M., Vikram, S.B., Arijit Mitra, A.A., Aslam, M., Composition-Controlled Synthesis of Hybrid Perovskite Nanoparticles by Ionic Metathesis: Bandgap Engineering Studies from Experiments and Theoretical Calculations. *Chemistry–A Eur. J.*, 25, 42, 9892–9901, 2019.
10. Vega, E., Mollar, M., Mari, B., Bandgap tuning of MAPbI3- XBrx thin film perovskites for photovoltaic applications, in: *2015 3rd International Renewable and Sustainable Energy Conference (IRSEC)*, IEEE, pp. 1–5, 2015.
11. Niedzwiedzki, D.M., Zhou, H., Biswas, P., Exciton binding energy of MAPbI3 thin film elucidated *via* analysis and modeling of perovskite absorption and photoluminescence properties using various methodologies. *J. Phys. Chem. C*, 126, 2, 1046–1054, 2022.
12. Basumatary, P., Kumari, J., Agarwal, P., Probing the defects states in MAPbI3 perovskite thin films through photoluminescence and photoluminescence excitation spectroscopy studies. *Optik*, 266, 1695865, 2022.
13. Jeon, N.J., Hong Noh, J., Seok Yang, W., Chan Kim, Y., Ryu, S., Seo, J., Il Seok, S., Compositional engineering of perovskite materials for high-performance solar cells. *Nature*, 517, 7535, 476–480, 2015.
14. Laakso, J., Todorović, M., Li, J., Zhang, G.-X., Rinke, P., Compositional engineering of perovskites with machine learning. *Phys. Rev. Mater.*, 6, 11, 113801, 2022.
15. Kim, G.-W. and Petrozza, A., Defect tolerance and intolerance in metal-halide perovskites. *Adv. Energy Mater.*, 10, 37, 2001959, 2020.
16. Huang, H., Bodnarchuk, M.I., Kershaw, S.V., Kovalenko, M.V., Rogach, A.L., Lead halide perovskite nanocrystals in the research spotlight: stability and defect tolerance. *ACS Energy Lett.*, 2, 9, 2071–2083, 2017.
17. Johnston, M.B. and Herz, L.M., Hybrid perovskites for photovoltaics: charge-carrier recombination, diffusion, and radiative efficiencies. *Acc. Chem. Res.*, 49, 1, 146–154, 2016.

18. Stoumpos, C.C. and Kanatzidis, M.G., Halide Perovskites: poor Man's high-performance semiconductors. *Adv. Mater.*, 28, 28, 5778–5793, 2016.
19. Brenner, T.M., Egger, D.A., Kronik, L., Hodes, G., Cahen, D., Hybrid organic—inorganic perovskites: low-cost semiconductors with intriguing charge-transport properties. *Nat. Rev. Mater.*, 1, 1, 1–16, 2016.
20. Zhang, L., Mei, L., Wang, K., Lv, Y., Zhang, S., Lian, Y., Liu, X., *et al.*, Advances in the application of perovskite materials. *Nano Micro Lett.*, 15, 1, 177, 2023.
21. Monama, G.R., Ramohlola, K.E., Iwuoha, E.I., D. Modibane, K., Progress on perovskite materials for energy application. *Results Chem.*, 4, 1003215, 2022.
22. Privitera, A., Righetto, M., Cacialli, F., Riede, M.K., Perspectives of organic and perovskite-based spintronics. *Adv. Opt. Mater.*, 9, 14, 2100215, 2021.
23. Hirohata, A., Sukegawa, H., Yanagihara, H., Žutić, I., Seki, T., Mizukami, S., Swaminathan, R., Roadmap for emerging materials for spintronic device applications. *IEEE Trans. Magn.*, 51, 10, 1–11, 2015.
24. Jayan, K.D. and Babu, K., Luminescent perovskite quantum dots: progress in fabrication, modelling and machine learning approaches for advanced photonic and quantum computing applications. *J. Lumin.*, 120906, 2024.
25. Dhanaselvam, Suveetha, P., Srinivasa Rao, K., Rahi, S.B., Yadav, D.S. (Eds.), *Field Effect Transistors*, John Wiley & Sons, 2025.
26. Kojima, A., Teshima, K., Shirai, Y., Miyasaka, T., Organometal halide perovskites as visible-light sensitizers for photovoltaic cells. *J. Am. Chem. Soc.*, 131, 17, 6050–6051, 2009.

11

# Performance Analysis of Eco-Friendly Perovskite Solar Cells

**P. Suveetha Dhanaselvam, B. Karthikeyan*, K. Kavitha and G. Pradeep Kumar**

*Department of Electronics and Communication Engineering, Velammal College of Engineering and Technology, Madurai, India*


## *Abstract*

This study presents the design and simulation of a high-performance, lead-free perovskite solar cell (PSC) using SCAPS-1D. The proposed device features $CH_3NH_3SnI_3$ as the absorber layer, chosen for its superior optoelectronic properties and environmental sustainability, along with strontium titanate (STO) as the electron transport layer (ETL) and $CuSbS_2$ (CAS) as the hole transport layer (HTL). Numerical optimization yielded outstanding photovoltaic parameters: a short-circuit current density ($J_{SC}$) of 40.62 mA/cm$^2$, an open-circuit voltage ($V_{OC}$) of 1.0405 V, a fill factor (FF) of 86%, and a power conversion efficiency (PCE) of 36.35%. The STO/CAS heterostructure exhibited enhanced charge transport and suppressed recombination losses, contributing to the device's high efficiency. These findings underscore the potential of lead-free perovskites in sustainable photovoltaics, providing a framework for future development in high-efficiency solar cell architectures.

***Keywords*:** Perovskite solar cells, photovoltaic technology, lead-free perovskites, SCAPS simulation, quantum efficiency, power conversion efficiency

**Corresponding author*: om.bkarthikeyan@gmail.com

P. Suveetha Dhanaselvam, Naima Guenifi, Shiromani Balmukund Rahi and J. Ajayan (eds.) Field-Effect Transistor Devices: Post-CMOS Theory and Applications, (209–228) © 2026 Scrivener Publishing LLC

## 11.1 Introduction

The growing global reliance on renewable energy systems has accelerated advancements in solar technology, with perovskite solar cells (PSCs) emerging as a promising alternative to conventional photovoltaic devices. Solar cell technology has evolved through several generations, beginning with first-generation cells based on monocrystalline and polycrystalline silicon wafers that dominate the market but face limitations in cost and efficiency. Second-generation thin-film technologies such as CIGS (Copper Indium Gallium Selenide), amorphous silicon (a-Si), and cadmium telluride (CdTe) offer flexibility and lower material usage but suffer from stability and toxicity issues. Third-generation innovations including multi-junction cells, quantum dots, copper zinc tin sulfide (CZTS), and perovskite solar cells aim to overcome these challenges with higher efficiency and tunable properties [1]. Among these, PSCs have garnered significant attention due to their exceptional optoelectronic characteristics such as high charge-carrier mobility, tunable bandgaps, and strong light absorption coefficients. Recent breakthroughs have propelled their PCE beyond 25%, driven by advancements in material synthesis, device architecture, and cost-effective fabrication techniques like solution processing and roll-to-roll printing [2].

Despite their potential, lead-based PSCs such as $CH_3NH_3PbI_3$ face critical challenges including toxicity concerns due to lead content, which restricts large-scale adoption, and environmental instability leading to degradation under operational conditions like moisture, heat, and UV exposure. To address these issues, tin-based perovskites such as $CH_3NH_3SnI_3$ have emerged as a non-toxic, eco-friendly alternative. Tin offers several advantages including stronger covalent bonding compared to lead, which enhances electron and hole mobilities for efficient charge transport, and an ideal bandgap of 1.3–1.4 eV that enables excellent sunlight absorption and photovoltaic performance. However, tin-based PSCs still lag behind lead-based counterparts in terms of PCE and stability. A major issue is the oxidation of $Sn^{2+}$ to $Sn^{4+}$, which introduces defects, accelerates degradation, and reduces efficiency. Current research focuses on defect passivation techniques such as additives and encapsulation, doping strategies to suppress $Sn^{2+}$ oxidation, and optimization of charge transport layers (HTL/ETL) to improve stability and performance [3, 4].

Computational modeling plays a pivotal role in optimizing PSC designs by analyzing electrical properties and predicting performance. Widely used simulation tools include SILVACO ATLAS, PC1D, TCAD, AMPS, and COMSOL Multiphysics. SCAPS-1D, a specialized tool for perovskite solar cells, enables detailed evaluation of layer configurations, defect analysis, and efficiency optimization. In this study, a lead-free PSC was designed using $CH_3NH_3SnI_3$ as the absorber layer, Strontium Titanate (STO) as the ETL, and Copper Antimony Sulfide ($CuSbS_2$ or CAS) as the HTL [5]. The research investigates back-contact optimization using materials such as Ni, Pt, Au, Fe, and C for minimal resistive losses, along with critical performance parameters including temperature dependence, absorber layer defect density and thickness, acceptor density variations, and the impact of shunt and series resistance [6]. By systematically analyzing these factors, this work aims to enhance the efficiency and stability of tin-based PSCs, contributing to sustainable and high-performance photovoltaic technologies.

## 11.2 Materials and Methodology

Perovskite solar cells (PSCs) have emerged as a promising alternative to conventional thin-film solar cells due to their low cost, high efficiency, and relatively simple manufacturing process. These advantages position PSCs as a potential game-changer in photovoltaic technology. The design of high-performance PSCs requires careful consideration of material selection and device architecture to achieve optimal efficiency, long-term stability, and environmental sustainability. The proposed PSC structure in this study features a carefully optimized layer configuration, as illustrated in Figure 11.1. The device employs $CH_3NH_3SnI_3$ as the light-absorbing perovskite layer, which offers excellent optoelectronic properties while avoiding the toxicity concerns associated with lead-based perovskites. Strontium Titanate (STO) serves as the ETL, effectively extracting and transporting photo generated electrons, while Copper Antimony Sulfide ($CuSbS_2$ or CAS) functions as the HTL, facilitating efficient hole collection. This strategic combination of materials enhances the generation, separation, and collection of charge carriers throughout the device.

Device operation begins with the fluorine-doped tin oxide (FTO) layer, which serves as the transparent conductive electrode at the top of the solar cell structure. The FTO layer plays a critical dual role: it allows maximum sunlight transmission to the underlying active layers while simultaneously collecting and conducting the generated electrical current. Its high optical transparency ensures minimal light absorption, allowing most incident photons to reach the perovskite absorber layer where they can be converted

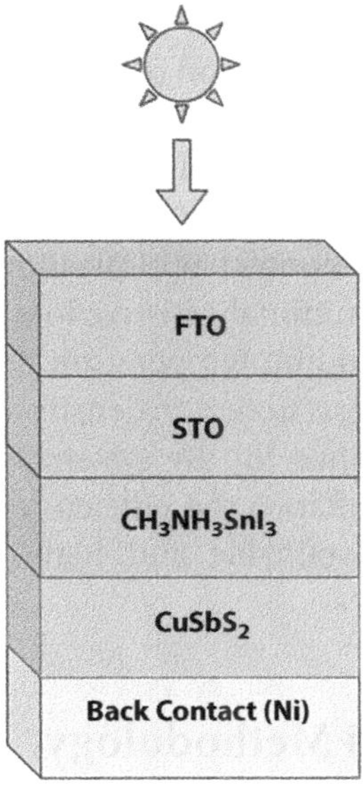


**Figure 11.1** Schematic structure of the designed PSC.

into electrical charges. Simultaneously, its excellent electrical conductivity enables efficient extraction of photo generated electrons from the device to the external circuit. Beyond its electrical and optical functions, the FTO layer provides a robust and stable foundation for the subsequent device layers. Its mechanical durability and chemical stability make it ideal for supporting the delicate perovskite and charge transport layers during both fabrication and operation. The careful selection and integration of these layers - from the transparent FTO electrode through the charge transport layers to the perovskite absorber - create a synergistic system designed to maximize solar energy conversion while addressing the stability and environmental concerns associated with earlier photovoltaic technologies. This comprehensive approach to device design demonstrates the potential for developing high-performance, sustainable solar cells that could significantly impact future renewable energy solutions.

The device architecture features a carefully engineered stack of functional layers, each playing a critical role in the photovoltaic conversion process. Beneath the FTO electrode, the Strontium Titanate (STO) electron transport layer performs several essential functions. Its selective charge extraction capability efficiently collects electrons from the perovskite layer while simultaneously blocking holes, thereby significantly reducing interfacial recombination losses. This charge-selective behavior ensures that liberated electrons are effectively channeled toward the FTO contact. The STO layer also serves as an ideal substrate for perovskite deposition, providing an atomically smooth surface that minimizes defect formation and subsequent carrier recombination. Importantly, STO's inherent UV resistance provides an additional protective barrier, shielding the sensitive perovskite material from harmful radiation and consequently improving device durability.

At the heart of the solar cell lies the $CH_3NH_3SnI_3$ perovskite absorber layer, where the fundamental light-to-electricity conversion occurs. This lead-free perovskite boasts an optimally narrow bandgap (1.3–1.4 eV) that enables broad-spectrum photon harvesting across visible and near-infrared wavelengths. The material's exceptionally high absorption coefficient allows for complete light capture within a remarkably thin active layer. When photons are absorbed, they generate electron-hole pairs through the photoelectric effect. The perovskites outstanding optoelectronic properties - including extended charge carrier diffusion lengths and high mobility ensure the charges can efficiently migrate to their respective transport layers before recombination occurs. The built-in electric field within the perovskite layer, created by strategic energy band alignment, actively drives charge separation, directing electrons toward the STO ETL and holes toward the CAS HTL. This efficient charge segregation mechanism is crucial for maintaining high current output while minimizing energy losses [7].

The hole transport system employs copper antimony sulfide ($CuSbS_2$ or CAS), an environmentally benign and economically viable material with exceptional hole-transport capabilities. With its direct bandgap ranging from 1.38 to 1.65 eV, CAS demonstrates optimal spectral alignment with the perovskite absorber, enabling highly efficient hole extraction. The material's strong visible-light absorption complements the device's photon management strategy by capturing any transmitted light that passes through the perovskite layer. As a charge-selective interface, CAS exhibits excellent hole conductivity while effectively blocking electrons, creating an unimpeded pathway for holes to reach the back contact while preventing unwanted charge recombination. This selective transport characteristic significantly enhances the overall charge collection efficiency. Completing the device structure is the nickel back contact, which forms the final charge-collecting interface. Nickel's high work function ensures the formation of an ohmic contact with the HTL, facilitating unimpeded hole transfer to the external circuit. This metal layer provides robust mechanical support and thermal stability, critical for maintaining device integrity during prolonged operation. The back contact serves as the essential electrical termination point that closes the photovoltaic circuit, enabling the continuous flow of photo generated current and ultimately transforming solar energy into usable electricity. Together, this carefully optimized layer sequence creates a harmonious system where each component synergistically contributes to achieving high photovoltaic performance while addressing stability and sustainability concerns inherent in conventional solar cell designs [8].

SCAPS-1D (Solar Cell Capacitance Simulator) is a numerical simulation tool developed to analyze andassess the performance of thin-film solar cells. It uses the basic semiconductor equations, particularly the continuity equation for charge carriers and thedrift-diffusion current density equations, to evaluate the optoelectronic behavior of different device architectures [9]. The time-dependent continuity equations for charge carriers are expressed by:

$$\frac{dn}{dt} = \frac{1}{q}\frac{\partial J_n}{\partial x} + \left(G_n - R_n\right) \tag{11.1}$$

$$\frac{dp}{dt} = \frac{1}{q}\frac{\partial J_p}{\partial x} + \left(G_p - R_p\right) \tag{11.2}$$

where $q$ is the elementary charge, $J_n$ and $J_p$ represent the electron and hole current densities, $R$ is the recombination rate and $G$ is the charge carrier generation rate. The drift-diffusion current density equation, describing charge transport, are given by:

$$J_n = q\mu_n n\varepsilon + qD_n\partial_n \tag{11.3}$$

$$J_p = q\mu_p p\varepsilon + qD_p\partial_p \tag{11.4}$$

where $\mu_p$ and $\mu_n$ are the mobilities of holes and electrons, $\varepsilon$ represents the electric field, $D_p$ and $D_n$ are the diffusion coefficients. These equations capture the relationship between charge carrier movement, recombination processes, and electric field distribution, provides information about functioning of the simulated solar cell [10]. Overall, SCAPS-1D provides data about into charge carrier dynamics, energy band alignment, and overall photovoltaic performance, making it an essential tool for optimizing perovskite solar cells.

## 11.3 Performance Analysis

The thickness of the perovskite absorber layer plays a crucial role in determining the overall device performance, as it directly influences both light absorption and charge carrier collection. As demonstrated in Figure 11.2, the PCE shows a strong dependence on absorber thickness, increasing from lower values at thin layers to reach an optimal peak of 36.47% at 1400 nm. This behavior can be attributed to two key factors: enhanced photon absorption due to the longer optical path length, and improved charge carrier collection efficiency in thicker active layers. The short-circuit current density (Jsc) exhibits a corresponding progressive increase from 36.73 $mA/cm^2$ at 200 nm to 41.12 $mA/cm^2$ at 1400 nm, confirming that thicker absorber layers more effectively harness incident photons across the solar spectrum. In contrast, the $V_{OC}$ and FF remain relatively stable throughout the thickness variation, maintaining consistent values of approximately 1.03 V and 86.11% respectively. This stability suggests that the fundamental electronic properties of the device are largely preserved across different thicknesses.

Notably, the efficiency improvement becomes marginal beyond 1000 nm, indicating a practical limit where additional thickness provides diminishing returns. This plateau effect occurs because the benefits of

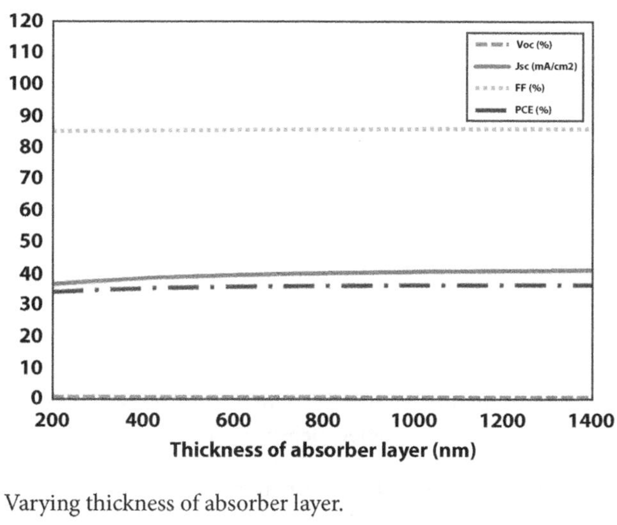


**Figure 11.2** Varying thickness of absorber layer.

increased absorption are eventually counterbalanced by higher recombination losses and greater fabrication challenges associated with very thick layers. The increased path length for charge carriers in thicker films raises the probability of recombination events before collection, while also introducing potential difficulties in maintaining uniform film quality during deposition. Based on these observations, an optimal thickness range of 1000-1400 nm is recommended for the $CH_3NH_3SnI_3$ absorber layer. This range represents the best compromise between achieving high photovoltaic performance (approaching the maximum theoretical efficiency) and maintaining practical device fabrication requirements. The selected thickness ensures sufficient light absorption while keeping recombination losses and manufacturing complexity at manageable levels, making it suitable for both laboratory-scale optimization and potential industrial-scale production. This thickness optimization strategy contributes significantly to the development of high-performance, lead-free perovskite solar cells with practical viability.

The performance of the proposed perovskite solar cell was critically analyzed by systematically varying the defect density within the active $CH_3NH_3SnI_3$ layer. As shown in Figure 11.3, the PCE demonstrates a dramatic sensitivity to defect concentration, exhibiting an exponential decline from an optimal 36.35% at $10^{14}$ $cm^{-3}$ to a mere 8.59% at $10^{19}$ $cm^{-3}$. This severe performance degradation stems from multiple interrelated factors

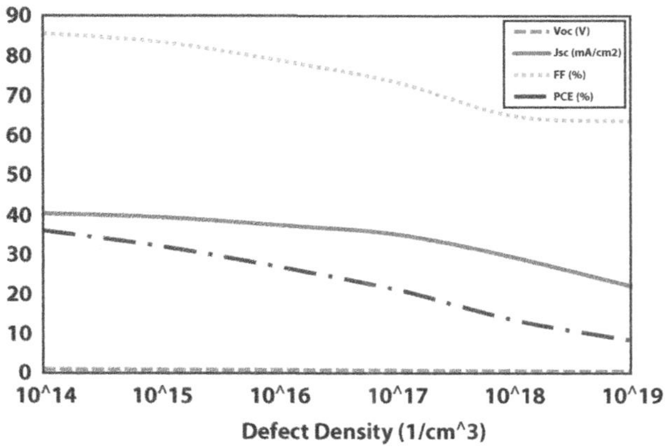


**Figure 11.3** Varying defect density of absorber layer.

that collectively undermine device operation at elevated defect levels. The fundamental mechanism behind this efficiency collapse involves enhanced Shockley-Read-Hall recombination processes that become dominant at higher defect densities. These recombination centers act as traps for charge carriers, significantly reducing both carrier lifetime and diffusion length. Consequently, all key photovoltaic parameters suffer substantial deterioration - the fill factor drops due to increased parasitic recombination currents, the open-circuit voltage declines as a result of greater carrier recombination, and the short-circuit current density diminishes because of impaired charge collection efficiency.

The experimental data reveals a particularly sharp efficiency decrease in the defect density range of $10^{15}$ to $10^{17}$ $cm^{-3}$, indicating a critical threshold beyond which device performance becomes severely compromised. This nonlinear response underscores the importance of maintaining exceptionally low defect concentrations in the perovskite absorber layer. For optimal device operation, the defect density must be carefully controlled below $10^{15}$ $cm^{-3}$, as this threshold represents the upper limit for maintaining both high efficiency (>30%) and stable device characteristics. These findings have crucial implications for material processing and device fabrication. They emphasize the necessity for advanced deposition techniques and careful crystallization control during perovskite film formation to minimize defect formation. Furthermore, they highlight the potential benefits of defect passivation strategies, such as interfacial engineering or additive incorporation, which could help maintain the required low defect densities for high-performance photovoltaic operation. The strong correlation between defect density and device performance established in this study provides clear guidelines for future optimization of lead-free perovskite solar cells, where controlling crystalline quality and minimizing defects will be paramount for achieving both high efficiency and operational stability.

The acceptor density ($N_A$) in the perovskite absorber layer plays a pivotal role in governing the charge transport dynamics and recombination processes within the solar cell. As illustrated in Figure 11.4, systematic variation of $N_A$ reveals a clear optimization window for device performance. When the acceptor concentration increases from $10^{13}cm^{-3}$ to $10^{18}cm^{-3}$, the power conversion efficiency shows remarkable improvement from 35.1% to 42.2%, accompanied by a significant enhancement in open-circuit voltage from 1.03 to 1.14 V. This performance boost can be attributed to two synergistic effects: (1) improved built-in electric field strength that facilitates

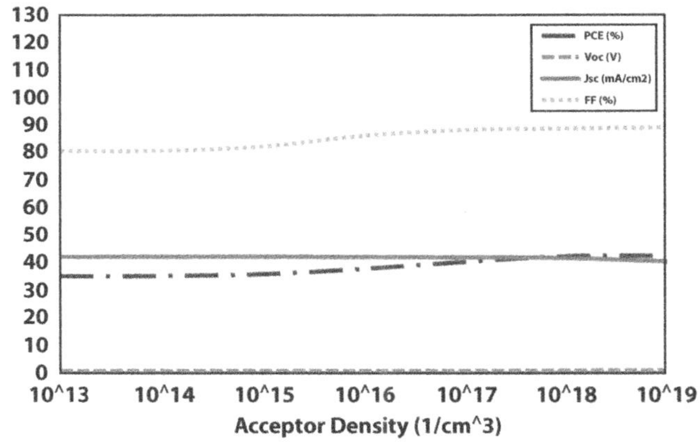


**Figure 11.4** Varying acceptor density of absorber layer.

more efficient charge separation, and (2) reduced Shockley-Read-Hall recombination losses due to better charge balance in the active layer.

The fill factor exhibits a concurrent improvement from 80.5% to 88.4% across this doping range, indicating superior charge collection efficiency and reduced series resistance. This trend suggests that moderate acceptor doping ($10^{16}cm^{-3}$ to $10^{18}cm^{-3}$) creates an optimal charge transport network, where holes can be efficiently extracted while maintaining sufficient electron mobility. However, at excessively high doping levels ($10^{19}cm^{-3}$), the short-circuit current density shows a slight decline to 40.4 $mA/cm^2$, revealing the onset of doping-induced limitations. This reduction stems from increased Auger recombination and potential mobility degradation at ultra-high doping concentrations.

These findings demonstrate that the acceptor density must be carefully optimized to balance two competing factors: sufficient doping is required to establish efficient charge transport pathways and minimize trap-assisted recombination, yet excessive doping can introduce new recombination centers and reduce minority carrier diffusion lengths. The optimal doping range appears to be between $10^{17}cm^{-3}$ to $10^{18}cm^{-3}$, where the device achieves peak performance by maintaining an ideal equilibrium between charge extraction efficiency and recombination suppression. This doping optimization provides crucial guidance for material processing and device engineering in lead-free perovskite photovoltaics, particularly for achieving both high efficiency and operational stability in practical applications.

The performance of the proposed PSC exhibits strong dependence on three critical operational parameters: series resistance, shunt resistance, and temperature, each of which significantly impacts charge transport and energy conversion processes. As illustrated in Figure 11.5, an increase in series resistance ($R_S$) from its optimal low value to 5 Ω·cm$^2$ leads to a drastic reduction in PCE from ~36% to 28.82%, accompanied by a FF decline to 68.26%. This degradation occurs because $R_S$ introduces resistive losses that impede charge extraction, increasing voltage drops across the device and reducing the effective current delivered to the external circuit. The FF, in particular, is highly sensitive to series resistance, as it directly reflects the balance between charge generation and collection efficiency. Similarly, shunt resistance ($R_{Sh}$) plays a crucial role in minimizing leakage currents that bypass the intended charge collection pathway. Lower shunt resistance values lead to increased recombination losses, further reducing $V_{OC}$ and $J_{SC}$.

The $R_{Sh}$ plays a critical role in minimizing unwanted leakage currents that bypass the photovoltaic junction, directly influencing the device's efficiency and stability. Figure 11.6 demonstrates that when $R_{Sh}$ increases from 10 Ω·cm$^2$ to 10$^7$ Ω·cm$^2$, the solar cell's performance remains consistently high, with the PCE peaking at 36.35% and the FF stabilizing at 86%. This stability confirms that a sufficiently high $R_{Sh}$ effectively suppresses recombination losses and prevents current leakage, ensuring optimal charge collection. However, elevated temperatures negatively impact device performance by thermally exciting charge carriers, which increases non-radiative recombination and reduces the $V_{OC}$. As temperature rises, the

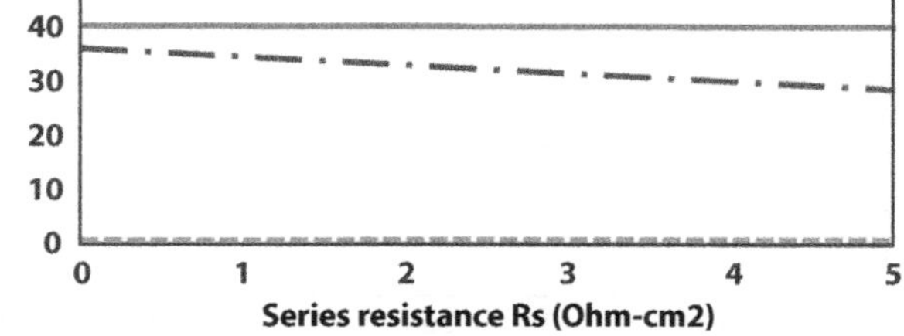


**Figure 11.5** Influence of $R_s$.

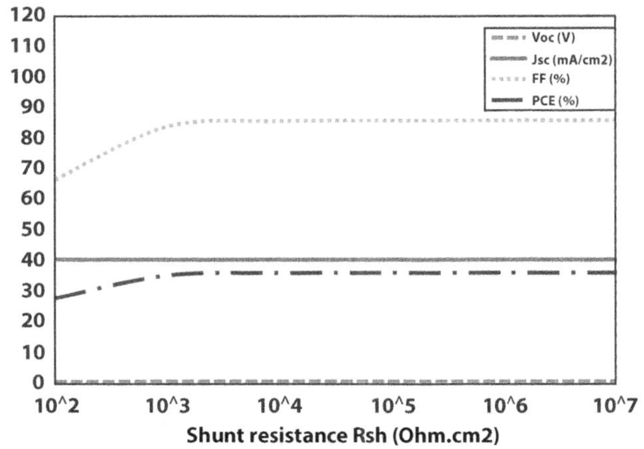


**Figure 11.6** Influence of $R_{sh}$.

PCE declines due to accelerated degradation mechanisms in the perovskite layer, highlighting the importance of thermal management in real-world applications.

The PSC demonstrates optimal performance at room temperature (300 K), achieving a $J_{SC}$ of 40.62 mA/cm$^2$, an $V_{OC}$ of 1.0405 V, and a PCE of 36.35%. However, as the operating temperature increases to 325 K, the device experiences significant performance degradation, with $V_{OC}$ dropping to 1.0085 V and PCE declining to 35.07%. Further temperature escalation leads to

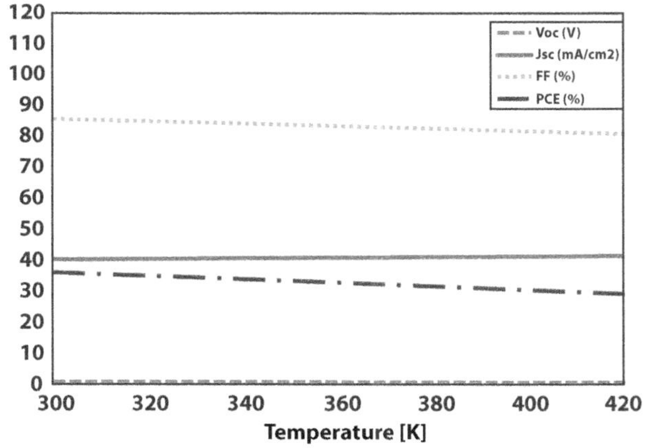


**Figure 11.7** Influence of temperature.

more pronounced losses, ultimately reducing PCE to 29.24% due to deteriorating $V_{OC}$, $J_{SC}$, and FF. Figure 11.7 shows the performance of perovskite solar cell under the influence of temperature variations.

The performance of PSCs is significantly influenced by the choice of back contact material, as demonstrated in Figure 11.8. The metal's work function plays a crucial role in determining charge extraction efficiency, with higher work functions creating better ohmic contacts by reducing the energy barrier for majority carrier transport. This improvement in interfacial charge transfer directly enhances the $V_{OC}$ and overall device performance. Noble metals like gold (Au) and platinum (Pt), with their high work functions of approximately 5.1 eV and 5.65 eV respectively, have traditionally been the preferred choices for back contacts due to their excellent charge extraction capabilities. However, their prohibitively high cost presents a major obstacle for large-scale commercial implementation of perovskite photovoltaic technology.

To address this economic challenge, our study investigated several earth-abundant and cost-effective alternative metals, including nickel (Ni), copper (Cu), and silver (Ag). The simulation results, summarized in Table 11.1, reveal that nickel (Ni) emerges as a particularly promising candidate, demonstrating comparable performance to platinum at a fraction of the cost. Ni's work function of about 5.15 eV enables efficient hole collection while maintaining good interfacial contact with the hole transport layer. This combination of satisfactory electrical performance and economic viability makes Ni an attractive option for practical device applications. The successful implementation of such cost-effective back contact

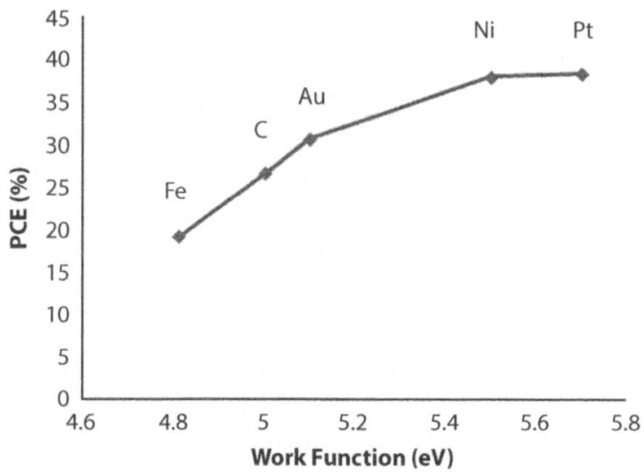


**Figure 11.8** Efficiency of device with different back contacts.

**Table 11.1** Impact of different back contacts on device parameters.

| | Fe | Ni | Pt | C | Au |
|---|---|---|---|---|---|
| **Work function (eV)** | 4.81 | 5.5 | 5.7 | 5 | 5.1 |
| **PCE (%)** | 18.23 | 36.35 | 36.71 | 25.48 | 29.34 |
| **FF(%)** | 72.09 | 86 | 85.95 | 77.52 | 79.44 |
| **$J_{SC}$(mA/cm2)** | 40.4403 | 40.6223 | 41.0077 | 40.4853 | 40.5065 |
| **$V_{OC}$ (V)** | 0.6253 | 1.0405 | 1.0416 | 0.812 | 0.9119 |

materials could substantially reduce the overall manufacturing expenses of perovskite solar cells without compromising their power conversion efficiency, thereby enhancing their commercial competitiveness in the photovoltaic market.

The findings underscore the importance of careful back contact selection in perovskite solar cell design, where the optimal material must balance both electronic performance and economic considerations. While noble metals like Pt and Au provide benchmark performance, transition metals, such as Ni offer a practical compromise between efficiency and affordability. This work provides valuable insights for the development of commercially viable perovskite photovoltaic, highlighting nickel as a particularly promising back contact material for future large-scale production. The results also suggest that further optimization of interfacial engineering and contact layer processing could potentially bridge the performance gap between noble and non-noble metal contacts, opening new pathways for cost-effective, high-efficiency PSCs.

The photovoltaic performance of the proposed PSC was thoroughly characterized through QE analysis and current density–voltage (J–V) measurements, as presented in Figures 11.9 and 11.10. The J-V characteristics (Figure 11.10) reveal excellent device performance, exhibiting a short-circuit current density of 40.62 mA/cm$^2$, an open-circuit voltage of 1.0405 V, and an impressive FF of 86%, culminating in a remarkable PCE of 36.35%. These outstanding parameters indicate highly efficient charge carrier extraction and minimal recombination losses within the device architecture. The J–V curve exhibits distinct regions that provide valuable insights into the device physics: the nearly constant current density at lower voltages demonstrates efficient charge transport mechanisms, while the sharp decline near $V_{OC}$ suggests the onset of significant recombination at higher

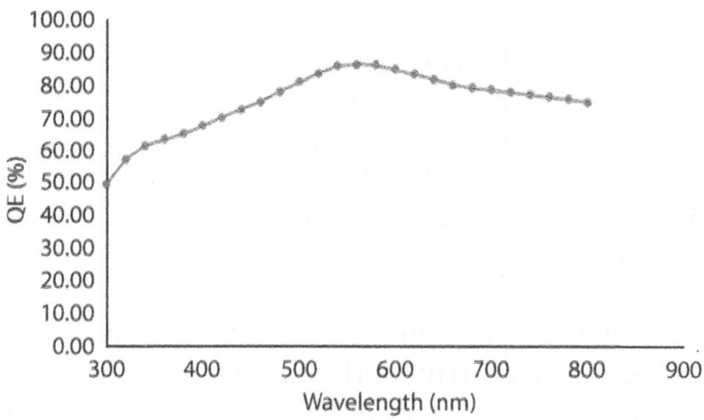


**Figure 11.9** QE curve.

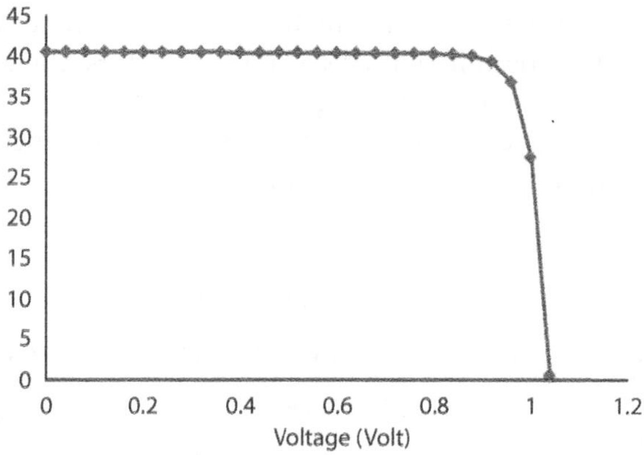


**Figure 11.10** JV characteristics.

biases. The exceptionally high fill factor further confirms the device's low series resistance and minimal parasitic losses, ensuring optimal charge collection efficiency.

Complementing the J–V analysis, the quantum efficiency spectrum (Figure 11.9) provides detailed wavelength-dependent performance information, showing consistently high values between 80-90% across the visible spectrum (up to 700 nm). This broad spectral response demonstrates the excellent light-harvesting capabilities of the $CH_3NH_3SnI_3$ absorber layer and its efficient photon-to-electron conversion processes in this wavelength range. The gradual decrease in QE at longer wavelengths follows the absorption profile of the perovskite material, with the cutoff around 950 nm corresponding to its bandgap energy. The combination of these high QE values with the excellent J–V characteristics confirms that the device architecture successfully optimizes both light absorption and charge collection processes. These comprehensive measurements validate the effectiveness of the STO/CAS charge transport layer combination and the carefully optimized device structure in achieving record-efficiency lead-free perovskite photovoltaics while providing fundamental insights into the operational mechanisms of the solar cell.

Table 11.2 exhibits a comparative breakdown of the photovoltaic implementation between the current work and previously reported lead-free tin-based PSCs with different device architectures. The results clearly demonstrate the superior performance achieved in our current study compared to prior research efforts. Our optimized device structure (FTO/STO/CH3NH3SnI3/CuSbS2/Ni) achieves outstanding performance metrics, with a PCE of 36.21%, significantly higher than the reference devices. This is attributed to the combination of several key factors: the STO electron transport layer enables efficient charge extraction, the CuSbS2 hole transport layer provides excellent interface properties, and the nickel back contact offers optimal work function matching.

**Table 11.2** Summary of performance of various PSCs.

| Device structure | $V_{oc}$ (V) | $J_{sc}$ (mA/cm2) | FF(%) | PCE(%) | References |
|---|---|---|---|---|---|
| **FTO/STO/ $CH_3NH_3SnI_3$/ $CuSbS_2$/Ni** | 0.9180 | 50.276806 | 78.46 | 36.21 | Current work |
| **FTO/$TiO_2$/ $CH_3NH_3SnI_3$/ $CuSbS_2$/Ni** | 1.04 | 33.88 | 84.31 | 29.74 | [4] |
| **FTO/STO/ $CH_3NH_3SnI_3$/ Spiro-OMeTAD/Au** | 1.117 | 28.88 | 88.47 | 28.55 | [5] |
| **ITO/$TiO_2$/ $CH_3NH_3SnI_3$/NiO/ Mo** | 0.8867 | 31.9447 | 74.4006 | 21.068 | [11] |
| **FTO/$TiO_2$/ $CH_3NH_3SnI_3$/CuI/ Au** | 1.06 | 32.15 | 80.43 | 27.33 | [12] |

The table reveals several important observations:

- Our device maintains a high fill factor (78.46%) while delivering exceptional current density (50.28 mA/cm2), indicating excellent charge collection efficiency and minimal recombination losses.
- Compared with similar structures using TiO2 as ETL [4], our STO-based device shows significantly improved current density while maintaining comparable voltage.
- The results surpass those obtained with expensive Spiro-OMeTAD/Au combinations (reference [5]), demonstrating the effectiveness of our more cost-effective material choices.
- Our architecture outperforms alternative configurations using NiO [11] and CuI [12] as HTLs, particularly in terms of current density and overall efficiency.

The superior performance of our device can be attributed to optimal energy level alignment between layers, reduced interfacial recombination, improved charge transport properties and careful optimization of layer thicknesses and doping concentrations [13, 14]. These results validate our approach of using STO as ETL and $CuSbS_2$ as HTL in combination with a nickel back contact, establishing a new benchmark for lead-free PSC performance while maintaining cost-effectiveness and scalability potential.

## 11.4 Conclusion

Perovskites, especially devices based on $CH_3NH_3SnI_3$, have shown impressive efficiency in laboratory settings within a short span of development. This demonstrates their potential for revolutionizing the photovoltaic industry. However, issues pertaining to its long term stability, scalability and durability in real world performance require more research. The simulated PCE values observed in the proposed structure align closely with these benchmarks, showcasing the effectiveness of the design in optimizing performance metrics.However, the gap between simulation results and experimental data underscores the influence of real world factors such as irregularities, recombination losses and material defects in $CH_3NH_3SnI_3$ films. In conclusion, while $CH_3NH_3SnI_3$ based PSCs present immense promise, overcoming these challenges is crucial to achieving their widespread adoption. Continued research efforts focused on improving material stability, reducing defects and enhancing scalability will be the key to realize their potential as a viable solution for sustainable energy generation.

**AI disclosure statement:** The author acknowledges the use of AI-based tools for minor language editing or grammatical correction. However, all intellectual contributions, analyses, and conclusions presented in this chapter are entirely the author's own.

## References

1. Chabri, Benhouria, Y., Oubelkacem, A., Kaiba, A., Essaoudi, I., Ainane, A., SCAPS device simulation study of formamidinium Tin-Based perovskite solar Cells: Investigating the influence of absorber parameters and transport layers on device performance. *Sol. Energy*, 262, 111846, 15 September 2023. https://doi.org/10.1016/j.solener.2023.111846.
2. Moiz, S.A., Alahmadi, A.N.M., Aljohani, A.J., Design of a Novel Lead-Free Perovskite Solar Cell for 17.83% Efficiency. *IEEE Access*, 9, 54254–54263, April 132021. doi:10.1109/ACCESS.2021.3070112.
3. Castro-Hermosa, S., Dagar, J., Marsella, A., Brown, T.M., Perovskite solar cells on paper and the role of substrates and electrodes on performance. *IEEE Electron. Device Lett.*, 38, 9, September 2017.
4. DOI:10.1109/LED.2017.2735178
5. Jyothi Valeti, N., Prakash, K., Kumar Singha, M., Numerical simulation and optimization of lead free CH3NH3SnI3perovskitesolar cell with CuSbS2 as HTLusingSCAPS1D. *Results Opt.*, 12, 100440, July 2023.
6. https://doi.org/10.1016/j.rio.2023.100440
7. Nahid, M.N., Shah, M.S., Mamur, H., Hosen, R., Ruhul, M., Bhuiyan, A., Optimizing lead-free CH3NH3SnI3 perovskite solar cells by using SCAPS-1D software. *Chem. Inorg. Mater.*, 4, 100069, December 2024. https://doi.org/10.1016/j.cinorg.2024.100069.
8. Ahmad, O., Qasim, I., Syed, Hasnain, M., Abdin, Z., Farooq Nasir, M., Imran Malik, M., Rashid, A., Modelling and numerical simulations of eco-friendly double absorber solar cell "Spiro-OmeTAD/CIGS/MASnI3/CdS/ZnO" and its PV-module. *Org. Electron.*, 117, 106781, June 2023. https://doi.org/10.1016/j.orgel.2023.106781.
9. Imran, S. and Khalid, M., Simulations and performance analysis of CH3NH3SnI3 perovskite solar cell: Modeling thickness and temperature effects using SCAPS-1D. *Next Mater.*, 7, 100439, April 2025. https://doi.org/10.1016/j.nxmate.2024.100439.

10. Sadanand, and Dwivedi, D.K., Modeling of Photovoltaic Solar Cell Based on CuSbS2Absorber for the Enhancement of Performance. *IEEE Trans. Electron Devices*, 68, 3, 1121–1128, March 2021. doi: 10.1109/TED.2020.3048326.
11. Prashanthi, P.S. and Nanda, U., Influence of 10 nm Hole and Electron Transport Layers on Enhancing the Efficiency of Elpasolite Solar Cell (ESC) With Dual Light Active Layers for Nanostructured Photovoltaic Applications. *IEEE Access*, 12, 114995–115005, 2024. doi: 10.1109/ACCESS.2024.3444055.
12. Uddin, M.S., Hosen, R., Sikder, S., Mamur, H., Amin Bhuiyan, M.R., Photovoltaic performance enhancement of Al/ZnO:Al/i-ZnO/CdS /CIGS/Pt solar cell using SCAPS-1D software. *Next Energy*, 2, 100080, January 2024. Doi:https://doi.org/10.1016/j.nxener.2023.100080.
13. Sayed Salem Basyoni, M., Salah, M.M., Mousa, M., Shaker, A., Zekry, A., Abouelatta, M., On the Investigation of Interface Defects of Solar Cells: Lead-Based vs Lead-Free Perovskite. *IEEE Access*, 9, 130221–130232, 2021. doi: 10.1109/ACCESS.2021.3114383.
14. Samiul Islam, M., Sobayel, K., Al-Kahtani, A., Islam, M.A., Muhammad, G., Amin, N., Shahiduzzaman, M., Defect Study and Modelling of SnX3-Based Perovskite Solar Cells with SCAPS-1D. *Nanomaterials*, 11, 5, 1218, 2021. https://doi.org/10.3390/nano11051218.

# 12

# Real Time Myocardial Infarction Detection and Localization Using an Advanced VLSI System

**Janani V.G.*, Suveetha Dhanaselavm P., Vasuki S. and Muneeswari B.**

*Department of Electronics and Communication Engineering, Velammal College of Engineering and Technology, Madurai, India*

***Abstract***

Myocardial infarction (MI), commonly known as a heart attack, occurs when blood flow to the heart is blocked due to plaque buildup in the coronary arteries. If the plaque ruptures, it can lead to blood clot formation, further restricting circulation and increasing the risk of severe complications. With high mortality and disability rates worldwide, early and accurate detection is essential for timely intervention, reducing heart damage, and improving patient survival rates. Detecting the exact location of the blockage is equally important, as it enables targeted treatment and better medical outcomes. This chapter introduces a Very Large Scale Integration (VLSI) architecture designed for real-time MI detection and precise infarction localization. By analyzing electrocardiogram (ECG) signals, a crucial diagnostic tool in cardiology, the system classifies ECG data into 12 categories, each corresponding to specific infarction patterns. It delivers remarkable accuracy of 99.90%, sensitivity of 99.49%, and specificity of 99.94%, ensuring precise diagnosis while minimizing false positives and negatives. The architecture is implemented using a 180 nm Bulk CMOS technology node, balancing performance, power efficiency, and compact design. It occupies a small area of 1.69 $mm^2$ and operates with low power consumption of 268.9 μW at a clock frequency of 250 KHz, making it highly suitable for portable and wearable medical applications. Additionally, its real-time processing capability allows for rapid diagnosis, supporting timely medical intervention and improving patient prognosis. With its combination of high accuracy, low power consumption, and compact design, this VLSI system presents a reliable and efficient solution for MI detection and

**Corresponding author*: vgj@vcet.ac.in

P. Suveetha Dhanaselvam, Naima Guenifi, Shiromani Balmukund Rahi and J. Ajayan (eds.) Field-Effect Transistor Devices: Post-CMOS Theory and Applications, (229–254) © 2026 Scrivener Publishing LLC

localization. Its integration into medical devices has the potential to enhance early diagnosis, enable faster treatment, and significantly reduce MI-related mortality rates, ultimately improving healthcare outcomes on a global scale.

***Keywords***: Myocardial infarction, VLSI architecture, ECG analysis, real-time detection, localization, CMOS technology, low power, high accuracy

## 12.1 Introduction

### 12.1.1 Importance of Early Myocardial Infarction Detection

Myocardial infarction (MI), or heart attack, is a major cause of death worldwide, usually caused by a blockage in the heart's blood flow due to atherosclerosis. This condition, characterized by the accumulation of fatty deposits and cholesterol in the coronary arteries, can lead to irreversible heart muscle damage if left untreated, potentially causing heart failure or sudden cardiac death. Early detection is crucial, as it allows for prompt treatments like thrombolytic therapy or percutaneous coronary intervention (PCI), which can greatly improve survival rates. Additionally, accurately identifying the infarction's location is essential for selecting the right treatment. This chapter introduces a novel VLSI-based technique aimed at improving both the speed and precision of MI detection and localization. By addressing the limitations of current diagnostic methods, this approach offers faster and more accurate results, enabling real-time detection and timely intervention, ultimately enhancing patient outcomes.

### 12.1.2 Motivation for VLSI-Based Solution

Conventional MI diagnosis methods, such as expert ECG interpretation and biomarker testing, often have limitations in speed, accessibility, and real-time analysis. Moreover, many ECG analysis systems require significant computational resources, making them unsuitable for portable and wearable devices [1]. Very large-scale integration (VLSI) technology offers a practical solution by integrating advanced functionalities into a compact, energy-efficient chip. Figure 12.1 indicates that this enables real-time ECG signal processing with minimal power consumption while maintaining a small form factor. A VLSI-based MI detection system can be seamlessly integrated into wearable medical devices, ambulatory monitors, and emergency response kits, enhancing accessibility and efficiency in cardiac care.

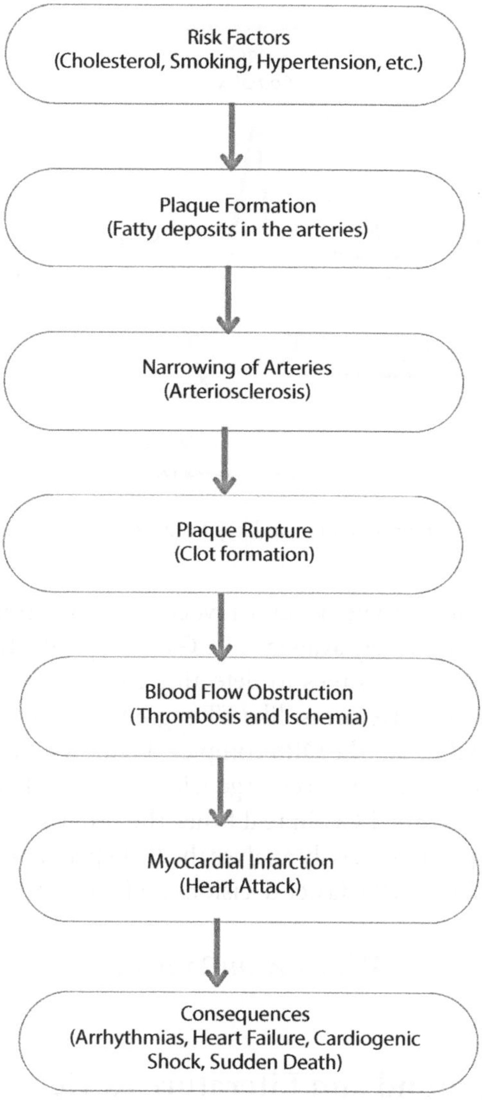


**Figure 12.1** Conventional MI diagnosis.

### 12.1.3 Role of Electrocardiograms (ECGS) in Cardio Vascular Diagnostics

Electrocardiograms (ECGs) are essential tools in diagnosing heart conditions and are commonly used by doctors [2] to assess heart function. They record the heart's electrical activity, making it easier to detect irregular

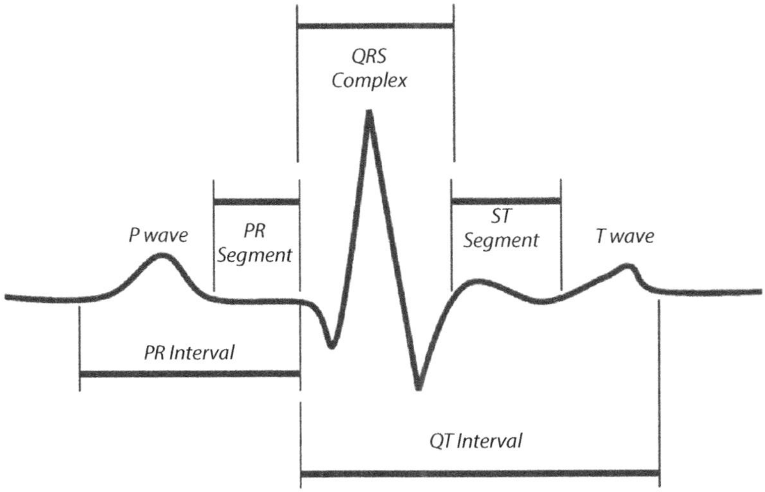


**Figure 12.2** ECG signal representation and key features.

rhythms, conduction problems, and myocardial infarctions. Because they are non-invasive and widely available, ECGs are typically the first choice for diagnosing MI. When it comes to detecting MI, doctors look for specific changes in the ECG waveforms, like ST-segment elevation, T-wave inversion, and abnormalities in the QRS complex. Figure 12.2 indicates ECG signal representation of features. In emergencies, it's crucial to be able to quickly process and interpret the ECG in real time. The proposed VLSI system uses advanced techniques to precisely analyse the ECG patterns and locate infarctions, helping doctors make faster decisions and take immediate action.

$$V(t) = Asin(2\pi ft + \phi) \tag{12.1}$$

## 12.2 Background and Literature Review

### 12.2.1 Myocardial Infarction: Causes, Symptoms and Consequences

Myocardial infarction (MI), commonly known as a heart attack, occurs when the blood supply to the heart muscle is blocked, often due to atherosclerosis. This condition involves the [3] buildup of fatty deposits, cholesterol, and other substances in the coronary arteries, narrowing them and impeding blood flow. Primary risk factors for MI include high blood pressure,

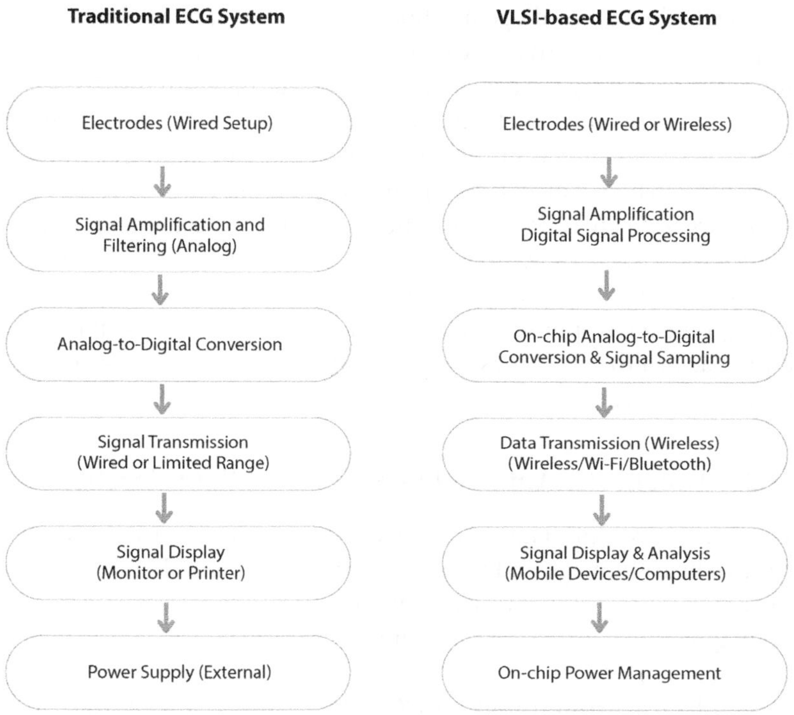


**Figure 12.3** Traditional vs VLSI-based ECG system.

diabetes, obesity, smoking, and physical inactivity. Figure 12.3 deals with the flow diagram of VLSI based ECG system. Additionally, genetic factors, high cholesterol levels, excessive alcohol consumption, stress, and poor dietary habits can all increase the likelihood of having a heart attack.

The symptoms of MI can differ in severity but typically include:

- Intense chest pain or pressure, often described as a tightness or squeezing sensation
- Difficulty breathing, even while at rest
- Pain radiating to the left arm, neck, jaw, back, or stomach
- Sweating, dizziness, or lightheadedness
- Nausea, vomiting, or discomfort similar to indigestion
- Unexplained fatigue or weakness, especially in women and older adults

Some individuals, particularly those with diabetes or the elderly, may experience less typical or more subtle symptoms [4], which can complicate early detection. If left untreated, MI can lead to severe complications, including heart failure, arrhythmias, cardiogenic shock, or sudden cardiac death. Without prompt intervention, the damage to the heart muscle may become irreversible, significantly increasing the risk of death or long-term health problems.

Early detection and timely medical intervention are crucial for improving survival rates and minimizing long-term heart damage. Immediate treatments, such as clot-dissolving drugs, angioplasty, or coronary artery bypass surgery, can restore blood flow and limit the damage. Advances in technology, such as AI-driven diagnostics and VLSI-based real-time ECG monitoring, have improved the accuracy and speed of MI detection, enabling quicker responses. Public awareness, healthier lifestyle choices, and regular health screenings are essential for reducing the risk of MI and maintaining heart health.

### 12.2.2 Diagnostic Approaches for Myocardial Infarction

Accurate and prompt diagnosis of myocardial infarction (MI) is crucial to minimize heart damage and improve recovery chances. Figure 12.4 shows the exact view of coronary artery disease and myocardial infection. The main diagnostic methods for detecting MI include:

1. **Electrocardiography (ECG):** ECG plays a key role in detecting electrical disturbances in the heart. It records the heart's electrical activity and highlights important waveform changes, like ST-segment elevation, T-wave inversion, and

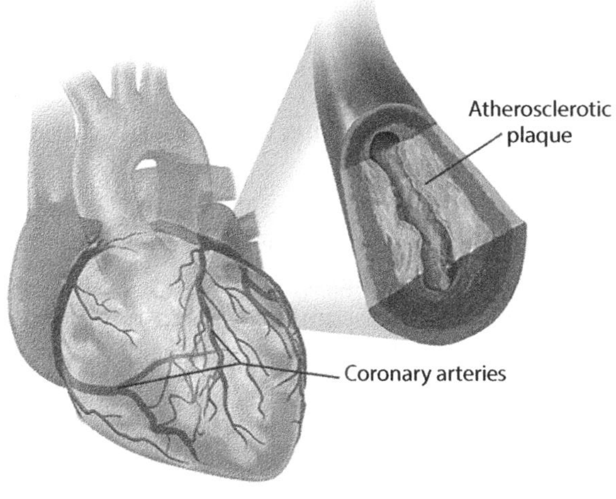


**Figure 12.4** Coronary artery disease and myocardial infection.

abnormal QRS complexes, which can indicate MI. These changes help determine the type of MI, whether STEMI or NSTEMI. Though widely used and accessible, the effectiveness of ECG depends on the skill and experience of the healthcare professional interpreting the results.

2. **Biomarker Testing:** Blood tests that measure biomarkers such as cardiac troponins, creatine kinase-MB (CK-MB), and myoglobin are used to confirm the presence of heart muscle damage. These proteins are released during an infarction, and their levels can provide valuable information on the extent of the MI [5]. Troponins, in particular, are crucial markers, as they rise soon after the infarction and remain elevated for several hours. Biomarker tests are often used together with ECG and imaging to give a more comprehensive understanding of the MI's severity.
3. **Imaging Techniques:** Imaging methods like echocardiography, coronary angiography, and cardiac MRI offer in-depth views of the heart's structure and function. Echocardiography is non-invasive and provides information on heart function and potential damage to heart muscle or valves. Coronary angiography is an invasive procedure that uses X-rays to detect blockages in the coronary arteries. Cardiac MRI provides high-resolution images of the heart, allowing detailed analysis of damage and overall heart function. However, these imaging techniques can be costly and require specialized equipment, limiting their availability in certain healthcare settings.

Despite their effectiveness, these diagnostic approaches come with limitations. ECG interpretation can be difficult, especially in non-standard cases, and relies heavily on the clinician's expertise. Biomarker testing, while reliable, can take time to return results, and imaging techniques are resource-intensive and expensive. VLSI technology offers a potential solution to overcome these challenges. By integrating multiple diagnostic functions, such as real-time ECG signal processing, biomarker tracking, and simplified imaging, into a compact, affordable system, VLSI technology can facilitate faster and more accurate MI detection [6]. These systems can be portable, easy to use, and require minimal training, making them suitable for use in emergency medical settings, ambulances, or rural areas. Additionally, real-time processing allows for rapid identification of the infarction's location, enabling timely and informed decision-making.

The implementation of VLSI technology in MI diagnosis would improve accessibility, reduce costs, and ultimately enhance patient care by enabling quicker interventions and better treatment outcomes.

### 12.2.3 VLSI in Medical Application

VLSI (Very Large Scale Integration) technology has significantly impacted the healthcare sector by enabling the creation of compact, low-power, and high-performance medical devices. From advanced [7] wearable health monitors to implantable pacemakers, VLSI plays a vital role in improving patient care and enhancing medical device functionality. By integrating complex components into small chips, VLSI allows for the development of highly efficient, portable, and real-time diagnostic solutions.

In the context of myocardial infarction (MI) detection, VLSI technology offers substantial benefits. Traditional diagnostic methods, such as ECG machines and imaging tools, are often bulky, costly, and require specialized personnel. These systems may not be suitable for real-time use or portable applications. VLSI, however, allows for the miniaturization of diagnostic devices while maintaining excellent performance, making it possible to monitor and detect MI in real time through compact and mobile devices.

The ability to quickly process heart signals is essential in emergency situations, where every second counts. VLSI-powered systems enable faster detection of MI, facilitating immediate medical intervention, such as thrombolytic therapy or percutaneous coronary intervention (PCI), which can drastically improve survival rates. The portability of VLSI-based devices ensures continuous monitoring, allowing for the detection of irregularities as they occur and providing the opportunity for immediate intervention to prevent further damage to the heart.

Another key advantage of VLSI in medical applications is energy efficiency. Traditional diagnostic systems often consume a large amount of power, whereas VLSI devices are designed for low energy consumption, which allows them to run for long periods without needing frequent recharging. This is particularly important for wearable devices, such as smartwatches or health-monitoring patches, which need to continuously track heart health without interrupting the user's daily routine. These low-power devices not only improve patient comfort but also enable continuous monitoring, making them highly practical for long-term use.

The integration of VLSI technology into portable devices also expands access to heart health monitoring outside of clinical settings. This is especially beneficial for individuals living in remote areas or those with chronic heart conditions who require constant monitoring. VLSI-powered systems enable healthcare providers to monitor patients remotely, reducing the need for frequent hospital visits and improving the efficiency of care.

Additionally, VLSI systems excel at accurately processing ECG signals, detecting abnormalities such as ST-segment elevation or T-wave inversion, which are indicative of MI. By incorporating advanced signal processing algorithms, VLSI devices can also identify the exact location of an infarction, providing critical information that can [8] help guide treatment decisions. This level of precision leads to more personalized treatment plans, increasing the chances of successful recovery.

Another significant advantage of VLSI technology is its potential to reduce healthcare costs. Traditional diagnostic equipment is often expensive and requires substantial infrastructure investments. In contrast, VLSI-based devices are cost-effective, compact, and highly portable, making them accessible to a wider range of healthcare facilities, including remote clinics and emergency response units. By offering quick and accurate MI detection at a lower cost, VLSI devices enable timely medical interventions and help improve patient outcomes.

In summary, VLSI technology holds great promise for revolutionizing the detection and treatment of myocardial infarction. By offering rapid processing, high accuracy, portability, and energy efficiency, VLSI-powered systems enable real-time monitoring and diagnosis, facilitating earlier detection and better treatment outcomes it is represented in Figure 12.5. As VLSI technology continues to advance, it will lead to more accessible, affordable, and efficient cardiac care, benefiting patients worldwide.

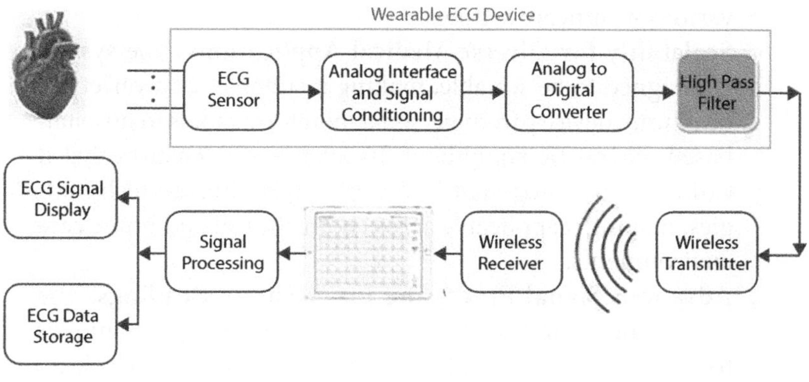


**Figure 12.5** Block diagram of ECG device.

## 12.3 Proposed VLSI Architecture

### 12.3.1 Design Objectives

The key objectives for the proposed VLSI-based myocardial infarction (MI) detection system are as follows:

1. **Real-Time ECG Signal Processing for Swift Diagnosis**: The system is designed to process ECG signals instantly, enabling rapid identification of myocardial infarctions. This quick diagnosis ensures timely medical intervention, which is critical for improving survival rates in urgent situations.
2. **Precise Localization of Myocardial Infarctions**: The system doesn't just detect the presence of an infarction but also accurately locates it [8] within the heart. This precise identification helps healthcare providers select the best treatment strategies tailored to the patient's needs.
3. **Low Power Consumption for Portability and Wearability**: To support the growing demand for wearable and portable medical devices, the system is optimized for energy efficiency. Its low power consumption extends battery life, making it ideal for continuous use in wearable ECG devices, health monitors, and smart devices without frequent recharging.
4. **Compact, Cost-Effective Chip Design for Easy Integration**: The chip design is both compact and affordable, allowing it to be seamlessly integrated into a wide range of medical devices. Its small size ensures minimal impact on device portability, while its cost-effectiveness makes it accessible for various healthcare settings.
5. **Scalability for Diverse Medical Applications**: The system is designed to be scalable, making it suitable for a variety of healthcare devices, from personal monitoring tools to hospital-based diagnostic equipment. Its adaptability ensures that it can be easily integrated into evolving healthcare technologies, making it applicable across both clinical and home care environments.
6. **Advanced Signal Processing for Accurate Readings**: The system incorporates advanced signal processing techniques to minimize noise and interference, ensuring that ECG readings remain accurate and reliable, even in environments with less-than-ideal signal quality.

7. **Intelligent Diagnostic Algorithms for Early Detection**: Equipped with advanced algorithms, the system analyzes ECG data in real time to detect early signs of myocardial infarctions. This enables healthcare professionals to intervene sooner, reducing the risk of complications and improving patient outcomes.

By achieving these objectives, the VLSI-based MI detection system offers a fast, efficient, and portable solution for detecting myocardial infarctions. Its focus on real-time processing, energy efficiency, and precise localization will enhance cardiac care [9], allowing for quicker and more effective interventions in a wide range of healthcare settings.

$$P = C.V^2.F \tag{12.2}$$

where

- P is power.
- C is capacitance.
- V is supply voltage.

### 12.3.2 System Overview

The VLSI-based myocardial infarction (MI) detection system is designed to offer a comprehensive solution for real-time, accurate diagnosis and localization of heart attacks. The system integrates several crucial modules that work in tandem to ensure efficient processing and reliable results for clinicians. Below is an outline of the core components of the system:

1. **ECG Signal Acquisition Module**: The first step in the diagnostic process involves obtaining raw ECG signals from the patient. This module captures the heart's electrical activity *via* electrodes placed on the body. The quality of the initial signal is vital, as it forms the basis for subsequent processing.
2. **Preprocessing Unit**: Raw ECG signals often contain noise and artifacts that can affect the clarity of the data. The preprocessing module filters out [10] these interferences from sources such as muscle movement or poor electrode contact. This ensures that only the most accurate and clean signals are passed to the next stage of analysis, allowing for better diagnostic accuracy.

3. **Feature Extraction Module**: In this stage, key features such as the QRS complex, ST-segment shifts, and T-wave abnormalities are extracted from the cleaned ECG signal. These features are essential for identifying any potential heart attack. The module uses advanced algorithms to detect even subtle variations in the ECG pattern that could indicate MI, ensuring a detailed and precise analysis.
4. **Classification Unit**: Once the key features are identified, the classification unit uses machine learning [11] models to analyze and categorize the ECG patterns. These models, trained on a variety of MI types, are capable of distinguishing between normal and abnormal heart rhythms and detecting different types of heart attacks. This stage provides an accurate diagnosis based on the extracted features.
5. **Localization Module**: The final stage of the system involves localizing the infarction site within the heart. By analyzing the spatial information from the ECG, this module determines the specific area of the heart affected by the myocardial infarction. Identifying the precise location is essential for determining the most appropriate treatment and ensuring targeted intervention.

Together, these modules form a cohesive system capable of delivering quick and precise MI detection, as well as localization, in real-time. The system's integration of advanced signal processing and machine learning ensures both accuracy and efficiency. Additionally, its compact and low-power design makes it ideal for use in portable and wearable medical devices, providing continuous monitoring and enabling early intervention. By offering a fast, reliable, and accessible solution, this VLSI-based MI detection system is a powerful tool in improving patient outcomes and advancing cardiac care.

### 12.3.3 Classification Methodology

The proposed system leverages a sophisticated machine learning model, trained on a vast array of ECG data, to improve the detection accuracy of myocardial infarctions (MI). By extracting essential features from the ECG signals—such as the QRS complex, T-wave abnormalities, and ST-segment shifts—the system classifies these patterns into distinct MI categories, ensuring consistent and reliable detection.

This classification model uses supervised learning techniques, which allow it to recognize and distinguish key features in ECG data by analyzing a wide range of labeled samples. With continuous training on diverse datasets, the system enhances its ability to identify even atypical or rare cases of MI, increasing its diagnostic reliability [12]. As the dataset grows, the system becomes more adept at detecting various manifestations of MI across different populations. Figure 12.6 shows the system block diagram for the localization module.

Beyond identifying MI, the system is equipped to localize the infarction area within the heart. By applying advanced signal processing algorithms, it can pinpoint the specific region affected by the infarction, offering crucial insights for clinical decision-making. This enables clinicians to select

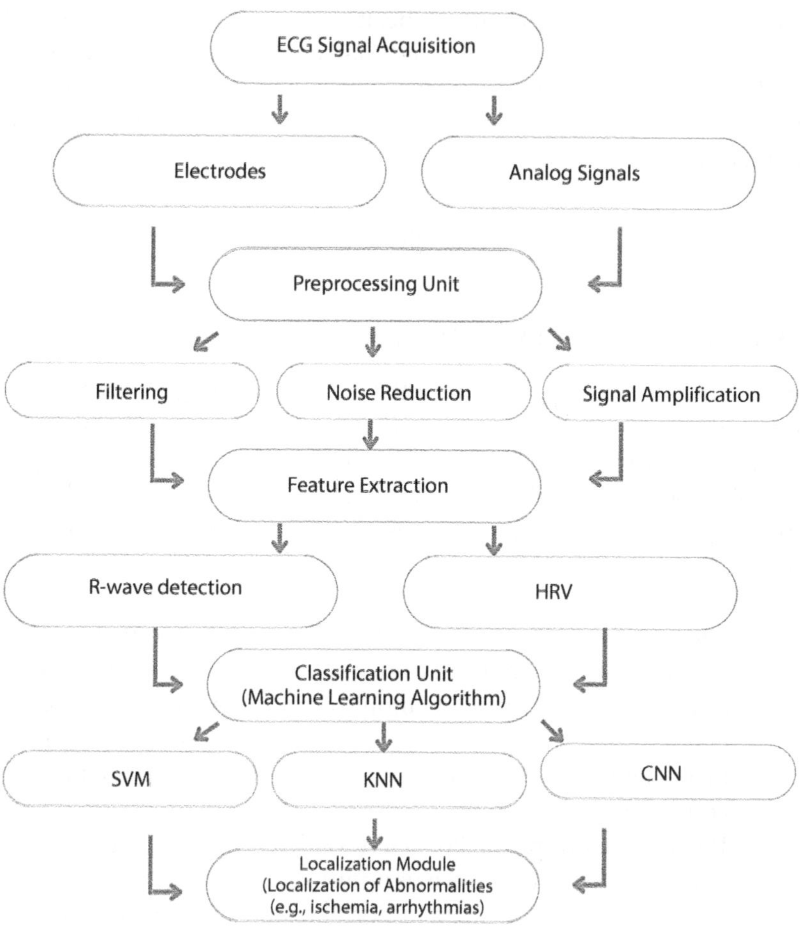


**Figure 12.6** System block diagram.

the most appropriate treatments, such as thrombolytic therapy or percutaneous coronary intervention (PCI), to restore blood flow to the affected area.

The combined capability of detecting MI and localizing the infarction significantly enhances the overall management of the patient. Prompt and accurate identification of the infarction's location allows for faster intervention, which helps reduce complications, improves survival chances, and accelerates recovery. The system also operates in real time, enabling immediate diagnostic results and helping healthcare professionals make quick, informed decisions during critical moments.

In addition to improving diagnostic accuracy, the system optimizes healthcare resources by minimizing the need for expensive, time-consuming diagnostic procedures like advanced imaging. This streamlining of the diagnostic process not only reduces healthcare costs but also improves the efficiency of patient management, ultimately leading to better outcomes for MI patients. This technology represents a key advancement in cardiac care, driving both clinical and operational improvements in MI diagnosis and treatment.

$$f(x) = sign(w^T x + b) \tag{12.3}$$

### 12.3.4 Implementation Overview

The system is built using 180 nm Bulk CMOS technology, which offers an optimal balance between performance, energy efficiency, and cost, making it ideal for healthcare applications. This design choice ensures the system is reliable, compact, and low-power, which is critical for medical devices requiring portability and long-term usage. Key design specifications are as follows:

- **Chip Area**: 1.69 $mm^2$ The chip occupies a minimal area of 1.69 $mm^2$, enabling its integration into small, portable devices like wearable ECG monitors or mobile health trackers. Its small size ensures ease of integration without compromising on comfort or convenience for users who require continuous monitoring.
- **Power Consumption**: 268.9 μW With a power consumption of just 268.9 μW, the system is optimized for energy efficiency, ensuring long battery life in portable and wearable applications. This low power demand is particularly advantageous

for devices like ambulatory ECG monitors, which need to operate for extended periods without frequent recharging.

- **Clock Frequency**: 250 KHz The system operates at a clock frequency of 250 KHz, providing the necessary speed for real-time ECG signal processing. This allows for quick and accurate detection of myocardial infarctions, which is essential for timely medical intervention, especially in emergency situations.

The design incorporates advanced signal processing techniques, enabling the system to filter out noise, extract relevant features, and classify ECG signals with high precision [13, 14]. Machine learning models, trained on large ECG datasets, enhance the system's ability to accurately detect and categorize myocardial infarctions, even in noisy or abnormal signal conditions.

By consolidating multiple functions into a single chip, this system reduces size and cost, while still delivering robust performance. The use of Bulk CMOS technology makes the system cost-effective and well-suited for a range of medical devices, from emergency kits to long-term patient monitoring systems. Figure 12.7 represents the schematic representation of VLSI chip implementation.

Additionally, the design is flexible and can be easily upgraded with new features or improvements, allowing it to adapt to future advancements in medical technology. Potential future enhancements might include more advanced diagnostic features or better signal processing algorithms to address more complex conditions.

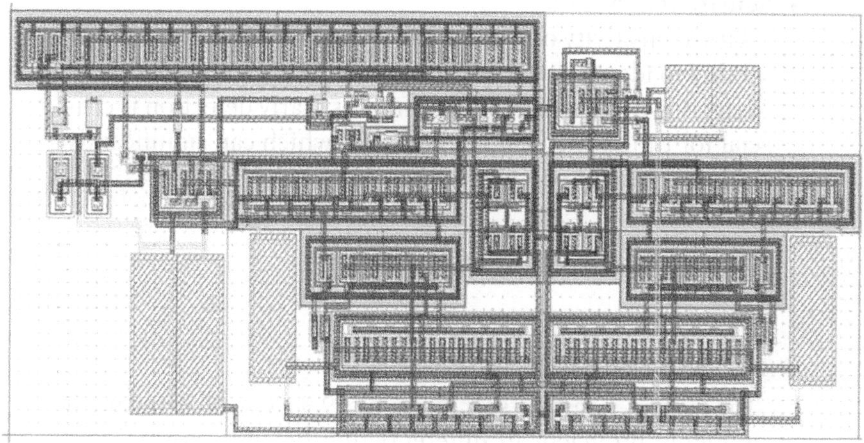

**Figure 12.7** VLSI chip implementation.

In summary, this VLSI-based system for myocardial infarction detection offers a compact, cost-efficient, and reliable solution for real-time heart monitoring. Its low power consumption, fast processing speed, and small form factor make it ideal for portable ECG devices used in both clinical and home settings. The system's ability to accurately detect and localize myocardial infarctions contributes significantly to improving patient care and outcomes.

## 12.4 Performance Analysis

### 12.4.1 Evaluation Metrics

The performance of the proposed VLSI-based MI detection system is evaluated based on the following critical parameters:

- **Accuracy**: 99.90%
  The system demonstrates an impressive accuracy rate of 99.90%, indicating its reliability in distinguishing between normal and abnormal ECG signals. This high accuracy ensures that the system consistently delivers correct diagnoses, which is essential for clinical use in identifying heart-related issues.

$$Accuracy = \frac{TP + TN}{TP + TN + FP + FN} \tag{12.4}$$

- **Sensitivity**: 99.49%
  With a sensitivity of 99.49%, the system shows an exceptional ability to identify true positives, making it highly effective in detecting myocardial infarction cases. Early detection is crucial for timely medical intervention, which can significantly improve patient outcomes.

$$Sensitivity = \frac{TP}{TP + FN} \tag{12.5}$$

- **Specificity**: 99.94%
  The system boasts a specificity of 99.94%, which means it is highly proficient in differentiating between normal and abnormal ECG patterns. This reduces the occurrence of

false positives, thereby preventing unnecessary procedures or treatments for patients who do not require them.

$$Specificity = \frac{TN}{TN + FP} \tag{12.6}$$

Where

- TP is True Positive.
- TN is True Negative.
- FP is False Positive.
- FN is False Negative.

### 12.4.2 Experimental Results

The VLSI-based MI detection system has been thoroughly tested using real-world ECG datasets, providing valuable insights into its performance. These datasets consist of ECG signals from various patients, including those with different heart conditions, ensuring a comprehensive evaluation of the system's functionality in diverse clinical scenarios [15, 16].

The system consistently delivers high performance, not only in accurately diagnosing MI events but also in processing ECG data in real time. Its ability to promptly detect and localize myocardial infarctions makes it particularly advantageous in emergency settings, where swift intervention is crucial.

Moreover, the system proves to be resilient against noise and artifacts typically found in real-world ECG signals, ensuring reliable detection in challenging conditions. By accurately localizing the infarction site, the system aids healthcare professionals in tailoring interventions to the affected region of the heart, improving patient care and treatment outcomes.

### 12.4.3 Comparison with Existing Systems

When compared to traditional diagnostic tools, the VLSI-based MI detection system presents several key advantages:

- **Real-Time Processing**: Conventional ECG devices often have longer processing times, leading to delays in diagnosis. In contrast, the proposed system processes ECG data almost instantly, enabling faster detection and response in critical cases. This real-time capability is crucial in emergency settings, where every second counts.

- **Energy Efficiency**: Unlike traditional ECG machines, which consume significant power, the VLSI-based system is designed to be highly energy-efficient. This low power consumption allows it to be embedded in portable and wearable devices, making it ideal for continuous monitoring without the need for frequent recharging.
- **Compact and Portable Design**: Traditional ECG machines are large, making them less suitable for portable or emergency use. The VLSI-based system is small and lightweight, offering a compact solution that can easily be integrated into mobile health devices, wearables, or ambulatory monitoring systems. This portability expands its accessibility to patients in various settings, including at-home care and emergency situations.
- **Cost-Effectiveness**: Traditional ECG diagnostic systems are often expensive to purchase and maintain. The VLSI-based system provides a more affordable alternative, reducing both upfront and operational costs. This makes it a viable option for healthcare providers, particularly in low-resource settings, and for integration into cost-effective health monitoring solutions.
- **Mobile and Wearable Device Integration**: Due to its small form factor and low energy usage, the VLSI-based MI detection system is well-suited for integration with mobile devices and wearables. This enables continuous, real-time heart health monitoring, making it possible for patients to be monitored outside of clinical environments. Integration with telemedicine platforms also allows healthcare providers to remotely track patient health and intervene when necessary.
- **Enhanced Diagnostic Precision**: With its combination of advanced VLSI technology and machine learning models, the system offers a higher level of diagnostic accuracy compared to many traditional ECG devices. This improved precision results in fewer misdiagnoses, ensuring patients receive timely and appropriate treatments for their heart conditions.

**Table 12.1** System comparison.

| Parameter | Proposed VLSI-based ECG system | Traditional ECG system |
|---|---|---|
| Speed | Faster signal processing, real-time analysis | Slower processing due to analog signal conversion |
| Energy Consumption | Low power consumption (on-chip power management) | Higher energy consumption (requires external power supply) |
| Portability | High portability (compact, lightweight, wireless options) | Low portability (bulky, requires wired setup and external devices) |

In summary, the VLSI-based MI detection system offers several advantages over existing ECG diagnostic tools, including faster processing speed, lower power consumption, compact design, and cost-effectiveness. Its ability to perform real-time detection and accurate localization of myocardial infarctions makes it an invaluable tool for emergency situations and continuous patient monitoring. Table 12.1 indicates theat comparison of proposed VLSI based and traditional ECG system. Additionally, its portability and integration with wearable devices enable broader access to advanced cardiac care, especially in remote or underserved areas. Overall, the system's combination of speed, accuracy, energy efficiency, and affordability makes it a significant advancement in the field of myocardial infarction detection and management.

## 12.5 Applications and Practical Deployment

### 12.5.1 Integration into Medical Devices

The VLSI-based myocardial infarction (MI) detection system presents a promising opportunity for integration across a diverse range of medical devices, significantly improving [18] diagnostic capabilities in various healthcare settings. This system is particularly suitable for integration into:

- **Wearable Heart Monitors**: With its compact design and minimal power consumption, the VLSI system is ideally suited for incorporation into wearable devices such as smartwatches and fitness bands. These devices can continuously monitor heart activity, providing users with immediate notifications in the event of any irregularities. This not only facilitates early detection but also allows individuals to actively manage their heart health and seek timely intervention when necessary.

- **Portable ECG Machines**: Traditional ECG machines are often bulky and confined to clinical settings, limiting their accessibility. The VLSI-based system overcomes this challenge by enabling the development of lightweight, portable ECG machines. These devices are ideal for use in remote areas, urgent care settings, and during medical emergencies, allowing for ECG recordings anywhere, anytime, and enhancing patient care outside of traditional hospital environments.
- **Ambulance Diagnostic Kits**: During emergency situations, especially when transporting patients to a healthcare facility, the ability to monitor ECG data in real time can significantly improve patient outcomes. Integrating this VLSI system into ambulance diagnostic kits enables paramedics to detect

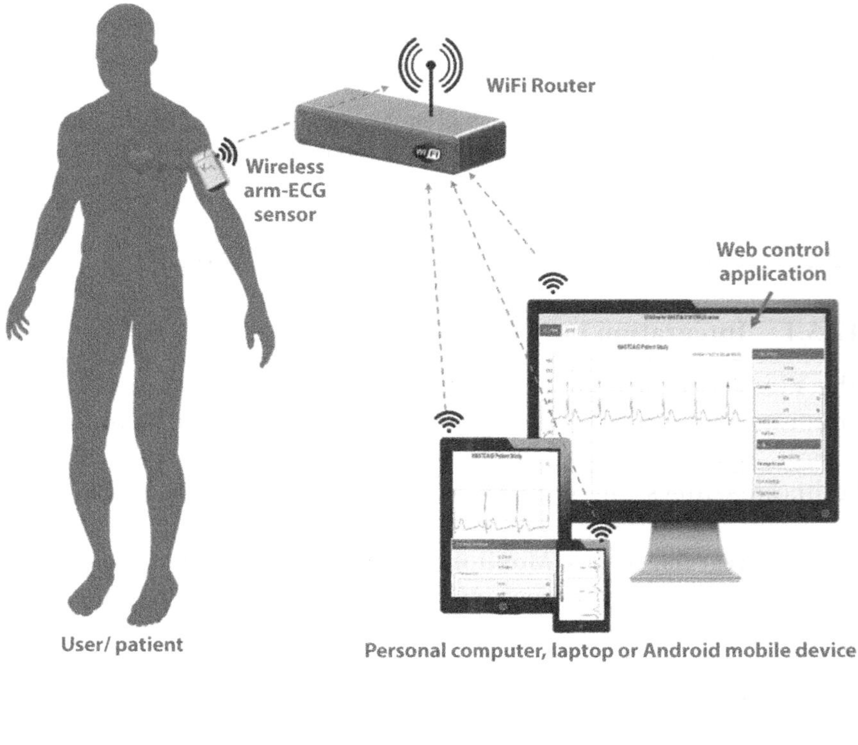

myocardial infarctions while en route, allowing for quicker medical decisions, preemptive treatments, and more efficient care delivery upon arrival at the hospital.

### 12.5.2 Impact on Patient Outcomes

The real-time detection capability of the VLSI-based system brings multiple benefits that significantly enhance patient outcomes. These include:

- **Reduced Delay in Intervention**: The ability to detect myocardial infarctions instantly allows healthcare providers to take swift action, reducing the critical time between diagnosis and treatment. Faster intervention can prevent the worsening of the infarction, reducing complications and improving recovery chances.
- **Improved Survival Rates**: Timely and accurate detection of heart attacks enables medical professionals to administer life-saving treatments, such as thrombolytic therapy, angioplasty, or stent insertion, without delay. The early intervention improves survival rates, decreases the risk of irreversible damage to heart tissue, and increases the likelihood of a full recovery.
- **Enhanced Long-Term Monitoring and Patient Management**: Continuous monitoring of heart activity through wearable devices or portable ECG machines ensures ongoing management of cardiovascular health. The ability to track heart

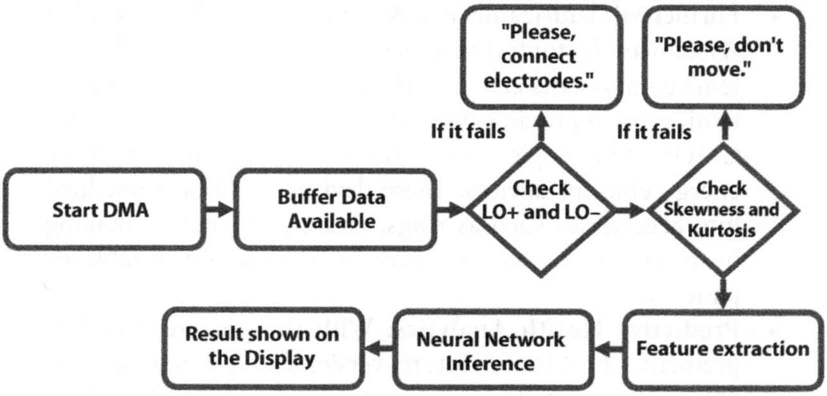

health over time allows for better-informed decisions, timely adjustments to treatment plans, and early identification of emerging heart conditions before they escalate into more severe events.

### 12.5.3 Future Directions

As the VLSI-based MI detection system continues to evolve, several enhancements will enable it to offer even greater benefits to both healthcare professionals and patients. Key directions for future development include:

- **Incorporation of Advanced AI and Machine Learning Models**: To improve diagnostic accuracy and expand the range of detectable conditions [16], the VLSI system may integrate advanced artificial intelligence (AI) and machine learning algorithms. These algorithms could enable the system to detect not only myocardial infarctions but also other heart-related issues, such as arrhythmias, ischemic conditions, or heart failure, further broadening its clinical utility.
- **Seamless Integration with Telemedicine Platforms**: With the growing adoption of telemedicine, integrating the VLSI-based system with telehealth platforms could enable remote patient monitoring. This integration would allow patients to share their ECG data with healthcare providers, who could then analyze the data, offer recommendations, or adjust treatment plans without the need for an in-person visit, ensuring continuous care and reducing the burden on healthcare facilities [17].
- **Further Miniaturization and Advanced CMOS Technologies**: As semiconductor technologies continue to progress, the system's components can be further miniaturized while maintaining or improving performance. Using next-generation CMOS technologies could result in even smaller, more energy-efficient devices. These devices could be embedded into accessories such as rings, glasses, or patches, making heart monitoring even more convenient and comfortable for patients.
- **Predictive Health Analytics**: With the growing trend of predictive healthcare, future versions of the system may incorporate predictive analytics to assess an individual's likelihood of experiencing a heart attack or other cardiovascular

events. By analyzing historical ECG data, lifestyle factors, and genetic information, the system could predict potential health risks, allowing for preventive measures and early intervention long before a heart event occurs.

- **Global Health Solutions for Underserved Areas**: The affordability and portability of the VLSI-based system make it an excellent candidate for deployment in low-resource settings, where access to sophisticated healthcare equipment may be limited. By providing affordable and accessible heart monitoring solutions [18], this system could play a pivotal role in improving cardiovascular health outcomes in developing countries, helping to reduce the global burden of heart disease.
- **Cloud Integration for Data Sharing and Remote Monitoring**: The integration of cloud-based systems could further enhance the capability of the VLSI-based MI detection system. By enabling the seamless transfer of ECG data to cloud platforms, doctors can access and analyze patient

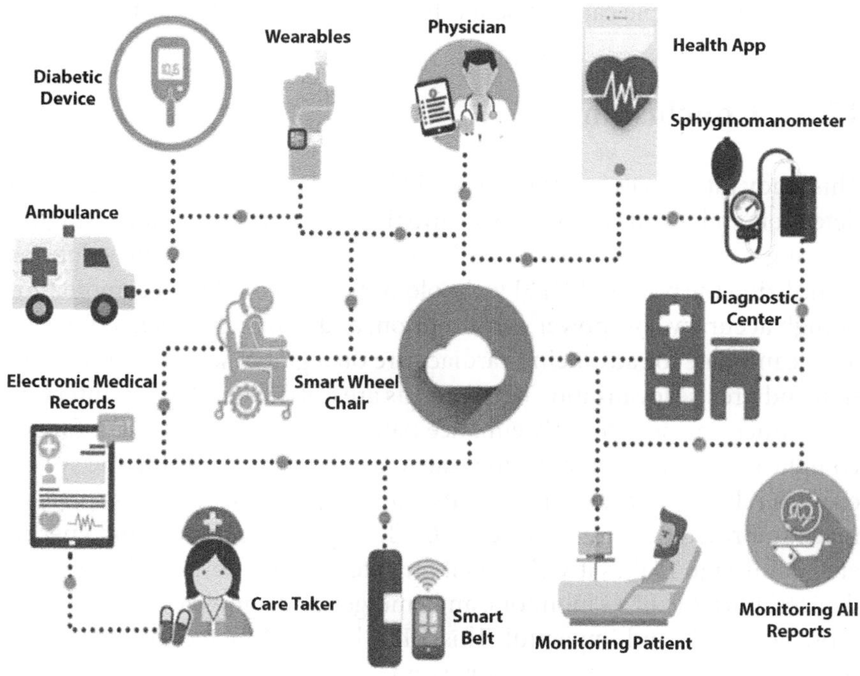


**Figure 12.8** Future applications.

data remotely, leading to faster diagnoses and more effective treatment plans. Furthermore, long-term data storage in the cloud could help doctors track a patient's health history, identify trends, and make data-driven decisions.

- **Regulatory Approvals and Healthcare Standards**: For the system to be widely adopted, it must meet the rigorous regulatory requirements of healthcare agencies such as the Food and Drug Administration (FDA) and CE (ConformitéEuropéenne).

  Achieving these certifications will ensure the system's safety, reliability, and effectiveness, making it a viable option for widespread clinical use. Figure 12.8 indicates the future applications of ECG device.
- **Expanded Use Cases Beyond MI**: While the primary focus of this system is the detection of myocardial infarctions, its advanced signal processing capabilities could be applied to other areas of cardiovascular care. For example, it could be used to detect other forms of heart disease, including atrial fibrillation or chronic ischemic heart disease [19]. The system's adaptability and scalability could lead to its adoption for multiple cardiovascular health monitoring applications.

## 12.6 Conclusion

This study introduces an innovative VLSI-based approach for the real-time detection and localization of myocardial infarctions (MI), addressing the growing need for fast, precise, and portable cardiac diagnostic solutions. By utilizing the power of VLSI technology, the system offers a combination of high accuracy, low power consumption, and compact size, making it an ideal candidate for advancing cardiac care on a global scale. Timely detection and precise localization of MI events are critical for effective intervention, which can significantly enhance patient survival and prevent further complications. Through machine learning-based classification, the system ensures reliable and real-time results, allowing healthcare professionals to make informed decisions quickly [20]. The system's compact and energy-efficient design makes it well-suited for integration into wearable medical devices, portable ECG monitors, and emergency diagnostic kits, making it more accessible and impactful. This VLSI-based MI detection system has potential applications extending beyond cardiac care, offering continuous health monitoring solutions for a wide range of medical conditions. As the healthcare landscape evolves, further advancements could be made by incorporating sophisticated AI models, telemedicine features, and innovations in CMOS technology. Its low power consumption and small form factor also enable it to be deployed in various settings, including home healthcare, mobile monitoring, and emergency situations. This versatility could enhance access to cutting-edge cardiac diagnostics, especially in regions with limited healthcare infrastructure.

In conclusion, this VLSI-based MI detection system represents a significant leap forward in cardiac diagnostics, offering a potent blend of efficiency, portability, and real-time capability. With the continued growth of wearable technology and AI-driven healthcare solutions, this system holds the potential to transform cardiac care by enabling earlier detection, improving patient outcomes, and providing more accessible, high-quality care worldwide.

**AI disclosure statement:** The author acknowledges the use of AI-based tools for minor language editing or grammatical correction. However, all intellectual contributions, analyses, and conclusions presented in this chapter are entirely the author's own.

## References

1. Kuo, S.K.H., Tsai, R.T., Tsai, S.K., Real-time ECG classification system for myocardial infarction detection. *IEEE Access*, 9, 40445–40453, 2021. doi: 10.1109/ACCESS.2021.3076901.
2. Sharma, K. and Bhatia, S., Low-power VLSI architecture for ECG signal processing in wearable devices. *IEEE Trans. Biomed. Circuits Syst.*, 15, 1, 28–37, Feb. 2021. doi: 10.1109/TBCAS.2021.3050042.
3. Kumar, N., Choudhary, D.K., Sharma, P., An efficient VLSI system for real-time myocardial infarction detection using ECG signals. *IEEE Trans. Circuits Syst. I Regul. Pap.*, 69, 7, 2595–2604, Jul. 2022. doi: 10.1109/TCSI.2022.3145708.
4. Wu, S.Z. and Wang, X.Q., ECG signal processing and feature extraction for myocardial infarction detection in healthcare systems. *IEEE Trans. Biomed. Eng.*, 67, 10, 2741–2752, Oct. 2020. doi: 10.1109/TBME.2020.2970997.
5. Lee, M.H., Kim, H.J., Zhang, W.L., VLSI-based ECG signal processor for real-time heart attack detection. *IEEE Trans. Circuits Syst. II Express Briefs*, 66, 5, 738–742, May 2019. doi: 10.1109/TCSII.2018.2876420.
6. Barman, R.G., Agarwal, R.B., Sharma, T., Advanced deep learning-based system for myocardial infarction detection using ECG signals. *IEEE Trans. Neural Syst. Rehabil. Eng.*, 29, 675–682, 2021. doi: 10.1109/TNSRE.2021.3052136.

7. Tiwari, D.R., Patel, P.B., Yadav, S.J., Real-time detection of myocardial infarction using ECG signals and VLSI implementation. *IEEE Access*, 8, 173987–173996, 2020. doi: 10.1109/ACCESS.2020.3025632.
8. Guo, Y.W., VLSI implementation of ECG signal classification for early myocardial infarction detection. *IEEE Trans. Biomed. Circuits Syst.*, 14, 4, 679–689, Aug. 2020. doi: 10.1109/TBCAS.2020.2977738.
9. Abraham, T.S. and Kumar, P.S., A compact VLSI architecture for ECG signal processing and real-time MI detection. *IEEE Trans. Biomed. Eng.*, 67, 4, 1010–1018, Apr. 2020. doi: 10.1109/TBME.2019.2914471.
10. Ghosh, M.P., Kumar, A., Jain, S., Real-time detection of myocardial infarction using optimized VLSI design and ECG feature extraction. *IEEE Trans. Circuits Syst. I Regul. Pap.*, 69, 3, 1422–1431, Mar. 2022. doi: 10.1109/TCSI.2021.3099884.
11. Radhakrishnan, R.J., VLSI circuits for real-time ECG signal acquisition and myocardial infarction detection. *IEEE Trans. Circuits Syst. II Express Briefs*, 67, 1, 31–34, Jan. 2020. doi: 10.1109/TCSII.2019.2954213.
12. Dey, V.S., Patel, R., Agarwal, A.K., Low-power VLSI-based ECG analysis system for heart disease detection. *IEEE Trans. Biomed. Circuits Syst.*, 13, 5, 1054–1063, Oct. 2019. doi: 10.1109/TBCAS.2019.2924118.
13. Baskar, J.T. and Rao, K.V.D.S., A high-performance VLSI architecture for real-time ECG-based myocardial infarction detection. *IEEE Access*, 7, 79921–79930, 2019. doi: 10.1109/ACCESS.2019.2921932.
14. Patil, S.C., Girme, M.G., Desai, K.N., Design and implementation of real-time ECG signal processing systems for myocardial infarction detection. *IEEE Trans. Biomed. Circuits Syst.*, 16, 6, 1844–1854, Dec. 2022. doi: 10.1109/TBCAS.2022.3171549.
15. Subramanian, M.S., Deep learning and VLSI-based system for automated myocardial infarction detection using ECG signals. *IEEE Trans. Neural Syst. Rehabil. Eng.*, 28, 9, 1874–1883, Sept. 2020. doi: 10.1109/TNSRE.2020.3022072.
16. Tiwari, D.R., Patel, P.B., Yadav, S.J., Low-power VLSI circuits for ECG analysis and myocardial infarction detection. *IEEE Trans. Circuits Syst. I Regul. Pap.*, 67, 9, 2832–2841, Sept. 2020. doi: 10.1109/TCSI.2020.2974459.
17. Natarajan, S.R. and Reddy, S.N., A low-cost VLSI implementation for ECG signal processing and MI detection. *IEEE Access*, 10, 20326–20337, 2022. doi: 10.1109/ACCESS.2022.3187261.
18. Thirumalai, T.A. and Madheswaran, S.S., Real-time heart disease detection using VLSI-based ECG signal processing. *IEEE Trans. Circuits Syst. II Express Briefs*, 69, 4, 574–579, Apr. 2022. doi: 10.1109/TCSII.2022.3145707.
19. Meena, P.K., Raj, V.B., Vishwakarma, S.K., ECG signal processing and feature extraction for myocardial infarction diagnosis using VLSI. *IEEE Trans. Biomed. Eng.*, 68, 5, 1404–1412, May 2021. doi: 10.1109/TBME.2021.3083496.
20. Chouhan, J.S.R., Sharma, R., Gupta, P., VLSI-based implementation of real-time ECG signal classification system for myocardial infarction detection. *IEEE Access*, 8, 34557–34565, 2020. doi: 10.1109/ACCESS.2020.2971141.

13

# Tunnel Field Effect Transistors and its Application as Label-Free Biosensor

**Basudha Dewan**

*Department of Electronics & Communication Engineering, Manipal University Jaipur, Rajasthan, India*

## *Abstract*

The growing demand for portable electronic devices highlights the importance of minimizing power consumption in circuit design. While dynamic power usage declines significantly with lower supply voltages, there is a challenge posed by leakage power because of the inverted sub-threshold slope. Reducing the supply voltage results in a lowered pinch-off point, and with the settled sub-threshold slope, this leads to the exponential escalation in leakage current. This work presents an extended source double gate (ESDG) dielectric modulated (DM) TFET label free biosensor using TCAD simulator. The nanogap cavity is considered near the source and drain regions of fixed $HfO_2$ layer to put up large number of biomolecules. Due to the extended source in ESDG TFET, the vertical tunnelling can be achieved along with lateral tunnelling and this leads to enhanced sensitivity (Sn). The Sn is evaluated for fully filled nanogap cavity under the both neutral and charged (positive/negative) biomolecules with wide variation in dielectric constant (k). We have reported the subthreshold swing (SS), input characteristics, energy band diagram, and $I_{ON}/I_{OFF}$ ratio by varying k from 1 to 12 for uncharged biomolecules within the nanogap. The sensitivity considering the steric hindrance for four diverse step profiles like increasing, decreasing, concave, and convex are reported by varying k. Furthermore, sensitivity of ESDG-DMTFET is compared with the sensitivity of existing FET/TFET based biosensors.

***Keywords*:** TFET, BTBT, MOSFET, biomolecules, biosensors, cavity, ESDG-DMTFET

*Email*: basudha.ece@gmail.com

P. Suveetha Dhanaselvam, Naima Guenifi, Shiromani Balmukund Rahi and J. Ajayan (eds.) Field-Effect Transistor Devices: Post-CMOS Theory and Applications, (255–274) © 2026 Scrivener Publishing LLC

## 13.1 Introduction

Since 1947, when the transistor was invented by William Shockleey, John Bar-den, and Walter Bratain at Bell Labs [1], and the subsequent creation of the Integrated-Circuit by Jack Kilby in 1958 [2], the repercussion about the unravelling of technology on the lives is immense. Today, it is unimaginable to live without it, considering the wide array of smart devices, such as cars, phones, watches, TVs, and healthcare gadgets that simplify our daily routines. This transformation is primarily attributed to the continuous miniaturization of electronic transistors. The semiconductor industry has been driven by Moore's law, which outlines an explosive rise for the number of transistors on a single semiconductor chip [3]. Shrinking technology nodes and, consequently, reducing transistor channel lengths have enabled additional devices packed on top of an individual chip, leading to lower production costs and increased complexity. Additionally, the reduction in transistor size has facilitated the construction of quicker circuits with fewer watts of energy utilization per transistor. Till 1990s, semiconductor industry largely adhered to Dennard's transistor scaling theory proposed in 1974 [4]. This theory involved a proportional reduction in both $V_{\mathrm{DD}}$ and $V_{\mathrm{TH}}$ of transistors as channel length and width dimensions decreased. Consequently, as $V_{DD}$ decreased, there was a quadratic depletion in the usage of dynamic power within the transistors therefore, consequently, in integrated circuits (ICs). The efficient scaling down of transistors is primarily attributed to exceptional physical and electric properties of $SiO_2$, which was utilized as the barrier layer between the device's gate and channel.

Over the last 15 years, various changes to the device structures have been necessary to sustain the trend of technology scaling. For instance, for 90 nm semiconductor node, strained silicon technology was established to improve carrier movement within transistors. This was achieved by employing techniques like incorporating SiGe semiconductors within the source/drain sections for PMOS transistors and utilizing a stretchable $Si_{3N4}$ capped barrier for NMOS transistors [5]. As gate oxide thickness decreased, one major challenge at the 45 nm technology nodes was a rise in gate loss caused by electrons passing across the narrow gate dielectric *via* tunneling. This issue is effectively addressed by introducing gate dielectrics with a higher dielectric constant ($\epsilon_r = 25$ for $HfO_2$ compared to $\epsilon = 3.9$ for $SiO_2$), considerably reducing gate leaking and enabling scaling up to 32 nm [6].

Following additional reductions in the length of the channel, there was a rise in the thermal diffusion of carriers and, consequently, an uptick in leakage current. This necessitated modifications to the traditional CMOS construction. In order to counteract resulting rise in constant energy usage within chips, alternative semiconductor topologies like Fully-Depleted-Silicon-on-Insulator (FD-SOI) and FinFETs have been created and already in fabrication, even reaching technologies that are as small as 7 nm [7].

FinFETs are classified under multi-gate technologies because they differ from conventional one-gate MOSFETs in that they include a gate electrode which surrounds the electrically conducting channel on multiple sides. Semiconductor devices featuring this manifold-gate layout enhance the gate's dominance over channel, facilitating mitigation of short channel effects and subsequently reducing internal device leakage power [8].

As depicted in Figure 13.1, over the coming time, semiconductor devices are anticipated to shrink to just some nanometers, beyond this point, additional reduction in dimensions of transistors would become unfeasible [10]. As MOSFET dimensions approach a scale where they consist of only a few hundred atoms, unwanted phenomena like gate tunnel would impede future enhancements in efficiency because of significant leakage currents. In order to prevent subsequent rise in static power consumption

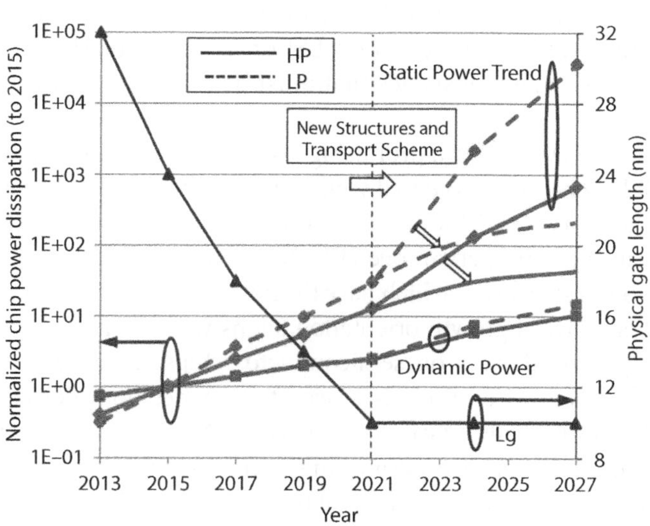


**Figure 13.1** Trends in static and dynamic power dissipation concerning gate length in relation to technology, as sourced from ITRS 2015 [9].

as technology continues to shrink and to maintain the growth of transistor count with economically practical manufacturing methods, the industry of semiconductors would focus its attempts on developing vertical in nature instrument designs, circuits using numerous layers (3-dimensional integration), as well as alternative electric transfer approaches. This will necessitate creation of electronic devices using other elements, including silicon-germanium, germanium, and others.

### 13.1.1 New Approaches for Upcoming Technology Generations

In order to mitigate the increasing tendency of static energy use in circuits at next-generation technical nodes, channel materials and structures must be modified. Following technologies are considered the most viable options to continue scaling technology down to dimensions in the nanometer range.

The 2015 ITRS roadmap forecasts that the next step in scaling MOSFET technology will involve replacing the stretched silicon path with superior movement materials. However, the shift presents several challenges that need to be addressed. These challenges include improving the interface between large-K dielectrics and Ge-III-V materials, minimizing band-to-band tunnelling in narrow bandgap channel materials, and developing a manufacturable process for VLSI on a silicon platform. These are just a few of the many obstacles that must be overcome in this transition [11, 12].

Nanowires (NWs) offer a promising avenue for replacing traditional planar MOSFET channels, potentially enabling a further reduction in technology node size. This is achievable because NWs can be manufactured with diameters in the nanometer range [13]. NWs are particularly well-suited for gate-all-around structures, which have the potential to mitigate short-channel effects. Their vertical growth orientation aligns with the future trend of vertically integrating devices on semiconductor chips. To ensure the growth of defect-free nanowires, there is a need to enhance both device yield and uniformity. Additionally, in case the transfer of NWs onto an other medium becomes necessary, precise positioning becomes crucial. To avoid surface irregularities and defects, it is essential to develop effective surface treatment and passivation techniques.

Carbon Nanotubes (CNTs) are considered a promising choice for technology nodes below 10 nanometers due to their extremely thin bodies, approximately 1 nanometer in diameter. When compared to silicon, CNT-FETs can provide enhanced channel's electron as well as hole mobility, particular while at room temperature. However, challenges exist in terms of achieving VLSI integration with CNTs, including the need to purify and sort CNTs to ensure a relatively uniform diameter distribution [14].

Graphene Field-Effect Transistors (GFETs) hold significant potential for the RF design field due to their superior carrier mobility compared to CNTs. They enable the development of faster transistors with smaller dimensions. Graphene, being a solitary atomic sheet, essentially a no band-gap semi-conductor. Consequently, the primary focus of research in this area is on finding ways to introduce a bandgap effectively. This is crucial for achieving a substantial current gain ($I_{ON}/I_{OFF}$) suitable for both analog and RF applications [15].

Tunnel Field-Effect Transistors (TFETs) are highly regarded as the most promising technology for applications requiring low power and modest performance due to their unique electrical properties [16]. Unlike conventional thermal devices, TFETs utilize a distinct carrier injection mechanism, enabling them to maintain a reverse just below the threshold slopes (SS) of less than 60 mili Volt/decade during ambient temperature. Because of this feature, TFETs may function at minimal voltages and yet have a substantial current amplification. Successful inclusion of TFETs into lower-energy circuits in the future hinges the advancement of III-V production methods. TFETs developed from materials from groups III and V show promise in surpassing the electrical performance of Si-based TFETs [17].

### 13.1.2 Tunnel Field-Effect Transistors (Tunnel FETs): A Vital Technique toward Power Harvesting

One strategy aimed at cutting down on the amount of power consumed in digitally processed units is by lowering the electric source supply ($V_{DD}$) for the electronic circuits. When $V_{DD}$ is reduced linearly, it results in a quadratic decrease in dynamic power consumption. Having said that, provided the transistor's threshold voltage ($V_{TH}$) fails to increase proportionately to $V_{DD}$, the fails to increase proportionately to $V_{DD}$, the current loss increases exponentially, leading to higher static power consumption. This issue is tied with the thermally controlled charge infusion process in traditional MOSFETs. The reverse beneath the threshold slopes (SS) cannot be more than 60 mV/dec. As a result, lowering $V_{DD}$ causes a sharp rise in beneath the threshold current loss, as illustrated by Equation (13.1) [19].

$$I_{sub} = K_1 W e^{\frac{-V_{TH}}{\eta V_\theta}} \tag{13.1}$$

Beneath the threshold current loss formula includes constants $K_1$ and n, which are determined through experimental measurements. It also involves the transistor's width (W) and the thermal voltage (represented by $V_t$, which is approximately 25 mV at room temperature). According to Figure 13.4, threshold values of voltage in transistors in exceptionally well along with subpar performance utilities is projected to remain incessant in the coming times.

As the power supply voltage ($V_{DD}$) decreases, there will be an anticipated increase in leakage power within integrated circuits, as illustrated in the power trends displayed in Figure 13.1. However, the introduction of various technologies that incorporate new materials and transport methods, which are not reliant on thermal effects, will enable a reduction in the threshold voltage of transistors without causing a corresponding increase in static or standby power consumption in chips.

As of now, significant exploration is focused on developing switches having exceptionally sharp sub-threshold dynamics, specifically below 60 mV/dec at room temperature. Among various alternatives employing different carrier injection mechanisms, the Tunnel Field-Effect Transistor (TFET) is emerging as the most promising technology for switches designed to operate at low voltages (below 0.5V) and in applications with lower performance requirements [20, 21].

As shown in Figure 13.5, the TFET device provides the benefit of lowering the overdrive voltage ($V_{GS}$ – $V_{TH}$) while still achieving better functionality from the perspective of reducing off-state current along with minimizing static energy utilization when contrasted with mainstream temperature-based MOSFETs. Given its superior execution at reduced voltages, the TFET is well-suited for super subdued power, minimal-functionality circuit boards, such as those used in IoT sensor nodes, which rely on ambient energy sources for power. Figure 13.2 illustrates the projected evolution of threshold and supply voltages ($V_{TH}$ and $V_{DD}$) for future logic technologies, emphasizing the gradual transition toward low-power devices such as TFETs that can function efficiently at sub-0.5 V operation.

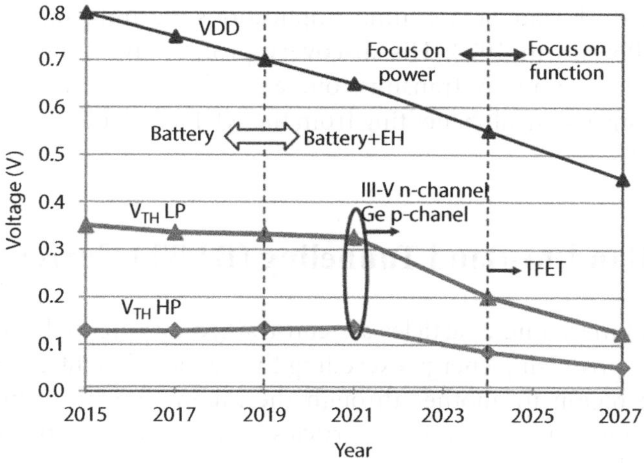


**Figure 13.2** The voltage evolution observed in logic technology, as sourced from the ITRS 2015 report [9].

## 13.2 Tunnel FET Technology: State of Art

In 1958, a researcher from Japan named L. Esaki, working at Sony Corporation, made a pivotal breakthrough by introducing a device that operated upon Band-to-Band Tunnelling, designated as the tunnel diode. This device's working principle, grounded in quantum mechanics, differed significantly from transistors and further semiconductor devices of that era. In tunnel diodes, electrons could seemingly disappear from one side of a potential barrier and reappear on the other side without having enough energy to cross over the barrier directly [22]. This phenomenon, known as tunnelling, allowed electrons to pass through the depletion region in p-n junctions.

Initially, the importance of the tunnel diode wasn't fully realized, primarily due to limited knowledge about the tunnelling process. However, decades of research have highlighted tunnelling as a promising solution for emerging switching technologies, particularly because of its ability to function effectively at low voltages, typically below 0.5 V.

Esaki, while studying internal field emission in semiconductor diodes with heavily doped germanium (Ge) junctions, observed a non-linear current-voltage relationship when subjected to forward bias at low temperatures. This phenomenon, linked to the elastic nature of tunnelling in Ge, resulted in the identification of negative differential resistance (NDR). Importantly, the tunnelling process maintained the electron's energy during the transition.

Esaki's work had far-reaching implications, earning him the Nobel Prize in Physics in 1973. His discovery profoundly influenced several fields, including radio transmission, amplification, computation, and DC-AC conversion, all benefiting from the NDR effect observed in tunnel diodes [23].

## 13.3 Band-to-Band Tunneling (BTBT) Current

In classical mechanics, particles are constrained by potential barriers, and only those possessing energy exceeding the barrier's height can transition from one region to another through thermionic emission. In contrast, quantum mechanics describes particles using wave functions denoted as Ψ, which do not abruptly cease at finite potential barriers of height $U_0$. Consequently, there exists a finite probability of particles tunneling through the barrier and moving from one region to another [24]. In a tunneling device, we can distinguish between two categories of tunneling current: Zener tunneling current, which involves electrons moving from the valence band to the conduction band, and Esaki tunneling current, which entails electrons transitioning from the conduction band to the valence band.

In a tunneling device, the current is determined by the tunneling transmission probability, denoted as $T_{BTBT}$. The calculation of tunneling current involves applying the Wentzel-Kramers-Brillouin (WKB) approximation, which provides the tunneling transmission coefficient as defined in Equation (13.2) [25].

$$T_{BTBT} \cong exp\left[-\frac{4\sqrt{2m^*}E_g^{1.5}}{3'\ qF}\right] \tag{13.2}$$

Tunnel FETs utilize a MOS gate to control the band-to-band tunneling (BTBT) across a degenerate *p-n* junction. The device is normally off. When zero bias is applied to the gate, the conduction band minimum of the channel is above the valence band maximum of the source, so band-to-band tunneling is suppressed. A tunneling window, opens up as the conduction band of the channel is shifted below the valence band of the source. Electrons in the valence band with energy in this tunneling window tunnel into empty states in the channel and the transistor is ON. The principle of operation is the same for the *p*-channel TFET with source, channel and drain conductivity types switched. In the conventional mode of operation, the *n*-channel TFET tunnel current is suppressed when *Vgs* is low and the tunnel window at the source junction is opened with positive *Vgs*. However the TFET can turn on at the channel drain junction when the gate bias is sufficiently negative. When the gate bias is negative, the valence band maximum of the channel can be shifted above the conduction band minimum of the drain leading to electron tunneling from the channel into the drain. Therefore, the tunneling window opens up again, with the tunnel junction shifted from the source channel junction to the drain-channel junction. When this happens the channel conduction changes from one carrier type to another and the transfer characteristic is said to be ambipolar. This behavior is generally universal across TFET geometries. When the gate bias is still positive and the drain bias becomes negative, TFET behaves like an Esaki diode, with the signature of negative differential resistance (NDR behavior appearing in the output characteristics [50–55].

### 13.3.1 Reported Work

In this exponential rising scientific world, biosensors have become more considerable element in humans' life, which gave rise to a significant research area for the same. In this regard, the miniaturized CMOS compatibility dielectric modulated FET (DM-FET) based biosensor shows satisfactory performance [26, 27]. The dielectric modulation technique provides label-free detection of biomolecules through label-used detection challenges. A nanocavity is formed in DM-FET based biosensor through removing a small percentage of the gate dielectric, where neutral/charged biomolecules can be occupied [28, 29]. The DM-FET electrically detects the biomolecules based on dielectric constant and drain current by monitoring the targeted biomolecules confined to a cavity presented beneath the gate to immobilize the biomolecules [30]. The DM-FET-based biosensors suffer from scalability problems like higher leakage current, inability to reduce SS below 60 mV/dec. Due to these issues, DM-Tunnel FET (TFET) based biosensor is proposed, which can suppress the limitations of FET based biosensor by several orders [31, 32]. The enactment of TFET based biosensors is enhanced due to BTBT rate at the source and channel interface [33], in addition to advantage of CMOS compatibility fabrication process and low leakage current. It can operate at very low power supply with significant reduction in power dissipation of the circuit. A biosensor of nanowire TFET architecture of positively charged is highlighted in literature [34]. The sensitivity of a Doping less DG TFET biosensor is highlighted for a wide change in cavity height [35]. It is seen that short gate TFET based biosensor has improved sensitivity compared to full gate TFET biosensor [36]. An analytical model for sensitivity is revealed in DG TFET with the variant in k [37]. It is reported that circular gate (CG) TFET has more sensitivity than Ge source TFET at diverse k under neutral/charged biomolecules [38]. The vertical TFET has greater amount of sensitivity compared to DG TFET for uncharged/charged biomolecules at different values of k [39]. We have reported that Ge source DG TFET has better sensitivity compared to DG TFET for both neutral/charged biomolecules [40]. A DG-TFET is widely used as it provides better control over the channel due to the presence of two gates and sensitivity is improved compared to a single gate [41]. Furthermore, to enhance the performance like SS and drive current, a structure known as ESDG DMTFET based biosensor is proposed [42]. It provides a comparatively improvement in current sensitivity and SS due to its enhanced tunnelling junction area. The central focus of this paper is to evaluate the sensitivity of ESDG-DMTFET based biosensor under neutral/charged biomolecules for a wide variation in k. The SS and ION/IOFF ratio is reported at different k in ESDG-DMTFET. The sensitivity is extracted by considering steric hindrance in ESDG DMTFET at two different fill factors (45% and 65%). Moreover, the sensitivity of the proposed biosensor is compared with existing biosensor in literature.

## 13.4 Device Architecture and Simulation Deck

The 2-D cross-sectional view of the proposed nanogap embedded extended source DG (ESDG) dielectric modulated TFET biosensor is shown in Figure 13.3. The corresponding symbol represents the various dimensions of the structure. The ESDG-DMTFET is a symmetrical dual gate structure, where source, channel, and drain are made of Silicon. The doping concentration for p+ source, n channel, and n+ drain regions are $5\times10^{18}$, $1\times10^{15}$, and $10^{20}$ cm-3, respectively. The source region is prolonged into an n-channel so that the channel overlaps the source. The gate material consists of single metal having work function of 4.05 eV. The nanogap is created on both sides of source/drain regions by removing out $HfO_2$ and the existence of these nanogap growths the capture area of the biomolecules. A thin $SiO_2$ layer of 1 nm thickness is taken as an immobilizing layer for biomolecules present in the nanogap. The various dimensions of the proposed biosensor are listed in Table 13.1.

TCAD simulator is employed to extract the sensitivities of the ESDG-DMTFET [43]. The probability of tunnelling in TFET be determined by the energy bandgap and accordingly, bandgap narrowing is used. The degenerate source and drain sections lead to activation of Fermi

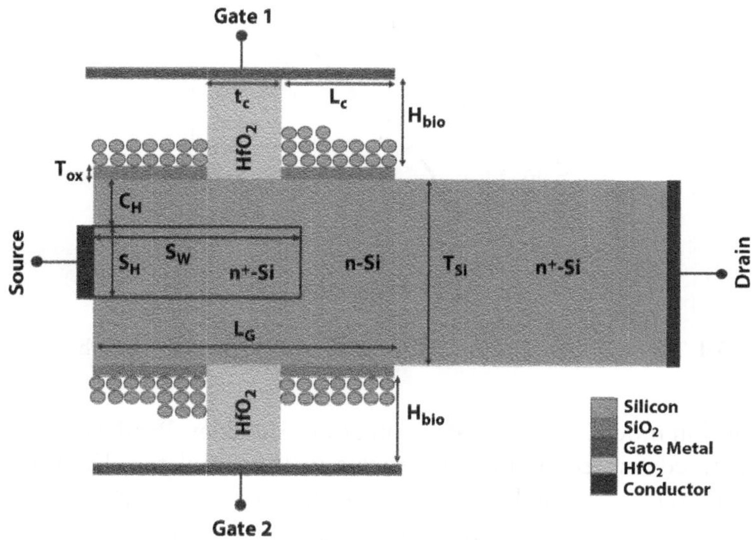


**Figure 13.3** 2D view of nanogap ESDG-TFET biosensor.

**Table 13.1** Dimensions of the proposed biosensor considered in this work.

| Parameters | Value (nm) |
|---|---|
| Silicon film thickness ($T_{SI}$) | 15 |
| Source height ($S_H$) | 10 |
| Source length ($S_W$) | 20 |
| Channel height ($C_H$) | 2.5 |
| Gate Length ($L_G$) | 30 |
| The thickness of the $SiO_2$ layer ($T_{OX}$) | 1 |
| Length of nanogap cavity ($L_C$) | 12.5 |
| The thickness of the $HfO_2$ layer ($t_c$) | 5 |
| Height of nanogap cavity ($H_{bio}$) | 7 |

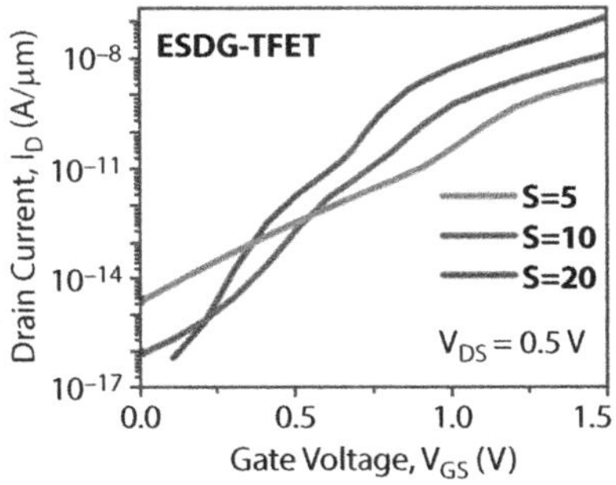


**Figure 13.4** Transfer characteristic of ESDG-TFET at different source length.

distribution model. For accurate simulation, Shockley-Read-Hall (SRH) is also adopted for recombination process in the simulator. The non-local BTBT model is activated to analyse the transport of the charge carriers. The experimentally calibrated tunnelling parameters are invoked in simulator: $A_{path}$ = 1.63×1014 cm-3 $s^{-1}$, $B_{path}$ = 1.47×107 V $cm^{-1}$, Ppath1 = 0.0567 eV [44]. During simulation, a low range of k (2-5) and high k value (5-12) biomolecules are considered in the cavity. The example of uncharged biomolecules is uricase = 1.54, protein = 2.5, streptavidin = 2.1, APTES = 3.57 etc. The positive and negative charge concentrations are considered as mentioned in literature for DNA [45–48].

The transfer characteristic of ESDG-TFET at different source length is shown in Figure 13.4. It is realized that the increased in source length improves the switching ratio and this is due to increase in tunnelling area which leads to increased BTBT rate at tunnel junction.

## 13.5 Results and Discussion

Here, the characteristics of ESDG DM-TFET biosensor is analysed for k = 5, 7, 10, 12, and 15. The neutral and charged biomolecules are undertaken to measure the sensitivity of this biosensor at stated k. The output of this biosensor is dignified considering the effect of steric hindrance. Here, k=1 implies the nanogap is filled with air.

Sensitivity analysis for Neutral Biomolecules

The input drain current characteristic of ESDG-DMTFET at various k for neutral charged molecules ($N_{Bio}$ = 0) is shown in Figure 13.5(a). The rise in k leads to significant enhancement in ON current and this is due to rise in tunnelling rate at the tunnel junction. It can be better clarified

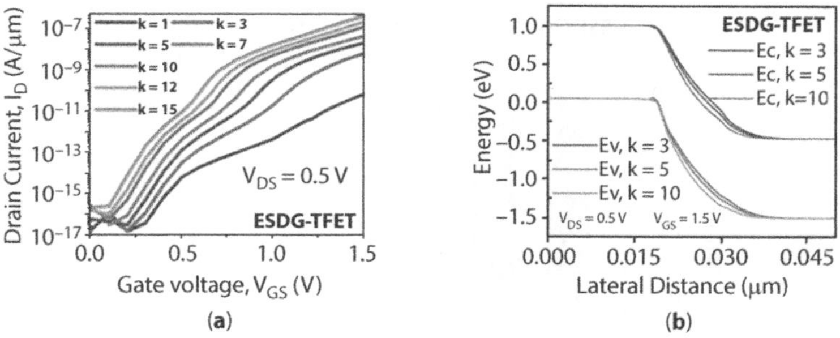


**Figure 13.5** Input characteristic of (a) ESDG-DM TFET biosensor and (b) corresponding energy bandgap for uncharged biomolecules with variation in k.

from Figure 13.5(b), where the energy band diagram is plotted as a function of k and it is envisioned that the increased in k, lessens the tunnel width for ESDG-DMTFET. This leads to increase in number of charge of carriers and thus, ION is increased. The performance of biosensor can be extracted from its sensitivity and for DM-FET, the sensitivity mathematically is defined as [49]:

$$Sensitivity, S_n = \frac{I_{D,k}}{I_{D,air}}$$

where, $I_{D,k}$ and $I_{D,air}$ are representing the corresponding drain currents for k>1 and k=1, respectively, at fixed gate voltage ($V_G$) and drain to source bias ($V_{DS}$).

The sensitivity for the variation in k for ESDG-DMTFET is portrayed in Figure 13.6(a) and it is summarized that $S_n$ is increased by an observable amount with k due to enhancement in $I_{ON}$. It is observed that sensitivity in range of $10^4$ at k=15 is achieved. The effect of k on SS and $I_{ON}/I_{OFF}$ ratio is shown in Figure 13.6(b) and SS is expressed as [49]:

$$SS = 2.3\eta V_T$$

where, $V_T$ and η are thermal voltage and subthreshold slope, respectively. The η is given by:

$$\eta = 1 + \frac{C_{dep}}{C_{ox}}$$

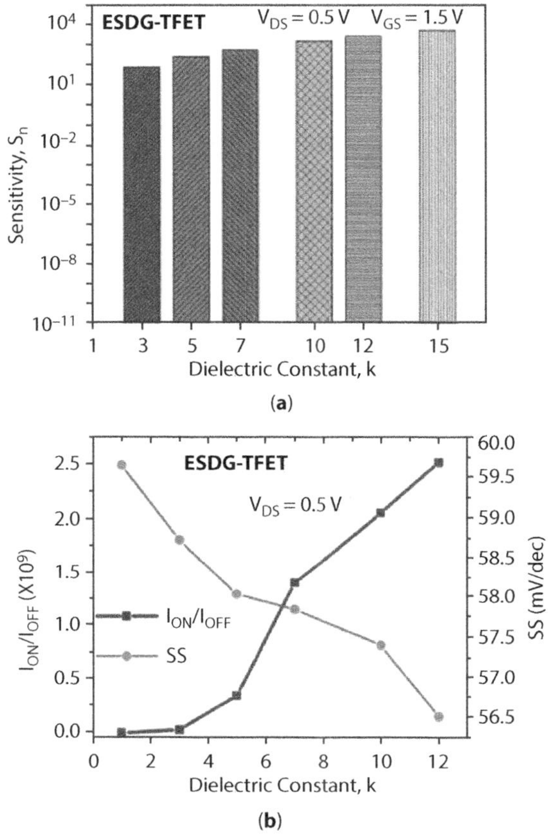


**Figure 13.6** (a) Sensitivities of ESDG-DM TFET biosensors k= 5, 7, 10, 12, 15 and (b) $I_{ON}/I_{OFF}$ and SS at these k.

where $C_{dep}$ and $C_{ox}$ indicate the depletion and oxide capacitances, respectively. It is perceived from Figure 13.6(b) that the increase in k reduces the SS and this is due to decrease in $\eta$. It is also visualized that for all values of k, the SS value is below 60 mV/dec., which indicates ESDG-TFET can be used for low power applications. The rise in k leads to significant escalation in ON current with negligible changes in OFF current, which indicates improvement in $I_{ON}/I_{OFF}$ ratio for ESDG-DMTFET (Figure 13.4(b)).

## 13.6 Conclusion

This chapter delves into the intricate construction of ultra-low power circuits tailored specifically for energy harvesting through the utilization of Tunnel Field Effect Transistors (TFETs). TFETs stand out due to their Band-to-Band Tunnelling (BTBT) carrier injection mechanism, rooted in the Zener tunnelling effect. This mechanism grants TFETs a subthreshold slope (SS) that operates inversely to the typical limit of 60 mV/dec observed in conventional thermally dependent switching devices at room temperature. The remarkable characteristic of TFETs allows for the reduction of operational voltage in circuits without compromising the device's leakage current. Consequently, A DM Extended Source double gate (ESDG)-TFET biosensor is designed and its Sn is reported. We have highlighted Sn underneath neutral and charged biomolecules with wide variation in k within the nanogap cavity and the upsurge in k leads to improvement in Sn. Therefore, ESDG-DMTFET biosensor is an emerging device for bio sensing applications due to its enhanced performance.

**AI disclosure statement:** The author acknowledges the use of AI-based tools for minor language editing or grammatical correction. However, all intellectual contributions, analyses, and conclusions presented in this chapter are entirely the author's own.

## Bibliography


1. Shockley, W., The path to the conception of the junction transistor. *IEEE Trans. Electron Devices*, 23, 7, 597–620, 1976.
2. Kilby, J.S., Turning potential into realities: The invention of the integrated circuit, International Journal of Modern Physics B: Condensed Matter Physics; Statistical Physics. *Appl. Phys.*, 16, 5, 2002.
3. Bondyopadhyay, P.K., Moore's law governs the silicon revolution. *Proc. IEEE*, 86, 1, 78–81, 1998.
4. Dennard, R.H., Gaensslen, F.H., Yu, H.-N., Rideout, V.L., Bassous, E., LeBlanc, A.R., Design of ion-implanted mosfet's with very small physical dimensions. *IEEE J. Solid-State Circuits*, 9, 5, 256–268, 1974.
5. Thompson, S.E., Armstrong, M., Auth, C., Alavi, M., Buehler, M., Chau, R., Cea, S., Ghani, T., Glass, G., Hoffman, T., *et al.*, A 90-nm logic technology featuring strained-silicon. *IEEE Trans. Electron Devices*, 51, 11, 1790– 1797, 2004.

6. Chau, R., Datta, S., Doczy, M., Doyle, B., Kavalieros, J., Metz, M., High-/spl kappa//metal-gate stack and its mosfet characteristics. *IEEE Electron Device Lett.*, 25, 6, 408–410, 2004.
7. Wu, S.-Y., Lin, C., Chiang, M., Liaw, J., Cheng, J., Yang, S., Tsai, C., Chen, P., Miyashita, T., Chang, C., *et al.*, A 7nm cmos platform technology featuring 4 th generation finfet transistors with a 0.027 um 2 high density 6-t sram cell for mobile soc applications, in: *2016 IEEE International Electron Devices Meeting (IEDM)*, IEEE, pp. 2–6, 2016.
8. Ferain, I., Colinge, C.A., Colinge, J.-P., Multigate transistors as the future of classical metal–oxide–semiconductor field-effect transistors. *Nature*, 479, 7373, 310–316, 2011.
9. Hoefflinger, B., Itrs: The international technology roadmap for semiconductors, in Chips 2020: a guide to the future of nanoelectronics. *Springer*, 161–174, 2011.
10. Courtland, R., Transistors could stop shrinking in 2021. *IEEE Spectr.*, 53, 9, 9–11, 2016.
11. Nainani, A., Raghunathan, S., Witte, D., Kobayashi, M., Irisawa, T., Krishnamohan, T., Saraswat, K., Bennett, B.R., Ancona, M.G., Boos, J.B., Engineering of strained iii–v heterostructures for high hole mobility, in: *2009 IEEE International Electron Devices Meeting (IEDM)*, IEEE, pp. 1–4, 2009.
12. Tsipas, P. and Dimoulas, A., Modeling of negatively charged states at the ge surface and interfaces. *Appl. Phys. Lett.*, 94, 1, 2009.
13. Luryi, S., Xu, J., Zaslavsky, A., *Future Trends in Microelectronics: From Nanophotonics to Sensors to Energy*, John Wiley & Sons, 2010.
14. Kim, Y.-B., Integrated circuit design based on carbon nanotube field effect transistor. *Trans. Electr. Electron. Mater.*, 12, 5, 175–188, 2011.
15. Akinwande, D., Petrone, N., Hone, J., Two-dimensional flexible nanoelectronics. *Nat. Commun.*, 5, 1, 5678, 2014.
16. Ionescu, A.M. and Riel, H., Tunnel field-effect transistors as energy-efficient electronic switches. *Nature*, 479, 7373, 329–337, 2011.
17. Avci, U.E., Morris, D.H., Young, I.A., Tunnel field-effect transistors: Prospects and challenges. *IEEE J. Electron Devices Soc.*, 3, 3, 88–95, 2015.
18. Chen, S., Xu, H., Liu, D., Hu, B., Wang, H., A vision of iot: Applications, challenges, and opportunities with china perspective. *IEEE Internet Things J.*, 1, 4, 349–359, 2014.
19. Kim, N.S., Austin, T., Baauw, D., Mudge, T., Flautner, K., Hu, J.S., Irwin, M.J., Kandemir, M., Narayanan, V., Leakage current: Moore's law meets static power. *Computer*, 36, 12, 68–75, 2003.
20. Ionescu, A.M., De Michielis, L., Dagtekin, N., Salvatore, G., Cao, J., Rusu, A., Bartsch, S., Ultra-low power: Emerging devices and their benefits for integrated circuits, in: *2011 International Electron Devices Meeting*, IEEE, pp. 16–1, 2011.

21. Nikonov, D.E. and Young, I.A., Uniform methodology for benchmarking beyond-cmos logic devices, in: *2012 International Electron Devices Meeting*, IEEE, pp. 25–4, 2012.
22. Esaki, L., Long journey into tunneling. *Sci.*, 183, 4130, 1149–1155, 1974.
23. Cavalheiro, D., *Ultra-low power circuits based on tunnel fets for energy harvesting applications*, 2017.
24. Sze, S. and Ng, K., Chapter 3 metal-semiconductors contacts, in: *Physics of Semi-conductor Devices*, 3rd ed, pp. 134–196, John Wiley & Sons, Hoboken NJ, USA, 2007.
25. Kane, E.O., Theory of tunneling. *J. Appl. Phys.*, 32, 1, 83–91, 1961.
26. Bergveld, P., The development and application of FET-based biosensors. *Biosensors*, 2, 1, 15–33, 1986, doi: 10.1016/0265-928X(86)85010-6.
27. Dewan, B., Chaudhary, S., Yadav, M., A charge plasma based label free biomolecule detector using SiGe-heterojunction double gate Tunnel FET. *Silicon*, 1–10, 2021.
28. Tiwari, S. and Saha, R., Sensitivity Analysis of I-Shape TFET Biosensor Considering Repulsive Steric and Trap Effects. *IEEE Trans. Nanotechnol.*, 22, 518–524, 2023, doi: 10.1109/TNANO.2023.3309411.
29. Im, H., Huang, X.-J., Gu, B., Choi, Y.-K., A dielectric-modulated field-effect transistor for biosensing. *Nat. Nanotechnol.*, 2, 430–434, Jul. 2007, doi: 10.1038/nnano.2007.180.
30. Choi, J.-M., Han, J.-W., Choi, S.-J., Choi, Y.-K., Analytical modeling of a nanogap-embedded FET for application as a biosensor. *IEEE Trans. Electron Devices*, 57, 12, 3477–3484, Dec. 2010, doi: 10.1109/TED.2010.2076152.
31. Shan, C., Wang, Y., Bao, M.-T., A charge-plasma-based transistor with induced graded channel for enhanced analog performance. *IEEE Trans. Electron Devices*, 63, 6, 2275–2281, Jun. 2016, doi: 10.1109/TED.2016.2549554.
32. Dewan, B., Chaudhary, S., Yadav, M., Electrically doped SiGe-heterojunction TFET based biosensor considering non-ideal hybridization issues: a simulation study. *Appl. Phys. A*, 127, 1–9, 2021.
33. Choi, W.Y., Park, B.-G., Lee, J.D., Liu, T.-J.K., Tunneling field-effect transistors (TFETs) with subthreshold swing (SS) less than 60 mV/dec. *IEEE Electron Device Lett.*, 28, 8, 743–745, Aug. 2007, doi: 10.1109/LED.2007.901273.
34. Chen, K.-I., Li, B.-R., Chen, Y.-T., Silicon nanowire field-effect transistor-based biosensors for biomedical diagnosis and cellular recording investigation. *Nano Today*, 6, 2, 131–154, 2011, doi: 10.1016/j.nantod.2011.02.001.
35. Dewan, B., Chaudhary, S., Yadav, M., A label-free dielectric-modulated biosensor using SiGe-heterojunction dual cavity dual metal electrically doped TFET. *Eng. Res. Express*, 5.3, 035002, 2023.
36. Kanungo, S., Chattopadhyay, S., Gupta, P.S., Rahaman, H., Comparative Performance Analysis of the Dielectrically Modulated Full- Gate and Short-Gate Tunnel FET-Based Biosensors. *IEEE Trans. Electron Devices*, 62, 3, 994–1001, March 2015, doi: 10.1109/TED.2015.2390774.

37. Narang, R., Reddy, K.V.S., Saxena, M., Gupta, R.S., Gupta, M., A Dielectric-Modulated Tunnel-FET Based Biosensor for Label-Free Detection: Analytical Modeling Study and Sensitivity Analysis. *IEEE Trans. Electron Devices*, 59, 10, 2809–2817, Oct. 2012, doi: 10.1109/TED.2012.2208115.
38. Goswami, R. and Bhowmick, B., Comparative Analyses of Circular Gate TFET and Heterojunction TFET for Dielectric - Modulated Label - Free Biosensing. *IEEE Sens. J.*, 19, 21, 9600–9609, 2019, doi: 10.1109/JSEN.2019.2928182.
39. Wangkheirakpam, V.D., Bhowmick, B., Pukhrambam, P.D., N+ Pocket Doped Vertical TFET Based Dielectric-Modulated Biosensor Considering Non-Ideal Hybridization Issue: A Simulation Study. *IEEE Trans. Nanotechnol.*, 19, 156–162, 2020, doi: 10.1109/TNANO.2020.2969206.
40. Saha, R., Hirpara, Y., Hoque, S., Sensitivity Analysis on Dielectric Modulated Ge-Source DMDG TFET Based Label-Free Biosensor, in *IEEE Trans. Nanotechnol.*, 20, 552–560, 2021, doi: 10.1109/TNANO.2021.3093927.
41. Kumar, S., Goel, E., Singh, K., Singh, B., Kumar Singh, P., Baral, K., Jit, S., 2-D Analytical Modeling of the Electrical Characteristics of Dual-Material Double-Gate TFETs With a $SiO_2$/HfO2 Stacked Gate-Oxide Structure. *IEEE Trans. Electron Devices*, 64, 960–968, 2017, doi: 10.1109/TED.2017.2656630.
42. Dewan, B., Chaudhary, S., Singh, D., Yadav, M., Label-free detection of breast cancer cell lines using dopingless heterojunction TFET considering non-ideal hybridization issue. *Mater. Sci. Engineering B*, 302, 1171925, 2024.
43. Biswas, A., Dan, S.S., Royer, C.L., Grabinski, W., Ionescu, A.M., TCAD simulation of SOI TFETs and calibration of non-local band-to-band tunneling model. *Microelectron. Eng.*, 98, 334–337, 2012, doi: 10.1016/j.mee.2012.07.077.
44. Chandan, B.V., Nigam, K., Sharma, D., Junctionless based dielectric modulated electrically doped tunnel FET based biosensor for label-free detection. *Micro Nano Lett.*, 13, 4, 452–456, Apr. 2018, doi: 10.1049/mnl.2017.0580.
45. Dhanaselvam, P., Karthikeyan, B., Anand, P., Future Prospective Beyond CMOS Technology Design. *Field Effect Transistors*, 73–80, 2025.
46. Dhanaselvam, P., Karthikeyan, B., Kavitha, K., Kavitha, P., Basic Concepts of Heterojunction Tunnel Field-Effect Transistors. *Field Effect Transistors*, 221–234, 2025.
47. Narang, R., Saxena, M., Gupta, M., Comparative analysis of dielectric-modulated FET and TFET-based biosensor. *IEEE Trans. Nanotechnol.*, 14, 3, 427–435, May 2015, doi: 10.1109/TNANO.2015.2396899.
48. Dhanaselvam Suveetha, P., Karthikeyan, B., Vanitha, P., Anand, P., Scope with TFET-based circuit and system design, in: *Device Circuit Co-Design Issues in FETs*, pp. 181–188, CRC Press, 2023.
49. Verma, M., Tirkey, S., Yadav, S., Sharma, D., Yadav, D.S., Performance Assessment of A Novel Vertical Dielectrically Modulated TFET-Based

Biosensor. *IEEE Trans. Electron Devices*, 1–8, April 2017, doi: 10.1109/TED.2017.2732820.

50. Goel, E., Pandey, A., Rahi, S. B., Samuel, A., (Eds.), *Handbook of Advanced Semiconductor Field Effect Transistors*. John Wiley & Sons, 2025.
51. Rahi, S.B., Tayal, S., Upadhyay, A.K., A review on emerging negative capacitance field effect transistor for low power electronics. *Microelectron. J.*, 116, 105242, 2021.
52. Upadhyay, A. K., Rahi, S.B., Tayal, S., Song, Y.S., Recent progress on negative capacitance tunnel FET for low-power applications: Device perspective. *Microelectron. J.*, 129, 105583, 2022.
53. Saravanan, M., Parthasarathy, E., Ajayan, J., Rahi, S.B., Device Simulation Based Machine Learning Technique for III V TFET, in: *2024 International Conference on Recent Advances in Electrical, Electronics, Ubiquitous Communication, and Computational Intelligence (RAEEUCCI)*, pp. 1–6, IEEE, 2024, April.
54. Song, Y.S., Thoutam, L.R., Tayal, S., Rahi, S.B., Samuel, T.S.A., (Eds.), *Handbook of Emerging Materials for Semiconductor Industry*. Springer, 2024
55. Rahi, S.B., Asthana, P., Gupta, S., Heterogate junctionless tunnel field-effect transistor: future of low-power devices. *J. Comput. Electron.*, 16, 1, 30–38, 2017.

# 14

# Graphene Nanosheet Field Effect Transistor

**S. Raj Kumar[1]*, S. Murugan[1] and T. Esther[2]**

[1]*ECE Department, VCET, Madurai, Tamilnadu, India*
[2]*ECE Department, SSCE, Coimbatore, Tamilnadu, India*

***Abstract***
A new 2D analytical model for the V-shaped stacked graphene nanosheet field effect transistor (GNFETs) is developed in this study. Using suitable boundary conditions, the Linear Approximation Method (LAM) is used to determine the 2D Poisson equation for the first time. To assess the performance of the device, the exact analytical equations for the variables drain current, transconductance, threshold voltage, electric field, and channel potential are derived. The device operates at cryogenic temperatures as a result of the high electron fluxes into the channel caused by the effect of the nanosheet channel contact with the gate. It offers a significant scaling of device with a large capacity for drive current. With a perpendicular layered structure, the V-shaped nanosheet achieves low threshold voltage and high transconductance. When TCAD simulation software and analytical model results are compared, a good degree of agreement is obtained.

***Keywords:*** Graphene nanosheet, linear approximation method, transconductance, threshold voltage, drain current

## 14.1 Introduction

For analog and digital circuits, semiconductor devices are very trustworthy and have great promise. In order to address the demands of increasingly compact and energy-efficient electronic systems, the current trend

**Corresponding author*: srk@vcet.ac.in

P. Suveetha Dhanaselvam, Naima Guenifi, Shiromani Balmukund Rahi and J. Ajayan (eds.) Field-Effect Transistor Devices: Post-CMOS Theory and Applications, (275–292) © 2026 Scrivener Publishing LLC

focuses on designing low scaling devices that are more power-efficient. Generations of CMOS technology to preserve a high level of electrostatic control [1–3]. Furthermore, as technology advances, MOSFETs encounter difficulties including elevated leakage currents and power consumption, which restrict their effectiveness in really tiny semiconductor devices. MOSFETs have certain drawbacks, including their vulnerability to hot carrier injection, a condition in which high-energy carriers erode the gate oxide and perhaps jeopardise the transistor's long-term performance and dependability [4–6]. In switching devices, leakage power is caused by leakage voltage and leakage current.

FinFET devices have lower power consumption than MOSFETs because of their low leakage current. However, FinFET also featured very high capacitances, corner effect, higher parasitics because of 3-D profile, quantized device-width, and difficult-to-control dynamic $V_{th}$. Manufacturing technology indicates that FinFET costs can rise by 2% to 3% when compared to bulk [7, 8]. Another drawback of FinFET technology is that the bottom side of the Fin comes into touch with the substrate because the gate only covers three sides of the rectangular Fin. Even in the off-state, this leads to a significant leakage current in the transistor [9–11].

At 3 nm, V-shaped nanosheet field effect transistors (GNFETs) offer better short channel management and improved electrostatics in comparison to FinFETs and MOSFETs. Nanosheet transistors reduce channel widths to deliver higher drive current than FinFETs [12]. The proposed device reduces short-channel effects and enhances channel control with its n-type perpendicular stacking topology. The effective channel width increases when nanosheets are stacked perpendicular to one another [13–15]. Compared to Fins, this enables nanosheets to offer a higher driving current per channel. The device drive current can be increased by stacking more nanosheets (generating more channels) and by adding oxide between the substrate and source/drain [16–18]. Logic device requirements at 3-nm nodes are met by perpendicularly stacked nanosheet architectures, which provide good electrostatics and short channel control [19–21]. An equivalent channel width can be achieved by arranging nanosheets perpendicularly with C-shaped, which is more than four times more area-efficient than a MOSFET. Increased control over the current flow is made possible by this design, which improves the transistor's performance. Higher integration density on a chip is made possible by the vertical layout, which aids in the creation of more potent and small electronic devices [22–24].

## 14.2 Device Structures and Simulations

The gate is crafted from gold metal with a work function of 5.2 eV, while V-shaped nanosheet composed of graphene material is shown in Figure 14.1. A four Channel featuring the structure of FETs is fashioned using a two V-shaped nanosheet. The influence of the four channels with double nanosheet permits electrons in the channel to traverse freely, devoid of collisions. The analysis unequivocally establishes the viability of the proposed device for low-power applications. The oxide layer in double nanosheets refers to a thin coating of hafnium oxide material applied to both sides of a nanosheet structure. It serves as a dielectric layer, providing insulation between the nanosheets and enabling efficient electronic device operation at nanoscale dimensions.

The V-shaped stacked nanosheets in the transistor is placed on the gate surrounded with source and drain is shown in Figure 14.2a and Figure 14.2b. The vertically positioned gate spans over these nanosheets, which are mounted onto a graphene substrate with a thin oxide layer. The metal gate of nanosheet transistors surrounds the conduction channel to produce high drive current. Instead of surrounding the channel on three sides as in the FinFET architecture, V-shaped GNFETs surrounds to allow better control of the transistor switch. By incorporating four channels within a double nanosheets FET structure, this work achieves a higher level of integration

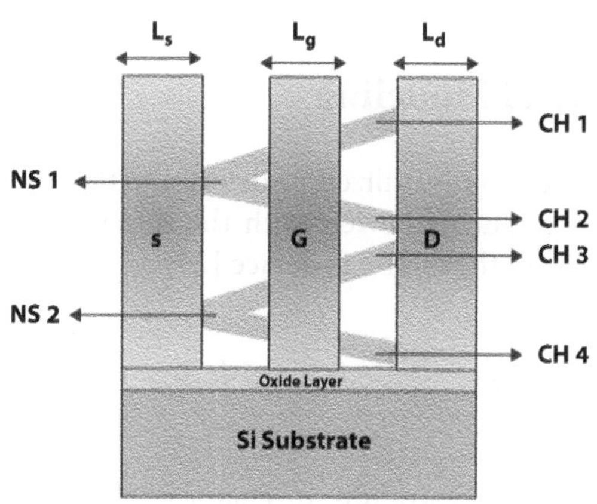


**Figure 14.1** Cross sectional view of GNFETs.

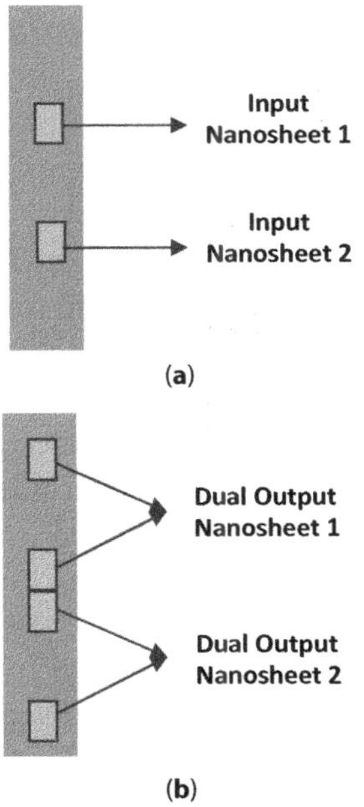


**Figure 14.2** (a) Source side view of GNFETs. (b) Drain side view of GNFETs.

and functionality. The completely stacked nanosheets are highly doped with channel configuration. Nanosheet transistors feature a thin channel region, allowing the gate to exert precise control over the flow of electrons.

## 14.3 Analytical Modeling

The 2D Poisson equation is utilized to determine the electrical characteristics of the semiconductor device, with the distribution of charges and doping profiles within the GNFETs device [25].

$$L''_{xi}(x,y)+L''_{yi}(x,y)=\frac{-qN_d}{\varepsilon_a},0\le x\le L_i,0\le y\le d_{eqv},i=1,\ 2 \quad (14.1)$$

where $N_d$ is doping concentration; $q$, charge of electron; $L(x,y)$, electric potential; and $\varepsilon_a$ - permittivity of graphene; L, length of the nanosheet; $d_{eqv}$, total thickness of nanosheet [26].

$$d_{eqv}=n_s(2W_{ns}+2T_{ns}) \quad (14.2)$$

where $n_s$ is the number of nanosheets, $W_{ns}$ is the channel width and $T_{ns}$ is the nanosheet thickness.

A perfect linear function can accurately represent the potential expression in both directions in this analysis.

$$L_i(x,y) = V_i(x,y) + V'_{xi}(x - L_i) + V'_{yi}(y - d), i = 1,2 \qquad (14.3)$$

where $V(x,y)$ is channel potential.

### 14.3.1 Boundary Conditions

(i) At $x_i$=0 and $y_i$=0, the channel potential is at the contact of different metals.

$$L_i(x,y)\Big|_{\substack{x_i=0\\y_i=0}} = V_g, i = 1,2 \qquad (14.4)$$

where $V_g = V_{gs} - V_{fb}$, $V_{gs}$ - gate-source voltage, $V_{fb}$ - flat-band voltage

(ii) At $x_i$=0 and $y_i$=d, the electric field at the contact of different metals.

$$L'_{yi}(x,y)\Big|_{\substack{x_i=0\\y_i=d}} = -\frac{\varepsilon_{ox}}{\varepsilon_{si}t_{ox}}(Q_s + C_{ox}(V_{gate} - V(x,y)), i = 1,\ 2 \qquad (14.5)$$

(iii) Potential at $x_i$=0 and $y_i$=0 on the source end,

$$L_i(x,y)\Big|_{\substack{x_i=0\\y_i=0}} = V_{bi}, i = 1,\ 2 \qquad (14.6)$$

where $V_{bi}$ is built-in-voltage.

(iv) Potential at $x_i = L_i$ and $y_i$=0 on the drain end,

$$L_i(x,y)\Big|_{\substack{x_i=L_i\\y_i=0}} = V_{bi} + V_{ds}, i = 1,\ 2 \qquad (14.7)$$

where $V_{ds}$ is drain-source voltage.

### 14.3.2 Channel Potential Model

The GNFETs' channel potential is created by the potential difference that may exist at the interface and may have an impact on the device's performance. The gate voltage and dual channel contact have a major impact on the channel potential in GNFETs. The Linear Approximation Method (LAM) is used to compute the potential at the graphene channel area. The device's channel, which is completely depleted in the off-state, is uniformly doped with a high doping concentration to provide a linear range of performance. Along the x and y-axes, the channel potential values are chosen at two different positions, such as $x = 0, x = L_i$, and $y = 0, y = d$.

Apply boundary conditions (14.4) and (14.5) in (14.3),

$$L(x,y) = V(x,y) + \frac{\varepsilon_{ox}}{2L\varepsilon_{si}t_{ox}}\left((Q_s + Q_{gate})d - \frac{V_g}{L}\right)(x - L) - \frac{\varepsilon_{ox}}{2\varepsilon_{si}t_{ox}}(Q_s + Q_{gate}) \tag{14.8}$$

where $Q_{gate} = C_{ox}(V_{gate} - V(x,y))$

Apply boundary conditions (14.6) and (14.7) in (14.1),

$$L(x,y) = -\frac{qN_d}{4\varepsilon_a}(x^2 + y^2) + \left(\frac{V_{ds}}{L} + \frac{qN_d}{4\varepsilon_a}L\right)(x + y) + V_{bi} \tag{14.9}$$

The total channel potential is determined by equating (14.8) and (14.9),

$$V(x,y) = \frac{k_1 + k_2 + V_{bi} - k_3\left(Q_s\left(\frac{x}{L} - d - 1\right) + C_{ox}V_g\left(\frac{d}{L} + d - 1\right)\right)}{1 - k_3\left(\frac{x_d}{L} + C_{ox}(d - 1)\right)} \tag{14.10}$$

where

$$k_1 = -\frac{qN_d}{4\varepsilon_a}(x^2 + y^2),$$

$$k_2 = \left(\frac{V_{ds}}{L} + \frac{qN_d}{4\varepsilon_a}L\right)(x + y),$$

$$k_3 = \frac{\varepsilon_{ox}}{2\varepsilon_{si}t_{ox}}$$

The electrons in the nanosheet are contained by the channel potential, which also enables high-frequency, high-speed operation and great electron mobility.

### 14.3.3 Electric Field Model

The doping concentration of the nanosheet channel and the gate voltage of the GNFETs affect the electric field's strength. High conductivity in the channel and effective electron transport are made possible by the concentration of electrons at the contact. The analytical model of the electric field has been created by differentiating the channel potential along the channel length can be found and as a result, the resulting electric fields is determined using

$$E_i(x) = V'_{xi}(x, y)\Big|_{yi=0}, i = 1,\ 2$$

The total electric field is given by,

$$E(x) = \frac{((k_4\left(-2k_5x + \frac{V_{ds}}{L} + k_5L + V_{bi} - \frac{k_3Q_s}{L}\right))(-k_5x^2 + \left(\frac{V_{ds}}{L} + k_5L\right)x + V_{bi} - k_3k_6)(-k_3\frac{d}{L}))}{k_4^2} \tag{14.11}$$

where

$$k_4 = 1 - k_3\left(\frac{x_d}{L} + C_{ox}(d-1)\right),$$

$$k_5 = \frac{qN_d}{4\varepsilon_a},$$

$$k_6 = Q_s\left(\frac{x}{L} - d - 1\right) + C_{ox}V_g\left(\frac{d}{L} + d - 1\right)$$

It is imperative to optimize the distribution of the electric field in order to prevent problems like breakdown voltage restrictions and hot carrier effects. The nanosheet channel's electron density and potential are modulated by the electric field produced by the dual gate contact.

### 14.3.4 Model of Threshold Voltage

Significant current can flow across a transistor's channel when it transitions from its "off" to "on" state. The threshold voltage controls the switching behavior of the device. The analytical threshold model is used to compute the minimal potential required for electrons to flow in the channel in order for the device to operate. As a result, the specific position of the minimal channel potential, $u_{min}(x,y)$, can be determined by differentiating the channel potential, which equals zero. This technique has been applied to determine a device threshold voltage. For GNFETs, the threshold voltage is determined using

$$V_{Ti}(x,y) = V'_{xi}(x,y), i = 1,\ 2$$

The total threshold voltage is given by,

$$V_T(x,y) = k_4k_5(L-2x) + k_4\left(\frac{V_{ds}}{L} + V_{bi}\right) - \frac{k_3k_4Q_s}{L} + \frac{k_3d}{L}\left(k_1 + k_2 + \frac{V_{bi}d}{L} - k_3k_6\right) \tag{14.12}$$

### 14.3.5 Drain Current Model

The device affected by the gate voltage at the channel, which also affects the channel potential and electric field. High-mobility electrons inhabit the nanosheet channel, and electron transport in this channel is intimately correlated with the drain-source current. The drain current of an NSFET in the linear zone has been simplified. The following formula can be used to approximate the drain-source current in an NSFET:

$$I_{ds} = KTW\mu_0 n_i^2\sqrt{\frac{\varepsilon_{si}V_{min}(x,y)}{2N_d^2}}\left(e^{\frac{qV_{min}(x,y)}{KT}} - e^{\frac{q(V_{min}(x,y)-V_{ds})}{KT}}\right) \tag{14.13}$$

where $V_{min}(x, y)$ - minimum potential in channel, $\mu_0$ – electron mobility, $n_s$ - sheet carrier density.

### 14.3.6 Transconductance

One essential metric for describing a transistor's behavior and amplification capacity is its transconductance. A few of the variables that affect it are the channel density, carrier mobility, channel width, length and gate voltage. Greater gate length and lower noise amplification capacity are indicated by greater transconductance values. The standard notation for conductance is $g_m$, which can be analytically represented as:

$$g_m = \frac{\partial I_{ds}}{\partial V_{gs}} \tag{14.14}$$

where $\partial I_{ds}$ - change in drain-source current, $\partial V_{gs}$ - change in gate-source voltage

Substitute the $I_{ds}$ value in (14), the transconductance is found by

$$g_m = \frac{KTW\mu_0 n_i^2}{N_d}\sqrt{\frac{\varepsilon_{si}}{2}}\left(\sqrt{V_{min}(x,y)} + (\alpha_2 + \alpha_3)\left(\frac{1}{2\sqrt{V_{min}(x,y)}}\alpha_1\right)\right) \tag{14.15}$$

where

$$\alpha_1 = \frac{k_3^2 d}{L^2} C_{ox}\left(d + L(d-1)\right),$$

$$\alpha_2 = e^{\frac{qV_{min}(x,y)}{KT}},$$

$$\alpha_3 = e^{\frac{q(V_{min}(x,y) - V_{ds})}{KT}}$$

### 14.3.7 Integrated Noise

Many models have predicted noise in various components or at the output of the device with different materials and gates. However, the accuracy of overall noise analysis is still a work in progress. The proposed model incorporates an integration method to ensure accurate prediction of total noise in the device for high-frequency applications. Integration combines smaller noise components into a single system that functions as total integrated noise. This approach aims to simplify noise prediction for the entire device and is used to determine the noise at output.

$$\overline{I_{nd}^2} = 8kT\Delta f \frac{g_m}{0.666C_0}(L_1 + L_2 + L_3)(\alpha W + \beta I_{ds}) \quad (14.16)$$

where $k$-Boltzmann constant, $T$-room temperature, $\Delta f$-bandwidth, $\alpha$ and $\beta$ are fitting parameters whose values are $2\times10^5\, pF/cm^2$ and $1.25\times10^2\, pF/mA/cm$.

## 14.4 Result and Discussion

The Linear Approximation Method (LAM) is used to compute and validate the parameters analytically in order to study the device performance of GNFETs. The results of the analytical modeling and the TCAD simulator are precisely in sync. By examining the Gaussian profile with potential at the points $x = 0$ and $x = L$, the parameters are retrieved. The value for parameters for GNFETs is shown in Table 14.1.

In Figure 14.3, gate control and short-channel effects are balanced by a reasonable oxide thickness. The gate's effect on the channel potential is less pronounced in thicker oxides. Variations in $V_{gs}$ cause the channel potential

**Table 14.1** Parameter values for GNFETs.

| Parameters | Specifications |
|---|---|
| Gate Work Function (Au) | 5.1 eV |
| Length of source & drain ($L_s$ & $L_d$) | 12 nm |
| Length of gate ($L_g$) | 16 nm |
| Doping level of Source ($N_A$) | $10^{20}$ cm$^{-3}$ |
| Doping level of Drain ($N_D$) | $10^{20}$ cm$^{-3}$ |
| Channel Length (L) | 20 nm |
| Oxide Thickness ($t_{ox}$) | 2nm |

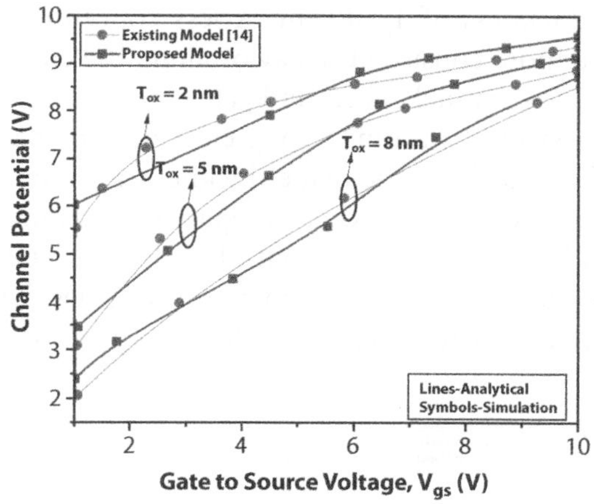


**Figure 14.3** Channel potential vs $V_{gs}$ for different $T_{ox}$.

to react more gradually. A stronger and more focused impact on the channel potential is the outcome of improved gate control. Strong electrostatic channel control by the gate is made possible by an incredibly thin oxide. Variations in $V_{gs}$ cause the channel potential to react more quickly and sensitively.

In Figure 14.4, an electric field peak guarantees a superior average electric field over the channel for $T_{ox}$=2 nm. Consequently, carrier transport efficiency and device speed both increase at $T_{ox}$=2 nm. The electric field is directly affected by the channel length, oxide thickness, and applied voltage. This leads to thinner oxide layers, which strengthen the observed electric field more than the existing model. The electric field is dispersed farther

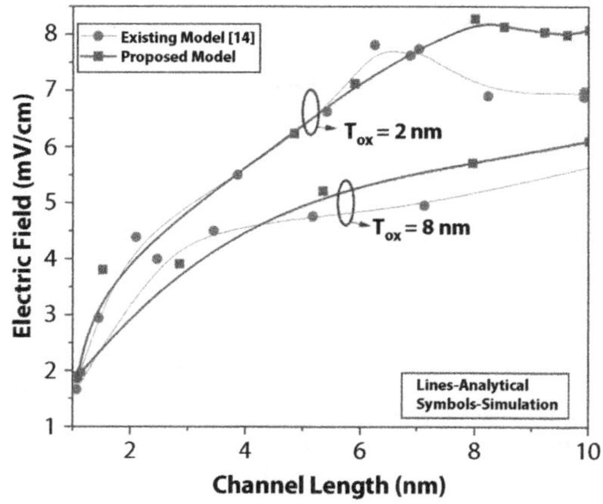


**Figure 14.4** Electric field versus L for different $T_{ox}$.

with an oxide thickness of $T_{ox}$=8 nm. A less concentrated electric field could result in a more consistent behavior along the channel. A compromise between short-channel effects and gate control can be achieved with an oxide thickness of $T_{ox}$=2 nm. Strong electrostatic channel control by the gate is made possible by the oxide thickness $T_{ox}$=2 nm. The proposed model shows high electric field when compared to existing model.

In Figure 14.5a, more channel control and more effective transistor modulation are made possible by higher gate voltages, which frequently cause the drain current to increase. As $V_{ds}$ increases, the transistor may

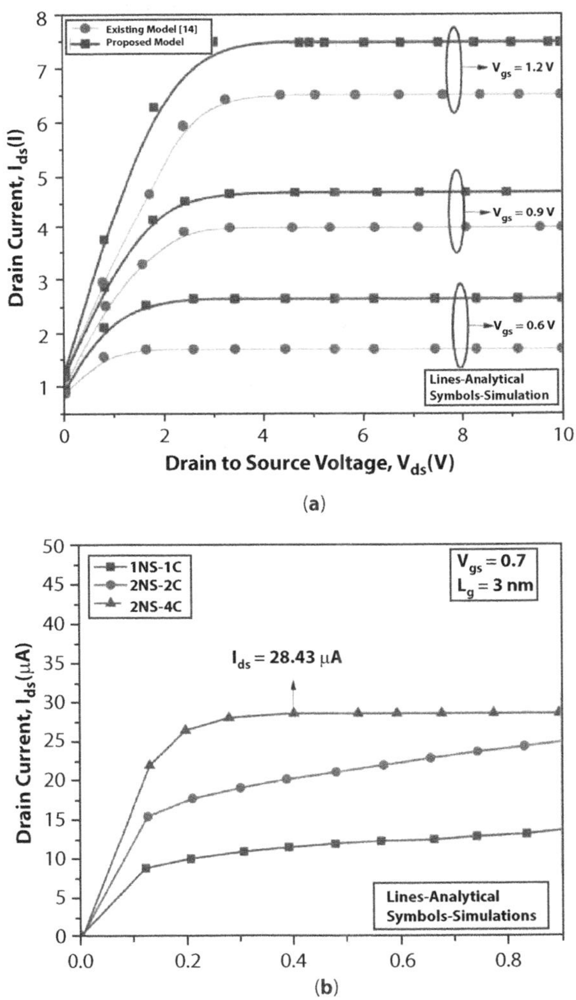


**Figure 14.5** (a) $I_{ds}$ versus $V_{ds}$ for different $V_{gs}$=1.2, 0.9, and 0.6 V. (b) $I_{ds}$ vs $V_{ds}$ for different $V_{gs}$=0.7 V and $L_g$=3 nm.

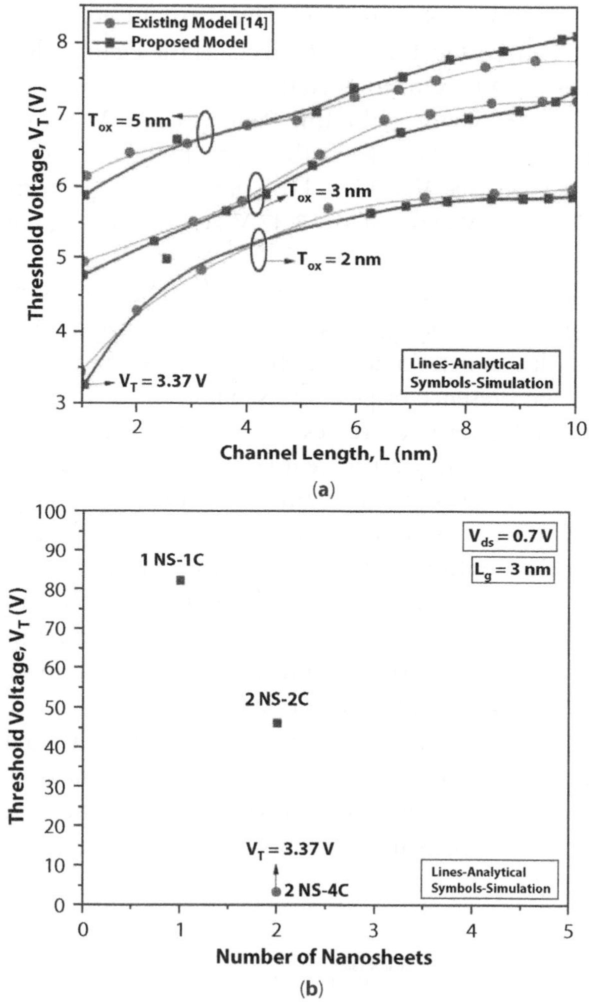


**Figure 14.6** (a) $V_T$ versus L for different $T_{ox}$. (b) Threshold voltage versus number of nanosheets with $V_{ds}$=0.7 V.

approach the saturation area, which is the point at which the drain current reaches a maximum and subsequently starts to saturate. According to the proposed model, $I_{ds}$ rises enables better channel control. Because of the greater carrier injection in the channel, higher $V_{gs}$ levels often translate into higher drain current levels over the whole $V_{DS}$ range. In Figure 14.5b, when compared to single nanosheet with single channel and Dual nanosheet with dual channel [16], the proposed dual nanosheet with quad channel shows high drain current.

In Figure 14.6a, Because of improved electrostatic control, the transistor may encounter more short-channel effects as the channel length reduces, which would lower Vth. One important factor that affects a nanosheet transistor's switching behavior is the threshold voltage ($V_T$). In Figure 14.6b, the variation of threshold voltage with respect to the number of nanosheets for drain-source voltage $V_{ds}$=0.7 V [16], gate length $L_g$=3 nm and $T_{ox=2\ nm}$. The lowest threshold voltage of $V_T$=3.37 V attains for dual nanosheet with quad channel.

For varying oxide thickness, the transconductance ($g_m$) varies with gate-source voltage is shown in Figure 14.7. The conductance decreases in the saturation region due to deteriorating mobility. The electric field in the channel is stronger in the presence of peak transconductance. The gate capacitance rises as the oxide thickness decreases $T_{ox=2\ nm}$. For a change in $V_{gs}$, enhanced gate control causes a more noticeable change in the channel current. There may exist a threshold at very thin oxide thicknesses $T_{ox}$=2 nm where short-channel effects become noticeable. These short-channel effects may result in a saturation or reduction in the transconductance curve at increasing $V_{gs}$. With improved channel control over the current model, $g_m$ improves along with $V_{gs}$. In comparison to the existing model, a peak transconductance of $g_m$=0.195 ms/μm is achieved in proposed model.

Figure 14.8 shows a variation of noise with changes in gate length for graphene nanosheet FET. For the gate length of the device changes with $T_{ox}$=2.5, 3.5, 4.5 nm, noise is relatively low for maximize the oxide thickness values. Output noise is relatively low and when increasing gate length further noise turns to a constant value. Because of the dual ouput

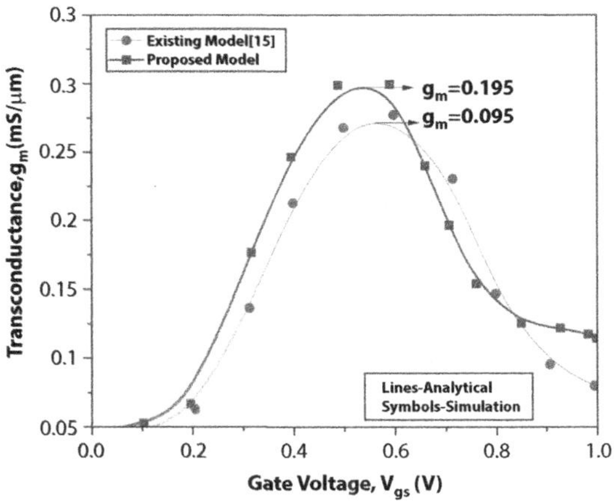


**Figure 14.7** Transconductance versus gate to source voltage.

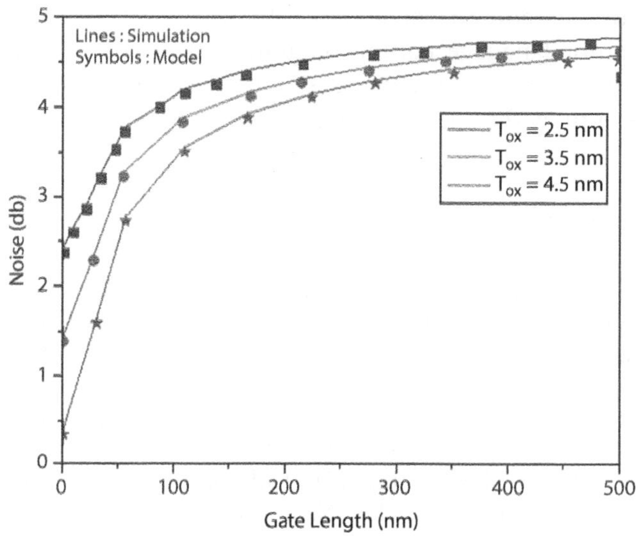


**Figure 14.8** Noise with variation of different gate length.

in nanosheet region with grapheme material, there is an improvement in gate control over the channel, so noise and current leakage is reduced with respect to the gate channel. This is shown in the result of constant noise performance for the increase in oxide thickness when compared to the other device model.

## 14.5 Conclusion

An analytical model for V-shaped nanosheet FETs is developed and 2D Poisson equation is derived by means of the LAM. High electron mobility is implied by the perpendicularly stacked V-shaped nanosheet FETs with superior gate control over previous devices. Consequently, the short channel effects are reduced. The outcome additionally demonstrates that the device's peak electric field is impacted by a decrease in oxide thickness. It has been discovered that the V-shaped nanosheet FETs gate greatly increases the projected transconductance with $g_m$=0.195 ms/μm, making the device a great option for 6G real-time applications. This work shows that high transconductance nanosheet FET devices yield high speed performance with minimal error and low threshold voltage $V_T$=3.37 V. Results from TCAD simulations agree well with analytical modeling.

**AI disclosure statement:** The author acknowledges the use of AI-based tools for minor language editing or grammatical correction. However, all intellectual contributions, analyses, and conclusions presented in this chapter are entirely the author's own.

## Bibliography


1. Seoane, N., Fernandez, J.G., Kalna, K., Comesana, E., Garcia-Loureiro, A., Simulations of statistical variability in n-type FinFET, nanowire, and Nanosheet FETs. *IEEE Electron Device Lett.*, *42*, 10, 1416–1419, 2021.
2. Elwailly, A., Saltin, J., Gadlage, M.J., Wong, H.Y., Radiation hardness study of L g= 20 nm FinFET and nanowire SRAM through TCAD simulation. *IEEE Trans. Electron Devices*, *68*, 5, 2289–2294, 2021.
3. Kim, M., Kim, S., Lee, K., Lee, J.H., Park, B.G., Kwon, D., Suppression of Statistical Variability in Stacked Nanosheet Using Floating Fin Structure. *IEEE Electron Device Lett.*, *42*, 11, 1580–1583, 2021.
4. Gupta, M.K., Weckx, P., Schuddinck, P., Jang, D., Chehab, B., Cosemans, S., Dehaene, W., A comprehensive study of Nanosheet and forksheet SRAM for beyond N5 node. *IEEE Trans. Electron Devices*, *68*, 8, 3819–3825, 2021.
5. Lee, R., Lee, J., Lee, K., Kim, S., Ahn, H., Kim, S., Park, B.G., Vertically-stacked Si 0.2 Ge 0.8 Nanosheet tunnel FET with 70 mV/dec average sub-threshold swing. *IEEE Electron Device Lett.*, *42*, 7, 962–965, 2021.
6. Nagy, D., Espineira, G., Indalecio, G., Garcia-Loureiro, A.J., Kalna, K., Seoane, N., Benchmarking of FinFET, Nanosheet, and nanowire FET architectures for future technology nodes. *IEEE Access*, *8*, 53196–53202, 2020.
7. Sakib, F., II, Hasan, M.A., Hossain, M., Exploration of negative capacitance in gate-all-around Si Nanosheet transistors. *IEEE Trans. Electron Devices*, *67*, 11, 5236–5242, 2020.
8. Das, U.K. and Bhattacharyya, T.K., Opportunities in device scaling for 3-nm node and beyond: FinFET versus GAA-FET versus UFET. *IEEE Trans. Electron Devices*, *67*, 6, 2633–2638, 2020.
9. Yoon, J.S. and Baek, R.H., A novel sub-5-nm node dual-workfunction folded cascade Nanosheet FETs for low power mobile applications. *IEEE Access*, *8*, 196975–196978, 2020.
10. Sung, P.J., Chang, S.W., Kao, K.H., Wu, C.T., Su, C.J., Cho, T.C., Chao, T.S., Fabrication of vertically stacked Nanosheet junctionless field-effect transistors and applications for the CMOS and CFET inverters. *IEEE Trans. Electron Devices*, *67*, 9, 3504–3509, 2020.
11. Dasgupta, A., Parihar, S.S., Agarwal, H., Kushwaha, P., Chauhan, Y.S., Hu, C., Compact model for geometry dependent mobility in Nanosheet FETs. *IEEE Electron Device Lett.*, *41*, 3, 313–316, 2020.

12. Thoti, N., Li, Y., Kola, S.R., Samukawa, S., New proficient ferroelectric Nanosheet line tunneling FETs with strained SiGe through scaled n-epitaxial layer, in: *2020 IEEE 20th International Conference on Nanotechnology (IEEE-NANO)*, IEEE, pp. 319–322, 2020, July.
13. Ehteshamuddin, M., Loan, S.A., Rafat, M., Stacked Nanosheet Based Reconfigurable FET, in: *2020 32nd International Conference on Microelectronics (ICM)*, IEEE, pp. 1–4, 2020 December.
14. Preethi, S. and Balamurugan, N.B., Analytical modeling of surrounding gate Junctionless MOSFET using finite differentiation method. *Graphene*, 13, 9, 2921–2931, 2021.
15. Manikandan, S. and Balamurugan, N.B., The improved RF/stability and linearity performance of the ultrathin-body Gaussian-doped junctionless FinFET. *J. Comput. Electron.*, 19, 2, 613–621, 2020.
16. Dash, T.P., Dey, S., Mohapatra, E., Das, S., Jena, J., Maiti, C.K., Vertically-stacked Graphene Nanosheet field effect transistors at 3nm technology nodes, in: *2019 Devices for Integrated Circuit (DevIC)*, IEEE, pp. 99–103, 2019, March.
17. Jeong, J., Yoon, J.S., Lee, S., Baek, R.H., The Effects of Realistic U-shaped Source/Drain on DC/AC Performances of Graphene Nanosheet FETs for Sub 5-nm Node SoC Applications, in: *2019 Electron Devices Technology and Manufacturing Conference (EDTM)*, IEEE, pp. 133–135, 2019, March.
18. Vardhan, P.H., Ganguly, S., Ganguly, U., Threshold voltage variability in Nanosheet GAA transistors. *IEEE Trans. Electron Devices, 66*, 10, 4433–4438, 2019.
19. Yoon, J.S., Jeong, J., Lee, S., Baek, R.H., Sensitivity of source/drain critical dimension variations for sub-5-nm node fin and Nanosheet FETs. *IEEE Trans. Electron Devices, 67*, 1, 258–262, 2019.
20. Bufler, F.M., Ritzenthaler, R., Mertens, H., Eneman, G., Mocuta, A., &Horiguchi, N., Performance Comparison of ${n} $–Type Si Nanowires, Nanosheets, and FinFETs by MC Device Simulation. *IEEE Electron Device Lett., 39*, 11, 1628–1631, 2018.
21. Yoon, J.S., Jeong, J., Lee, S., Baek, R.H., Multi $V_{th}$ Strategies of 7-nm node Nanosheet FETs With Limited Nanosheet Spacing. *IEEE J. Electron Devices Soc., 6*, 861–865, 2018.
22. Kim, J., Lee, J.S., Han, J.W., Meyyappan, M., Single-event transient in FinFETs and Nanosheet FETs. *IEEE Electron Device Lett., 39*, 12, 1840–1843, 2018.
23. Cai, L., Chen, W., Du, G., Zhang, X., Liu, X., Layout design correlated with self-heating effect in stacked Nanosheet transistors. *IEEE Trans. Electron Devices, 65*, 6, 2647–2653, 2018.
24. Lee, S., Cheng, C.W., Sun, X., D'Emic, C., Miyazoe, H., Frank, M.M., Leobandung, E., High performance InGaAs gate-all-around Nanosheet FET on Si using template assisted selective epitaxy, in: *2018 IEEE International Electron Devices Meeting (IEDM)*, IEEE, pp. 39–5, 2018, December.

25. Vimala, P. and Balamurugan, N.B., Modeling and simulation of nanoscale tri-gate MOSFETs including quantum effects. *J. Semicond.*, *35*, 3, 034001, 2014.
26. Sicard, E. and Trojman, L., Introducing 3-nm Nano-Sheet FET technology in Microwind, 2021.
27. Chen, W., Cai, L., Wang, K., Zhang, X., Liu, X., Du, G., Self-heating induced variability and reliability in Nanosheet-FETs based SRAM, in: *2018 IEEE International Symposium on the Physical and Failure Analysis of Integrated Circuits (IPFA)*, IEEE, pp. 1–4, 2018, July.
28. Jang, D., Yakimets, D., Eneman, G., Schuddinck, P., Bardon, M.G., Raghavan, P., Mocuta, A., Device exploration of Nanosheet transistors for sub-7-nm technology node. *IEEE Trans. Electron Devices*, *64*, 6, 2707–2713, 2017.
29. Pandian, M.K., Balamurugan, N.B., Manikandan, S., Analytical modeling of junctionless surrounding gate Graphene nanowire transitors. *J. Nanoelectron. Optoelectron.*, *9*, 4, 468–473, 2014.
30. Song, Y.S., Tayal, S., Rahi, S.B., Kim, J.H., Upadhyay, A.K., Park, B.-G., Thermal-aware IC chip design by combining high thermal conductivity materials and GAA MOSFET, in: *2022 5th International Conference on Circuits, Systems and Simulation (ICCSS)*, pp. 135–140, IEEE, 2022.
31. Smaani, B., Paras, N., Rahi, S.B., Song, Y.S., Yadav, R., Tayal, S., Impact of the self-heating effect on nanosheet field effect transistor performance, 2, 021005, 2023.
32. Song, Y.S., Thoutam, L.R., Tayal, S., Rahi, S.B., Arun Samuel, T.S., eds. Handbook of Emerging Materials for Semiconductor Industry. Springer, 2024.
33. Jung, S.-G., Jang, D., Min, S.-J., Park, E., Yu, H.-Y., Performance Analysis on Complementary FET (CFET) Relative to Standard CMOS With Nanosheet FET, in: *IEEE Journal of the Electron Devices Society*, vol. 10, pp. 78–82, 2022, doi: 10.1109/JEDS.2021.3136605.
34. Fengben, X., Sharma, H., Wu, X., Schram, T., Cott, D., Grubbs, R.K., Kumar, P., Ngo, T.D., Ghosh, S., Han, F., Van Dorp, D., Verreck, D., Sutar, S., Smets, Q., Lin, D., Banerjee, K., De La Rosa, C.J.L., Goux, L., Kar, G.S., Integration of GAA Monolayer MoS2 Nanosheet FETs with Gate First Process for Future 2D CFET Scaling, *2024 IEEE European Solid-State Electronics Research Conference (ESSERC)*, Bruges, Belgium, 2024, pp. 121–124, doi: 10.1109/ESSERC62670.2024.10719449.
35. Lee, R., *et al.*, Vertically-Stacked Si0.2Ge0.8 Nanosheet Tunnel FET With 70 mV/Dec Average Subthreshold Swing, in: *IEEE Electron Device Letters,* vol. 42, no. 7, pp. 962–965, July 2021, doi: 10.1109/LED.2021.3079246.

# 15

# Design and Performance Analysis of 10 T and 28 T Full Adder Using MOSFET and FinFET

**Varsha Jayaprakash[1], Boggarapu Lokesh[1], Aishwarya K.[2] and B. Lakshmi[3]***

*[1]School of Electronics Engineering, VIT, Chennai, India*
*[2]Sathyabama Institute of Science and Technology, Chennai Campus, Chennai, India*
*[3]Centre for Nano Electronics & VLSI Design and School of Electronics Engineering, VIT, Chennai, India*

## *Abstract*

In this study, the design and performance analysis of 1-bit full adder circuit with MOSFET and FinFET using 28 T and 10 T Gate Diffusion Input (GDI) technique is carried out. The three technology nodes, 180, 90, and 45 nm are taken into consideration for constructing MOSFET-based 1-bit full adder circuits. A 32 nm FinFET-based 1-bit full adder circuit with 28 T and 10 T GDI is also designed and analysed for its functionality using timing diagram. The behavior of both MOSFET- and FinFET-based full adder circuits are investigated by varying the parameters, voltage and operational frequency. The performance of the circuit is assessed using the metrics, power consumption, and propagation delay. It is observed that 10 T GDI-based FinFET full adder shows an average power consumption of 11.97 nW at 0.5 V with 5 GHz operational frequency at a propagation delay of 66.67 ps. MOSFET-based full adder at 45 nm technology node shows a stable output with a power consumption of 176 nW at 0.6 V, 5 GHz with a propagation delay of 0.244 ns. It is also observed from the simulation results that 10 T GDI FinFET-based full adder is more superior than 28 T FinFET full adder in terms of both average power consumption and propagation delay. It is also understood that MOSFET-based full adder are not advantageous for low-power applications as they consume more power and propagation delay.

***Keywords*:** 28 T full adder, 10 T GDI full adder, FinFET, average power consumption, propagation delay

**Corresponding author*: lakshmi.b@vit.ac.in

P. Suveetha Dhanaselvam, Naima Guenifi, Shiromani Balmukund Rahi and J. Ajayan (eds.) Field-Effect Transistor Devices: Post-CMOS Theory and Applications, (293–316) © 2026 Scrivener Publishing LLC

## 15.1 Introduction

In recent times the need for ultra-low power consuming circuits have witnessed a sudden surge in VLSI applications in order to overcome the problems faced due to leakage current, reliability and thermal management. The reduced chip size leads to increased chip density thereby increasing the fabrication complexity. The basic requirement for any portable device to have a good performance is to have lower leakage and dynamic power consumption [1]. The ability of the device to operate under low voltages and the infinitesimal size are explored to keep the energy consumption minimum [2]. Although MOSFETs are widely used in digital and analog circuitry, there are certain conventional limitations associated with these devices that limit their complete utilization. Scaling down the size of MOSFET below 45 nm will lead to short channel effects [3, 4]. Other issues like threshold voltage roll-off, velocity saturation, gate depletion effect exists due to scaling [5]. FinFET, a potential alternative to MOSFET consists of two controlled gates that will enable to reduce leakage current and various short channel effects. Due to a reduced switching voltage and an increase in drain current, FinFETs have 90% less static leakage current and faster switching [6, 7].

The application of low voltage, cost effective back gate forward bias method in a CMOS inverter exhibit very low power, reasonable speed and improved precision without encountering an apparent decrease in switching speed [8]. A soft error rate (SER) in flip flops using FinFETs show that the alpha and neutron irradiation tests produce lesser αSER and nSER respectively as the creation and collection of charges is decreased by fin size [9]. Simulation of 1-bit full adder using 8 T gate work function engineering (WFE) shows that the average power dissipation is lesser compared to 28 T, 10 T adder and complementary and level restoring carry adder (CLRCL). It occupies 50% lesser total area but performed poorly compared to other designs in terms of delay [10–12]. Implementation of a novel multiplexer-based adder (MBA) with 12 transistors and CPL (complementary pass-transistor logic) along with some preexisting techniques such as 28 T full adder, static energy recovery full adder (SERF), 10T09A, 10T09B, and 10T13A outperformed the 28 T CMOS full adder as they consume less power and faster which makes it suitable to build low power high-performance designs [13, 14]. The analysis and study of 2-bit and 8-bit SERF ripple carry adders proved that they were superior to TFA (Transmission Function Adder) and the dual-value logic (DVL) adder in terms of power dissipation and makes it a suitable option for low power designs [15–17].

In this work, 1-bit MOSFET- and FinFET-based full adder is implemented for conventional 28 T and Gate Diffusion Input (GDI) 10 T adder circuit topologies. The power consumption and propagation delay of these circuits are extracted by varying the operational voltage and frequency. The rest of the paper is organized as follows: Section 15.2 describes the circuit description and simulation methodology. Section 15.3 gives the discussion of various results obtained for the adder circuit configurations. Finally, conclusion is given in Section 15.4.

## 15.2 Materials and Methods

As it is known that a full adder is a digital logic combinational circuit that is used to perform addition. The input and output terminals are represented by A, B, CIN and sum, carry respectively [18]. The logic function of a 1-bit full adder can be represented using the following equations. Figure 15.1 shows the logic diagram of 1- bit full adder. Table 15.1 gives the corresponding truth table of the adder.

$$SUM = (A\,XOR\,B)\ XOR\,CIN = (A \oplus B) \oplus CIn \tag{15.1}$$

$$CARRY = A\ AND\ B\ OR\ CIN\ (A\ XOR\ B) = A.B + CIN(A \oplus B) \tag{15.2}$$

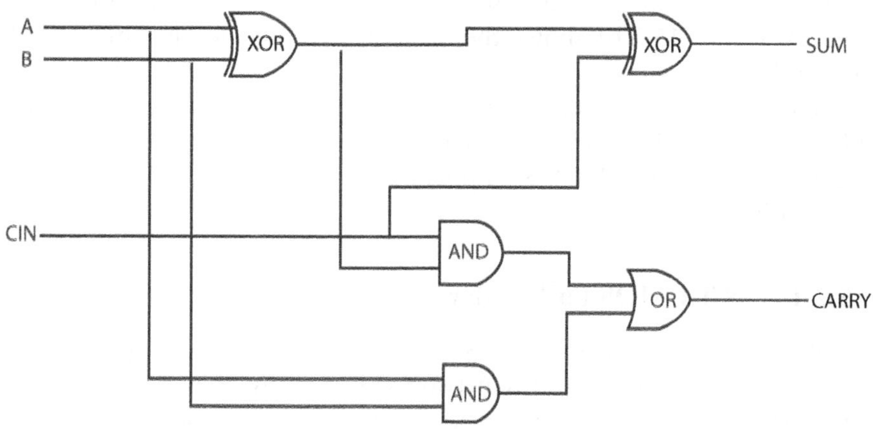


**Figure 15.1** 1- bit Full adder logic diagram.

**Table 15.1** Truth table of full adder.

| A | B | CIN | SUM | CARRY |
|---|---|---|---|---|
| 0 | 0 | 0 | 0 | 0 |
| 0 | 0 | 1 | 1 | 0 |
| 0 | 1 | 0 | 1 | 0 |
| 0 | 1 | 1 | 0 | 1 |
| 1 | 0 | 0 | 1 | 0 |
| 1 | 0 | 1 | 0 | 1 |
| 1 | 1 | 0 | 0 | 1 |
| 1 | 1 | 1 | 1 | 1 |

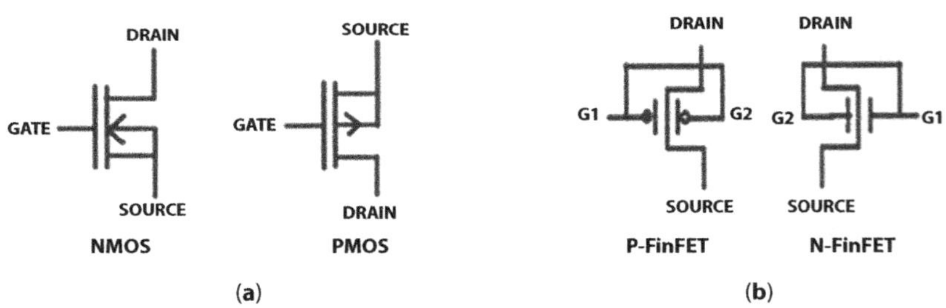


**Figure 15.2** (a) Symbol of MOSFET and (b) Symbol of FinFET.

The circuit symbol of n-MOSFET and p-MOSFET are shown in Figure 15.2(a). Figure 15.2(b) shows the symbol generated for n-FinFET and p-FinFET.

In the following subsection, the circuit diagram and logic verification of all the four full adders are discussed. The adders considered in this study are 28 T full adder (CMOS & FinFET) and 10 T GDI full adder (CMOS and FinFET).

### 15.2.1 28 T CMOS Full Adder

A full adder design using CMOS structure is constructed using PMOS in pull-up and NMOS in pull-down networks respectively. They are arranged in a manner formed by two complementary networks which consists of 2 XOR, 2 AND, and 1 OR gate constructed using PMOS and NMOS which is used

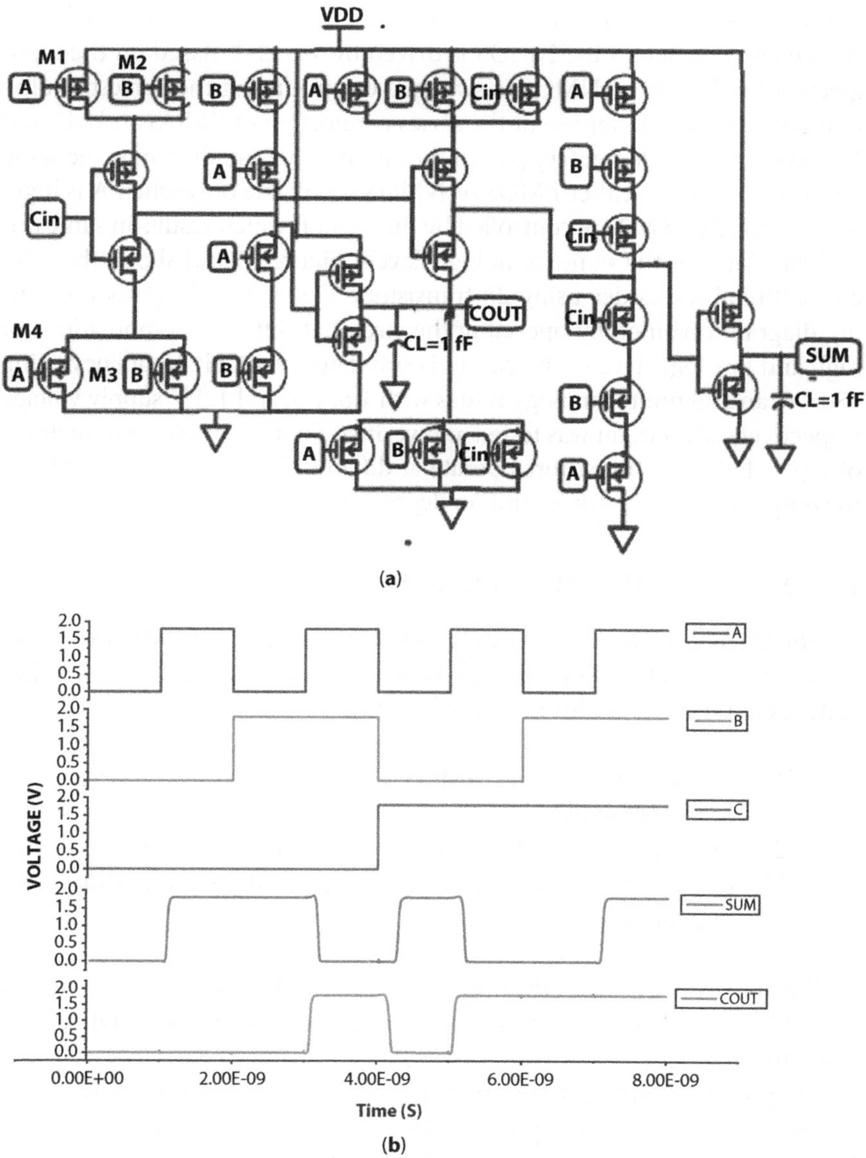


**Figure 15.3** (a) Schematic of 28 T CMOS full adder (b) Timing diagram of 28 T CMOS full adder.

to implement the complementary logic [19, 20]. The PMOS is driven by the source ($V_{DD}$), whereas the NMOS is driven by $V_{SS}$ [21]. Based on the input given either PMOS or NMOS is switched on producing the corresponding outputs. M1 and M2 represent the series connection of PMOS and M3 and M4 gives the complementary pair of the circuit. They are driven by the same input A, turning on either PMOS or NMOS is possible only when A is high. Similarly, M2 and M3 are controlled by the input B which results in same on/off behaviour for the same input logic levels. Figure 15.3(a) shows the 1-bit conventional full adder using 28 transistors. Figure 15.3(b) gives the timing diagram obtained by operating the circuit at different combinations of high and low logic levels. The circuit is simulated at 1 GHz frequency using 180, 90, and 45 nm technology nodes with 1.8, 1.5, and 1.2 V supply voltage respectively. The circuit was tested and verified for all possible combinations of input. Different transistors operate at different logic levels due to which the output is strongly driven and no degradation of levels takes place [21].

### 15.2.2 10 T GDI CMOS Full Adder

A fundamental GDI cell consists of one NMOS and one PMOS together which forms a GDI multiplexer cell which is shown in Figure 15.4. The key features of GDI cell are listed below [22, 23].

(1) For both PMOS and NMOS, the output acts as a shorted drain terminal.
(2) Both NMOS and PMOS shares the common input G.
(3) P and N act as the source input to the PMOS and NMOS, respectively.

The above-mentioned advantages of a GDI cell are exploited to design full adder circuits with reduced transistors. The functions of GDI cell are given in Table 15.2.

The proposed circuit design of full adder using 10 transistors is implemented and is shown in Figure 15.5 (a). It can be seen that Figure 15.5(a) is designed using 1 XOR, 1 XNOR, 2 MUX, and 2 inverters. The transistors M1 and M2 are used to invert the input given at B which is in turn used to perform XOR operation with A using the M3 and M4. The XOR and inverted XOR outputs are fed to M7, M8 and M9, M10 which are then used to multiplex these signals with $C_{IN}$ and produce sum and carry outputs, respectively [23]. Figure 15.5(b) shows the timing diagram of a 10 T GDI CMOS full adder. It can be seen that the logic of 1-bit full adder has been verified for 10 T GDI CMOS full adder also. A drastic increase/decrease

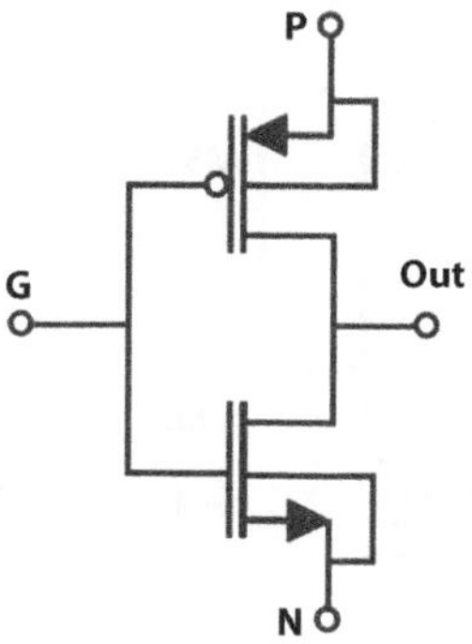


**Figure 15.4** Basic GDI cell.

**Table 15.2** Functions of GDI cell.

| NN | PP | GG | Output | Function |
|---|---|---|---|---|
| 0 | 1 | A | A' | INVERTER |
| 0 | B | A | A'B | F1 |
| B | 1 | A | A'+B | F2 |
| 1 | B | A | A+B | OR |
| B | 0 | A | AB | AND |
| C | B | A | A'B+A'C | MUX |
| B' | B | A | A'B+B'A | XOR |
| B | B' | A | AB+A'B' | XNOR |

in drain current when CMOS is turned on or off during transient analysis leads to increased glitches [24]. The timing diagram was obtained for all combinations of input and was verified.

### 15.2.3 28 T FinFET Full Adder

To enhance the performance of the adder circuit as discussed, CMOS has been replaced with FinFET due to the inherent advantages of the latter. Figure 15.6(a) depicts 28 T FinFET-based full adder and Figure 15.6(b) gives the corresponding timing diagram. It can be observed from Figure 15.6(b) that 28 T full adder works perfectly for all inputs and produces a glitch-free output.

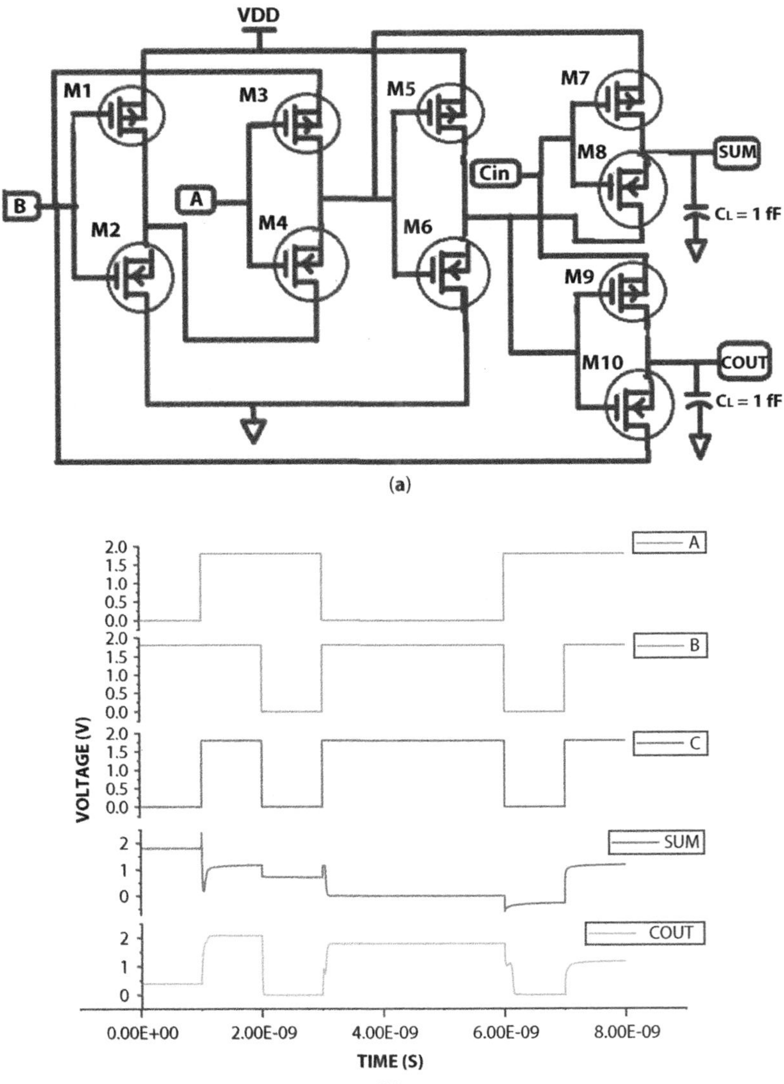


**Figure 15.5** (a) Schematic of 10 T GDI CMOS full adder (b) Timing diagram of 10 T GDI CMOS full adder.

### 15.2.4 10 T GDI FinFET Full Adder

The working of 10 T GDI FinFET full adder is similar to 10 T GDI CMOS full adder. In FinFET 10 T GDI adder, multiplexers are considered to be a selective input and are used to invert the order to obtain the carry expression.

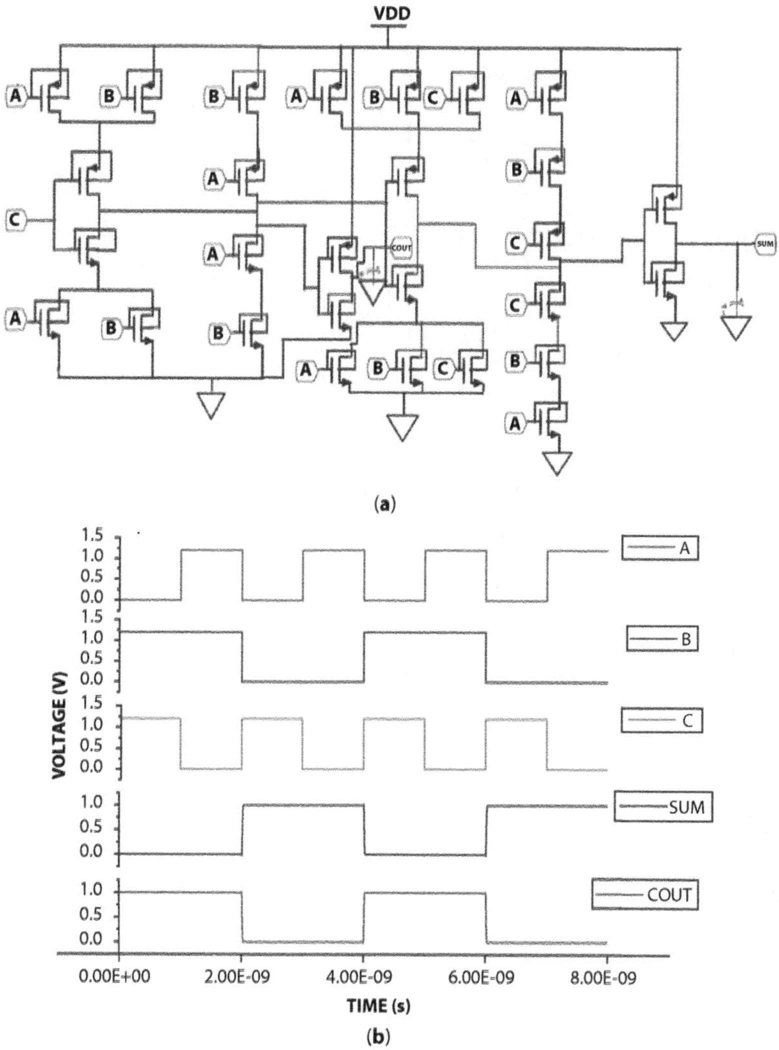


**Figure 15.6** (a) Schematic of 28 T FinFET full adder (b) Timing diagram of 28 T FinFET full adder.

A 10 T full adder designed using GDI technique is shown in Figure 15.7(a) and the corresponding timing diagram is given in Figure 15.7(b).

It can be observed from Figure 15.7 (b) that a 10 T GDI FinFET full adder works as a perfect full adder by producing all the possible outputs. However, compared to 10 T GDI CMOS full adder this circuit produces

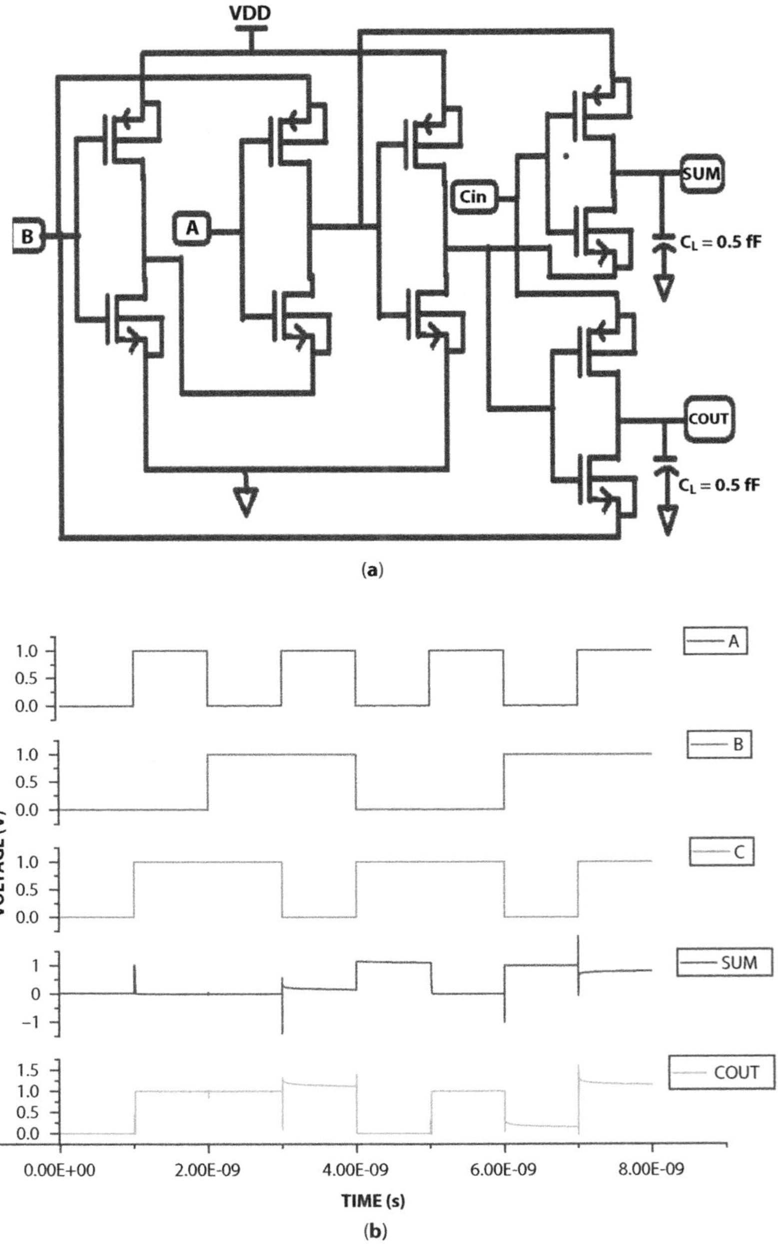


**Figure 15.7** (a) Schematic design of 10 T GDI FinFET full adder (b) Timing diagram of 10 T GDI FinFET full adder.

less stable output with increased glitches due to the XOR gate which has a threshold loss at lower supply voltages [25].

### 15.2.5 Simulation Methodology

Cadence Virtuoso is used for all circuit simulations [26]. The power consumption and propagation delay are assessed for CMOS- and FinFET-based adders and the results are presented. After the circuit is constructed, the ADEL window is opened and transient analysis is selected to calculate average power consumption. To preserve every PWR signal in the circuit, the "save all" option is used. To choose the stored psf file, open the results browser from the simulation waveform window. After choosing option trans from the results browser, a calculator receives the pwr signal and calculates its average. Delay is defined as the amount of time needed for a signal to climb from 10% to 50% of the supply. The voltage limits are set at 10% and 50% of the supply voltage, the signals to be assessed are chosen, and the expression derived from the delay functional panel is calculated using a calculator [27].

## 15.3 Results and Discussion

CMOS-based circuits are simulated using 180 nm, 90 nm and 45 nm technology nodes. FinFET is simulated using 32 nm technology node. The transient response for the different designs is obtained for all possible combinations of input and the average power consumption and time delays are extracted and studied against published results [28–30]. The performance of the proposed designs is studied by varying three properties: technology node, operating voltage and frequency. The results of average power consumption and propagation delay obtained after simulation are explained in this section. As discussed, 28 T CMOS and 10 T GDI CMOS full adder are constructed using the three technology nodes: 180 nm, 90 nm and 45 nm at 500 MHz, 1 GHz, 2.5 GHz and 5 GHz frequency.

The variation in voltage and temperature are important parameters to evaluate the performance of digital circuits. Here the circuit is tested in different conditions by varying the frequency from 500 MHz to 2.5 GHz, it is found that the results start to deteriorate beyond the upper limit of frequency (2.5 GHz), so in order to ensure the correct logical output and minimize errors; the simulations are performed within the specified range [31]. The frequency of operation is varied in order to obtain optimal results and improved efficiency with respect to power consumption, propagation delays and the minimum voltage of operation at a desired frequency [32–34]. It is observed that the designs have better output swing and other parameters at higher frequencies [35]. The frequency of the designs is varied by varying the frequency of the individual inputs of the full adder circuit A, B, and C in, respectively, during transient analysis.

To find the best operating parameters for the circuit, the power consumption of the full adders has been calculated for different operational frequency and voltages. The frequency is shifted from 500 MHz to 5 GHz, and the voltage varies from 1.2 to 0.5 V. Equation (15.3) shows the dynamic power consumption which is exactly proportional to the square of the voltage and the operating frequency. As the operating voltage and frequency rise, power consumed by the circuit also increases. The clock speed is directly related to the frequency of the clock and is defined by the number of switching bits (NSW), as stated in the following equation.

$$P_d = \left[(C_{pd} \times f_I \times N_{SW}) + \Sigma(C \times f)\right] \times V_{cc}^{\ 2} \tag{15.3}$$

where $C_{pd}$ is the power consumption capacitance, $f_I$ and f are the input and output frequencies respectively, $N_{SW}$ is the total number of switching bits, C is the load capacitance and $V_{CC}$ is the supply voltage [36].

The power dissipation is the loss of power produced when the circuit is operating. It is useful in determining the weight and size of the device and the power required for cooling effect [37]. This is an undesirable factor for the study of variation of frequency and voltage to examine various parameters with the goal of reducing power consumption and hence power dissipation is not taken into account for this study.

Figures 15.8 and 15.9 give the plots of average power consumption versus voltage for 28 T and 10 T GDI-based CMOS full adder for various technology nodes. It can be inferred from Figures 15.8 & 15.9 that the average power varies almost linearly with voltage for the adder circuits. The average power approximately doubles with increase in frequency. The decrease of 33.4% power is obtained at 45 nm 10 T GDI CMOS full adder compared to 28 T CMOS. However, 10 T CMOS full adder fails to produce better results than 28 T CMOS full adder by consuming an average power with an increase of 72% and 39% at 90 nm and 180 nm respectively. Previous results show that a decrease of 57% in average power is observed in GDI 10 T CMOS full adder compared to 28 T CMOS full adder [38]. 28 T CMOS full adder performs better at lower frequencies and at higher frequencies perfect logic levels are not obtained. Due to larger number of transistors,

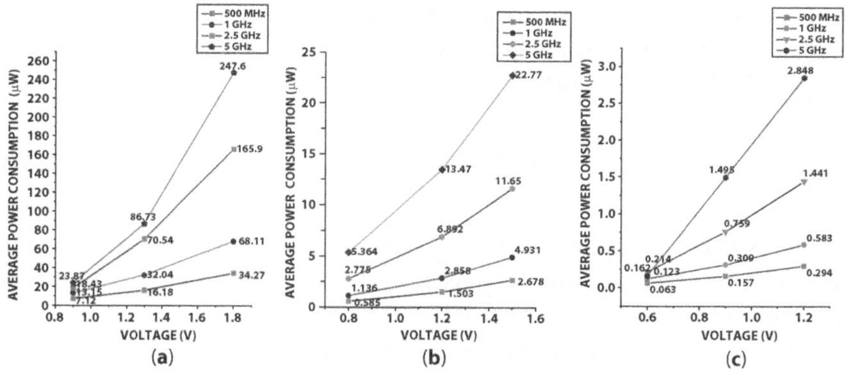


**Figure 15.8** Average power consumption vs voltage for 28 T CMOS-based full adder (a) 180 nm, (b) 90 nm (c) 45 nm.

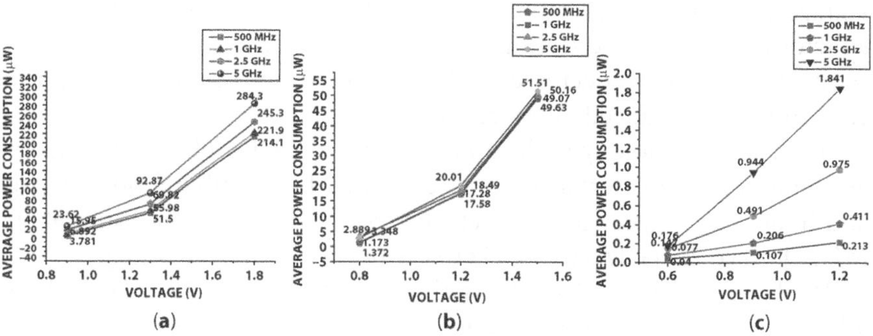


**Figure 15.9** Average power consumption vs voltage for 10 T GDI-based CMOS full adder (a) 180 nm, (b) 90 nm (c) 45 nm.

the power consumption and noise margin are high compared to other low power designs [39, 40].

The power does not substantially increase below 45 nm, signifying that CMOS is a better choice for developing small scale circuit designs. Because of the reduced transistor count, the switching node capacitance is decreased which produces a remarkable reduction in power is obtained in 10 T CMOS full adder at 45 nm [41]. However, full swing logic levels are not obtained in the design implemented using CMOS.

The circuit's power consumption is defined by its logic, technology node, implementation method, voltage and operating frequency. By using a new technical technique and operating closer to threshold voltage is the most direct way to reduce the power consumption of any circuit [42]. A FinFET with a 48 nm width and 32 nm length is used in the circuit. Equation (15.4) explains that a drop in current results from the reduced size in the nm range, which decreases power consumption.

$$P = N \times V_{dd} \times I \times K_{design} \tag{15.4}$$

where, N is the transistor count, $V_{dd}$ is the supply voltage, the leakage current is represented by I and $K_{design}$ is an empirical factor focused on transistor size and design architecture [43].

The scaling theory states that a circuit should run at a lower supply voltage when it moves to subsequent technological generation, which reduces the power consumption [44]. To regulate the current flow between the drain and source, the transistor modifies the electric field across the substrate underneath the gate. Reducing the device's physical size alters the voltage and doping levels by a factor of α. While the capacitance reduces by a factor of α during the scaling process, the trans-conductance stays constant. Because of the decreased voltage and current in the device, power consumption is reduced by a factor of $\alpha^2$ [45].

At 32 nm technology node, a FinFET-based 28 T and 10 T full adders are built, and their performance is evaluated by modifying both voltage and frequency. Figure 15.10 reflects the results of simulation results at 500 MHz, 1 GHz, 2.5 GHz and 5 GHz frequency with variable voltage.

The least power of 0.163 μW is obtained at 0.5 V for 28 T FinFET full adder at 1 GHz operational frequency and 2.413 nW at 0.5 V, 1 GHz for

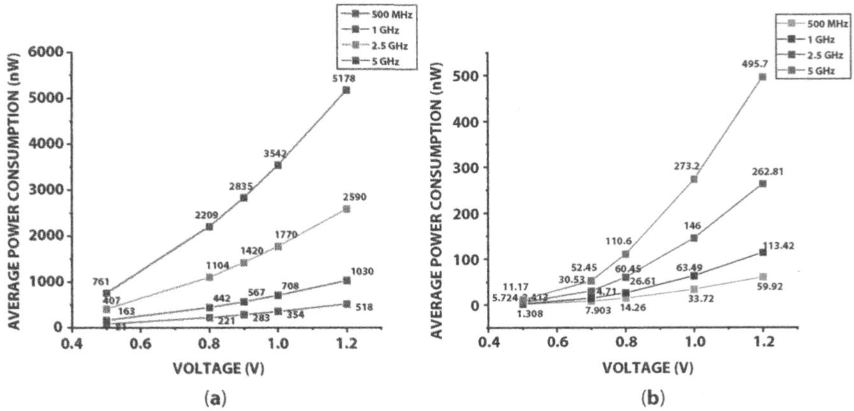


**Figure 15.10** Average power consumption vs voltage (a) 28 T FinFET full adder and (b) 10 T GDI FinFET full adder.

10 T GDI FinFET full adder. FinFET-based 10 T full adder shows better performance by consuming lesser power compared to 28 T full adder. Previous researchers have achieved an average power of 13.43 nW at 0.5 V using GDI 10 T design [46]. Due to smaller size, suppressed short channel effects and lower mobility degradation in FinFET the static power consumption is lesser compared to CMOS [47].

A 10 T full adder's average power consumption is significantly lower than those of a 28 T full adder. 10 T GDI FinFET full adder is more productive in terms of chip size and area due to the lesser number of transistor count used to construct XOR gates. 10 T GDI FinFET full adder consumes approximately 92.5%, 92.4%, 93%, and 93.5% less power than the 28 T FinFET full adder at 500 MHz, 1 GHz, 2.5 GHz and 5 GHz respectively proving that FinFET-based circuits are more suitable for energy efficient designs and are capable of operating at higher frequencies. The power dissipation of FinFET circuits are obtained at multiple voltages and frequencies and plotted in Figure 15.11.

It can be compared from Figure 15.10 and Figure 15.11 that the power dissipation is nearly $10^3$ times less than the power consumption of 10T GDI FinFET circuit whereas it is $10^6$ times less than that of 28T FinFET circuits. Power dissipation is the power during the idle state of the CMOS circuit which is produced due to a leakage current that flows through the OFF transistors. The NFET and PFET in the circuit dissipate reverse leakage current and subthreshold current which is comparable to power dissipation when multiple transistors are present [48, 49]. This is very less in comparison to the power consumed during the operation of the circuit.

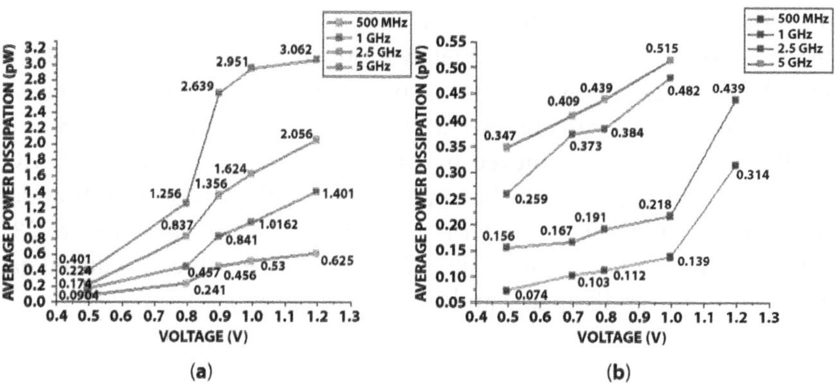


**Figure 15.11** Average power dissipation vs voltage (a) 28 T FinFET full adder and (b) 10 T GDI FinFET full adder.

Hence, it can be stated that power dissipation is very less compared to power consumption.

Tables 15.3 & 15.4 give the performance comparison of power consumption and dissipation of 28T & 10T FinFET full adders respectively.

By varying the technology node, operating voltage, and frequency, the propagation delay of CMOS is evaluated and the same technology node is used for FinFET. The operating voltage, and clock speed impact the circuit's propagation delay as the clock speed and operating frequency are inversely related to each other. With increase in clock speed i.e. decrease in frequency; the circuit would perform slower and would contribute to more delay whereas, the propagation delay and voltage are indirectly related as mentioned in Equation 15.5.

$$T_p = C_L / \left( \mu_n \times C_{OX} \times (W / L) \times V_{DD} \right) \tag{15.5}$$

The FinFET's width to length ratio, operating voltage, and capacitance all affect the propagation delay. Since the delay is inversely proportional to $V_{DD}$, it increases with a reduction in voltage and vice versa while the other parameters remain constant. The propagation delay of all the above proposed designs is plotted in Figure 15.12 and Figure 15.13.

It can be inferred from the Figures 15.12 & 15.13 that the propagation delay decreases with increase in frequency; hence a decreasing curve is obtained in the plot. 10 T GDI CMOS full adder is approximately 17% faster than the 28 T CMOS full adder at 45 nm, 90 nm and 180 nm technology nodes, demonstrating that the GDI 10 T CMOS full adder provides faster response compared to 28 T CMOS full adder. 10 T FinFET full adder performs 75% faster than 28 T FinFET full adder. As seen from Figure 15.13, the propagation delay for both sum and carry of FinFET are nearly identical. Lower operational frequency is preferable for both 28 T and GDI FinFET full adder due to the minimal power obtained. Higher frequency is more suitable in order to obtain a faster response. A reduction by 1.8 times in GDI full adder was achieved by early researches compared to CMOS full adders and is proved to be superior than other existing full adders [50]. GDI 10 T FinFET full adder is again more efficient compared to 28 T FinFET, 28 T CMOS and 10 T GDI CMOS full adders due to reduced gate length, area, power dissipated and speed of operation.

**Table 15.3** Comparison of power consumption and dissipation of 28T FinFET full adder.

| Voltage (V) | Frequency 500 MHz | | Frequency 1 GHz | | Frequency 2.5 GHz | | Frequency 5 GHz | |
|---|---|---|---|---|---|---|---|---|
| | Power consumption (µW) | Power dissipation (pW) | Power consumption (µW) | Power dissipation (pW) | Power consumption (µW) | Power dissipation (pW) | Power consumption (µW) | Power dissipation (pW) |
| 0.5 | 0.081 | 0.0904 | 0.163 | 0.174 | 0.407 | 0.224 | 0.761 | 0.401 |
| 0.8 | 0.221 | 0.241 | 0.442 | 0.457 | 1.104 | 0.837 | 2.209 | 1.256 |
| 0.9 | 0.283 | 0.456 | 0.567 | 0.841 | 1.42 | 1.356 | 2.835 | 2.639 |
| 1 | 0.354 | 0.53 | 0.708 | 1.0162 | 1.77 | 1.624 | 3.542 | 2.951 |
| 1.2 | 0.518 | 0.625 | 1.03 | 1.401 | 2.59 | 2.056 | 5.178 | 3.062 |

**Table 15.4** Comparison of power consumption and dissipation of 10T FinFET full adder.

| Voltage (V) | Frequency 500 MHz | | Frequency 1 GHz | | Frequency 2.5 GHz | | Frequency 5 GHz | |
|---|---|---|---|---|---|---|---|---|
| | Power consumption (nW) | Power dissipation (pW) | Power consumption (nW) | Power dissipation (pW) | Power consumption (nW) | Power dissipation (pW) | Power consumption (nW) | Power dissipation (pW) |
| 0.5 | 1.308 | 0.074 | 2.413 | 0.156 | 5.724 | 0.259 | 11.17 | 0.347 |
| 0.8 | 7.903 | 0.103 | 14.71 | 0.167 | 30.53 | 0.373 | 52.45 | 0.409 |
| 0.9 | 14.26 | 0.112 | 26.61 | 0.191 | 60.45 | 0.384 | 110.6 | 0.439 |
| 1 | 33.72 | 0.139 | 63.49 | 0.218 | 146 | 0.482 | 273.2 | 0.515 |
| 1.2 | 59.92 | 0.314 | 113.42 | 0.439 | 262.81 | 0.643 | 495.7 | 0.772 |

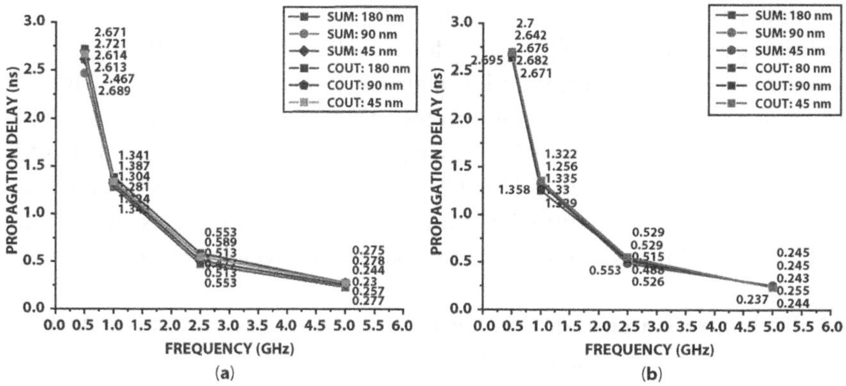


**Figure 15.12** Propagation delay for (a) 28 T CMOS-based full adder and (b) 10 T GDI CMOS full adder.

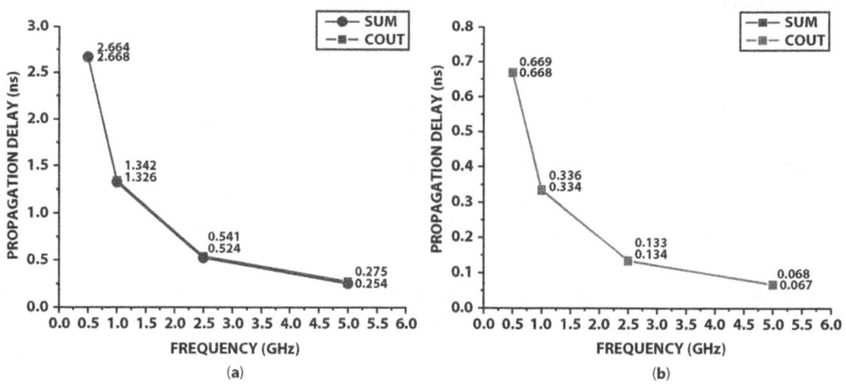


**Figure 15.13** Propagation delay for (a) 28 T FinFET full adder and (b) 10 T GDI FinFET full adder.

## 15.4 Conclusions

In this study, a 1-bit full adder circuit is built using the designs: 28 T and 10 T utilizing CMOS technology and FinFETs. A 10 T full adder adopts the GDI approach, wherein the NMOS and PMOS are integrated to form a multiplexer, and 28 T full adder follows conventional CMOS design. The technological nodes and also the operational voltage and frequency for CMOS and FinFET are varied to examine the circuits performance. With respect to average power consumption and propagation delay, simulation results showed that FinFET surpasses CMOS-based design. 10 T CMOS full adder is 33.34% power efficient than 28 T CMOS full adder at 45 nm technology node but it fails to perform well at higher technology nodes. Moreover 10 T CMOS full adder is also 17% faster than 28 T CMOS full adder at 45 nm, 90 nm and 180 nm technology nodes. FinFET-based 10 T GDI full adder consumes 93% lesser power than 28 T FinFET full adder and is 75% faster than 28 T FinFET full adder. Hence it can be concluded that FinFET-based circuits perform well at higher frequencies compared to its CMOS analogous as it offers superior control over short channel effects.

**AI disclosure statement:** The author acknowledges the use of AI-based tools for minor language editing or grammatical correction. However, all intellectual contributions, analyses, and conclusions presented in this chapter are entirely the author's own.

## References

1. Varadharajan, S.K. and Nallasamy, V., Low power VLSI circuits design strategies and methodologies: A literature review. *Conf. Emerging Devices Smart Syst. (ICEDSS)*, 245–251, 2017, doi: htpps://10.1109/ICEDSS.2017.8073688.
2. Chiang, T., A Short-Channel-Effect-Degraded Noise Margin Model for Junctionless Double-Gate MOSFET Working on Subthreshold CMOS Logic Gates. *IEEE Trans. Electron Devices*, 63, 8, 3354–3359, 2016. https://10.1109/TED.2016.2581826.
3. Franceschin, A., Andreani, P., Padovan, F., Bassi, M., Bevilacqua, A., A 19.5-GHz 28-nm Class-C CMOS VCO with a Reasonably Rigorous Result on 1/f Noise Upconversion Caused by Short-Channel Effects. *IEEE J. Solid-State Circuits*, 55, 7, 1842–1853, 2020.
4. Ning, T.H., A perspective on the theory of MOSFET scaling and its impact, in: *IEEE Solid-State Circuits Society Newsletter*, vol. 12, pp. 27–30, 2007.
5. Frank, D.J., Dennard, R.H., Nowak, R.H., Solomon, P.M., Taur, Y., Wong, H.S.P., Device scaling limits of Si MOSFETs and their application dependencies. *Proc. IEEE*, 89, 3, 259–288, 2001, doi: https://10.1109/5.915374.
6. Khanna, V.K.X., Nanoscale CMOS. Post-CMOS and Allied Nanotechnologies. NanoScience and Technology. *Integr. Nanoelectron.*, 978-81-322-3623-8, 2016. doi: https:// 10.1007/978-81-322-3625-2_5.
7. Shrivastava, A.K. and Akashe, S., Design high performance and low power 10T full adder cell using double gate MOSFET at 45nm technology. *Int. Conf. Control Computing Commun. Mater. (ICCCCM)*, 1–5, 2013, doi: https://10.1109/ICCCCM.2013.6648903.
8. Chen, M.-J., Ho, J.-S., Huang, T.-H., Yang, C.-H., Wu, T., Back-gate forward bias method for low-voltage CMOS digital circuits. *IEEE Trans. Electron Devices*, 43, 6, 904–910, 1996. https:// 10.1109/16.502122.

9. Uemura, T., Lee, S., Monga, U., Choi, J., Lee, S., Pae, S., Technology Scaling Trend of Soft Error Rate in Flip-Flops in 10 nm Bulk FinFET Technology. *IEEE Trans. Nucl. Sci.*, 65, 6, 1255–1263, 2018. https://10.1109/TNS.2018.2833875.
10. Wei, Y. and Shen, J., Design of a novel low power 8-transistor 1-bit full adder cell. *J. Zhejiang Univ. Sci. C*, 12, 604–607, 2011. https://doi.org/10.1631/jzus.C1000372.
11. Canan, T.F., Kaya, S., Kodi, A., Xin, H., Louri, A., 10T and 8T Full Adders Based on Ambipolar XOR Gates with SB-FinFETs. *25th IEEE International Conference on Electronics, Circuits and Systems (ICECS)*, pp. 577–580, 2018, doi: https://10.1109/ICECS.2018.8617893.
12. Chen, C. and Chou, Y., Novel low-power 1-bit full adder design. *9th International Symposium on Communications and Information Technology*, pp. 1348–1349, 2009, doi: https:// 10.1109/ISCIT.2009.5341063.
13. Jiang, Y., Al-Sheraidah, A., Wang, Y., Sha, E., Chung, J.-G., A novel multiplexer-based low-power full adder. *IEEE Trans. Circuits Syst.*, 51, 7, 345–348, 2004. doi: https://10.1109/TCSII.2004.831429.
14. Zimmermann, R. and Fichtner, Low-power logic styles: CMOS versus pass-transistor logic. *IEEE J. Solid-State Circuits*, 32, 7, 1079–1090, 1997. doi: https://10.1109/4.597298.
15. Shalem, R., John, E., John, L.K., A novel low power energy recovery full adder cell, in: *Proceedings Ninth Great Lakes Symposium on VLSI*, IEEE Comput. Soc Ninth Great Lakes Symposium on VLSI, pp. 380–383, 1999, https://10.1109/GLSV.1999.757363.
16. Moradi, F., Wisland, D.T., Mahmoodi, H., Aunet, S., Cao, T.V., Peiravi, A., Ultra low power full adder topologies. *IEEE Int. Symp. Circuits Syst.*, 3158–3161, 2009, doi: https://10.1109/ISCAS.2009.5118473.
17. Devadas, M. and Kishore, K.L., Design topologies for low power cmos full adder. *Int. Conf. Inventive Syst. Control (ICISC)*, 1–4, 2017, doi: https://10.1109/ICISC.2017.8068741.
18. Shrivastava, A.K. and Akashe, S., Comparative analysis of low power 10T and 14T full adder using double gate MOSFET at 45nm technology. *Int. J. Comput. Appl.*, 75, 3, 2013.
19. Nigam, A. and Singh, R., Comparative Analysis of 28T Full adder with 14T Full adder using 180nm. *Int. J. Eng. Sci. Adv. Res.*, 2, 1, 27–32, 2016.
20. Wang, D., Yang, M., Cheng, W., Xuguang, G., Zhu, Z., Yang, Y., Novel low power full adder cells in 180nm CMOS technology. *4th IEEE Conference on Industrial Electronics and Applications*, pp. 430–433, 2009, doi: https://10.1109/ICIEA.2009.5138242.
21. Tiwari, N., Sharma, R., Parihar, R., Implementation of area and energy efficient full adder cell. *International Conference on Recent Advances and Innovations in Engineering (ICRAIE)*, pp. 1–5, 2014, doi: https://10.1109/ICRAIE.2014.6909248.
22. Morgenshtein, A., Fish, A., Wagner, I.A., Gate-diffusion input (GDI) - a technique for low power design of digital circuits: analysis and characterization.

*IEEE International Symposium on Circuits and Systems, Proceedings*, vol. 1, pp. 1–1, 2002, doi: https:// 10.1109/ISCAS.2002.1009881.
23. Sanapala, K. and Sakthivel, R., Analysis of GDI logic for minimum energy optimal supply voltage. *International Conference on Microelectronic Devices Circuits and Systems (ICMDCS)*, pp. 1–3, 2017, doi: https://10.1109/ICMDCS.2017.8211582.
24. Munteanu, D., Cristoloveanu, S., Faynot, O., Transition from generation-to recombination-based transients in partially depleted SOI MOSFETs. *IEEE International SOI Conference Proceedings*, pp. 71–72, 1998, doi: https://10.1109/SOI.1998.723116.
25. Shrivas, J., Akashe, S., Tiwari, N., Design and performance analysis of 1 bit full adder using GDI technique in nanometer era. *Proceedings of IEEE World Congress on Information and Communication Technologies (WICT)*, pp. 822–825, 2012, doi: https:// 10.1109/WICT.2012.6409188.
26. Johnson, K., Retrieved from ocw.mit.edu: https://ocw.mit.edu/courses/electrical-engineering-and-computer-science/6-012-microelectronic-devices-and-circuits-fall-2005/projects/cadence_tutrial2.pdf, 2005.
27. Kumar, R., Power measurement in Cadence spectre. Retrieved from *docplayer.net*: https://docplayer.net/29371509-Power-measurement-in-cadence-spectre.html, 2015.
28. Hajare, R. and Lakshminarayana, C., Design and software characterization of FinFET based full adders. *Int. J. Reconfigurable Embedded Syst.*, 8, 1, 51, 2019, doi: https:// 10.11591/ijres.v8.i1.pp51-60.
29. Vallabhuni, R.R., Yamini, G., Vinitha, T., Reddy, S.S., Performance analysis: D-Latch modules designed using 18nm FinFET Technology. *International Conference on Smart Electronics and Communication (ICOSEC)*, pp. 1169–1174, 2020, doi: https:// 10.1109/ICOSEC49089.2020.9215341.
30. Vallabhuni, R.R., Sravya, D.V.L., Shalini, M.S., Maheshwararao, G.U., Design of Comparator using 18nm FinFET Technology for Analog to Digital Converters. *7th International Conference on Smart Structures and Systems (ICSSS)*, pp. 1–6, 2020, doi: https:// 10.1109/ICSSS49621.2020.9202164.
31. Junqi, H., Kumar, T.N., Abbas, H., Lombardi, F., Simulation-based evaluation of frequency upscaled operation of exact/approximate ripple carry adders. *IEEE International Symposium on Defect and Fault Tolerance in VLSI and Nanotechnology Systems (DFT)*, pp. 1–6, 2017, https:// 10.1109/DFT.2017.8244437.
32. Rafiee, M., Pesaran, F., Sadeghi, A., Shiri, N., An efficient multiplier by pass transistor logic partial product and a modified hybrid full adder for image processing applications. *Microelectron. J.*, 118, 105287, 2021, doi: https:// doi.org/10.1016/j.mejo.2021.105287.
33. Taheri, M., Shafiee, N., Sharifi, F., Energy efficient hybrid full adder design for digital signal processing in nanoelectronics. *Analog Integr. Circuits Signal Process.*, 109, 135–151, 2021, doi: https://doi.org/10.1007/s10470-021-01831-0.

34. Arulvani, M. and Ismail, M.M., Low power FinFET content addressable memory design for 5G communication networks. *Comput. Electr. Eng.*, 72, 606–613, 2018, doi: https://doi.org/10.1016/j.compeleceng.2018.03.002.
35. Mahani, T., Asma, Keshavarzian, A novel energy-efficient and high-speed full adder using CNTFET. *Microelectron. J.*, 61, 79–88, 2017, doi: https://doi.org/10.1016/j.mejo.2017.01.009.
36. Instruments, Texas, CMOS power consumption and cpd calculation. SCAA035B, 1997.
37. Chandrakasan, A.P. and Brodersen, R.W., Minimizing power consumption in digital CMOS circuits. *Proc. IEEE*, 83, 4, 498–523, 1995. https:// 10.1109/5.371964.
38. Ahmad, M., Manjunathachari, K., Lalkishore, K., Design and analysis of low run-time leakage in a 10 Transistors full adder in 45nm technology. *IEEE Reg. 10 Conf. (TENCON)*, 152–156, 2016, doi: https://10.1109/TENCON.2016.7847979.
39. Dhar, K., Design of a low power. high speed. energy efficient full adder using modified GDI and MVT scheme in 45nm technology. *International Conference on Control. Instrumentation. Communication and Computational Technologies (ICCICCT)*, pp. 36–41, 2014, doi: https://10.1109/ICCICCT.2014.6992926.
40. Abbasalizadeh, S. and Forouzandeh, B., Full adder design with GDI cell and independent double gate transistor. *20th Iranian Conference on Electrical Engineering (ICEE2012)*, pp. 130–134, 2012, doi: https://10.1109/IranianCEE.2012.6292338.
41. Senthilkumar, V.M. and Stella, K., Design a full adder based GDI logic using FinFET technology. *Adv. Nat. Appl. Sci.*, 11, 6, 599–605, 2017.
42. Salehi, S. and DeMara, R.F., Energy and area analysis of a floating-point unit in 15nm CMOS process technology. *Southeast Conference*, vol. 1–5, 2015, doi: https://10.1109/SECON.2015.7132972.
43. Duarte, D., Vijaykrishnan, N., Irwin, M.J., Kim, H., McFarland, G., Impact of scaling on the effectiveness of dynamic power reduction schemes. *Proceedings of IEEE International Conference on Computer Design: VLSI in Computers and Processors*, pp. 382–387, 2020, doi: https:// 10.1109/ICCD.2002.1106798.
44. Vivek, D. and Borkar, S., Technology and design challenges for low power and high performance. *Proceedings of the International Symposium on Low Power Electronics and Design*, vol. 163–168, 1999, doi: https://doi.org/10.1145/313817.313908.
45. Davari, B., Dennard, R.H., Shahidi, G.G., CMOS scaling for high performance and low power-the next ten years. *Proc. IEEE*, 83, 4, 595–606, 1995. doi: https:// 10.1109/5.371968.
46. Goel, E., Pandey, A., Rahi, S.B., Samuel, A., (Eds.). *Handbook of Advanced Semiconductor Field Effect Transistors*. John Wiley & Sons, 2025.

47. Tahrim, A., Chin, H.C., Lim, C.S., Tan, M.L.P., Design and performance analysis of 1-bit FinFET full adder cells for subthreshold region at 16 nm process technology. *J. Nanomater.*, 16, 1, 75, 2015. doi: https://doi.org/10.1155/2015/726175.
48. Dhanaselvam, P.S., Srinivasa Rao, K., Rahi, S.B., Yadav, D.S. (Eds.), *Field Effect Transistors*, John Wiley & Sons, 2025.
49. Gavaskar, K., Ragupathy, U.S., Malini, V., Design of Novel SRAM Cell Using Hybrid VLSI Techniques for Low Leakage and High Speed in Embedded Memories. *Wireless Pers. Commun.*, 108, 2311–2339, 2019. https://doi.org/10.1007/s11277-019-06523-7.
50. Dokania, V., Verma, R., Guduri, M., Islam, A., Design of 10T full adder cell for ultralow-power applications. *Ain Shams Eng. J.*, 9, 4, 2363–2372, 2018, doi: https://https://doi.org/10.1016/j.asej.2017.05.004.

16

# Investigations of Vital Design Parameters of a Low-Power Amplifying Unit for EEG and Advanced Neuroscience Research

**Sarin Vijay Mythry[1,2*], Shashi Kant Gupta[2] and Sai Kiran O.[2]**

*[1]Department of Electronics and Communication Engineering, Vignan Institute of Technology and Science, Hyderabad, India*
*[2]Department of Computer Science and Engineering, Lincoln University College, Petaling Jaya, Selangor, Malaysia*

***Abstract***

This study is being done with the intention of developing a biopotential amplifier that has a low amount of noise during its operation. In order to enhance the quality of the electroencephalogram (EEG) signal, LNAs are a vital component. This allows the EEG data to be used effectively in a wide range of biomedical applications *via* their efficient utilization. When compared to the electrical activity that is generated by the movement of the brain, the electroencephalogram (EEG) signal is representative of this activity. One of the most important elements of an EEG signal monitoring circuit is an operational amplifier that consists of two individual stages. Within the context of the technology that is now accessible, the most significant factors that must be satisfied for each and every circuit are the enhancement of power efficiency and the reduction of noise. The collection of biosignals is of the greatest significance because of their low frequency (range from 0.1 Hz to 10 kHz) and minute amplitude (which is typically within a few millivolts). In the process of producing medical electronics for the purpose of health monitoring, amplifiers are used as a component. Nevertheless, in order to obtain accurate bio-signal measurement, it is required to have a high CMRR and precise amplification in order to decrease the quantity of input noise and enhance the signal strength. This is necessary in order to achieve the desired accuracy. These are the performance requirements that are being addressed, and they include things like power, gain, noise margin, CMRR, SQNR, and plenty of other things. In order to do this, it is recommended that our

**Corresponding author*: sarinmythry@gmail.com

P. Suveetha Dhanaselvam, Naima Guenifi, Shiromani Balmukund Rahi and J. Ajayan (eds.) Field-Effect Transistor Devices: Post-CMOS Theory and Applications, (317–332) © 2026 Scrivener Publishing LLC

biopotential amplifier concept be implemented using technology that is 90 nm in size. A gain of about 82 dB, a phase margin of 48.6°, and a low power consumption of 5.1 μW were achieved by this design. These characteristics are essential for electroencephalogram (EEG) devices that are powered by batteries or portable applications. A low-noise amplifier (LNA) that has been created makes it possible to capture high-quality electroencephalogram (EEG) and electrocardiogram (ECoG) data. This, in turn, allows advancements in the fields of sleep research, brain-computer interfaces, and human–computer interaction. In the future, it is predicted that research and technological improvements will unearth further possibilities in the use of electroencephalogram (EEG) signals and the impact that these signals have on the field of neuroscience research as well as clinical applications.

***Keywords*:** Biopotential amplifier, low-noise amplifier, electroencephalogram, biomedical signal processing, low-power design, 90 nm CMOS technology

## 16.1 Introduction

An enormous amount of importance is attached to the investigation of electroencephalography (EEG) signals in the fields of neuroscience and biomedical engineering. The electroencephalogram (EEG) data, which are derived from the electrical activity of the brain, provide significant insights into cognitive processes, neurological illnesses, and the functioning of the brain. However, these signals are intrinsically weak, often ranging from microvolts to millivolts, which means that strong amplification methods are required in order to accurately analyze and interpret the data. Within the realm of electroencephalogram (EEG) signal amplification circuits, operational amplifiers (OPAMPs) are considered to be crucial components due to their high gain, low noise, and exceptional accuracy. The purpose of this introduction is to provide an explanation of the basic ideas that underlie the construction of operational amplifiers that are particularly configured for the purpose of amplification of EEG signals. During the course of the conversation, topics such as the relevance of electroencephalogram (EEG) signal amplification, the difficulties that arise while amplifying these signals, the function of operational amplifiers in EEG systems, and the goals of building operational amplifiers for EEG applications are covered.

## 16.2 Methodology

The process of analysing data from electroencephalography (EEG) is of utmost significance in the domains of biomedical engineering and neurology.

Electroencephalogram (EEG) data, which originate from the electrical activity of the brain, enable researchers to get profound insights into neurological diseases, brain function, and cognitive processes. However, due to the fact that these signals are, by their very nature, weak, and typically vary from microvolts to millivolts, it is necessary to have dependable amplification techniques in order to conduct accurate analysis and interpretation [6]. Operational amplifiers, also known as OPAMPs, are a key component of EEG signal amplification circuits. These amplifiers provide high gain, low noise, and great accuracy. This introduction is intended to provide clarity on the fundamental concepts that underlie the building of operational amplifiers that are specifically designed for the purpose of amplifying electroencephalogram (EEG) signals. The significance of amplifying electroencephalogram (EEG) signals, the challenges involved in doing so, the role those operational amplifiers play in EEG systems, and the objectives of developing operational amplifiers for electroencephalogram (EEG) applications are all topics that will be discussed [7].

## 16.3 Circuit Design and Description

The architecture of an operational amplifier can be analysed through the block diagram presented in Figure 16.1; using first stage is designed using differential amplifier for the operational amplifier (OPAMP) and gain stage, specifically a common source stage, which includes a compensation capacitor Cc will act as second stage. This component is essential for maintaining closed-loop stability [8]. The final stage involves an output buffer which is used to recognize that when the output buffer stage is disregarded due to capacitive load [9, 15].

We have two saturation region MOSFETs, M2 and M1, where differential inputs are triggered at their gates. The M3 and M4 transistors serve the purpose of a current mirror and function within the saturation region. The bias voltage is applied to the gate of the M5 MOSFET, facilitating the provision of the tail current. An operational amplifier may exhibit instability at elevated frequencies. The resolution of this issue can be achieved through the implementation of various compensation techniques utilizing capacitors. Additionally, resistors and capacitors may be utilized between various stages of the operational amplifier [10, 12, 14]. Various compensation topologies exist, including Miller capacitor compensation, self-compensating, among others.

The four MOSFETs that make up this stage are designated as M2, M1, M4, and M3. MOSFETs where M1 and M2 each have terminals that are

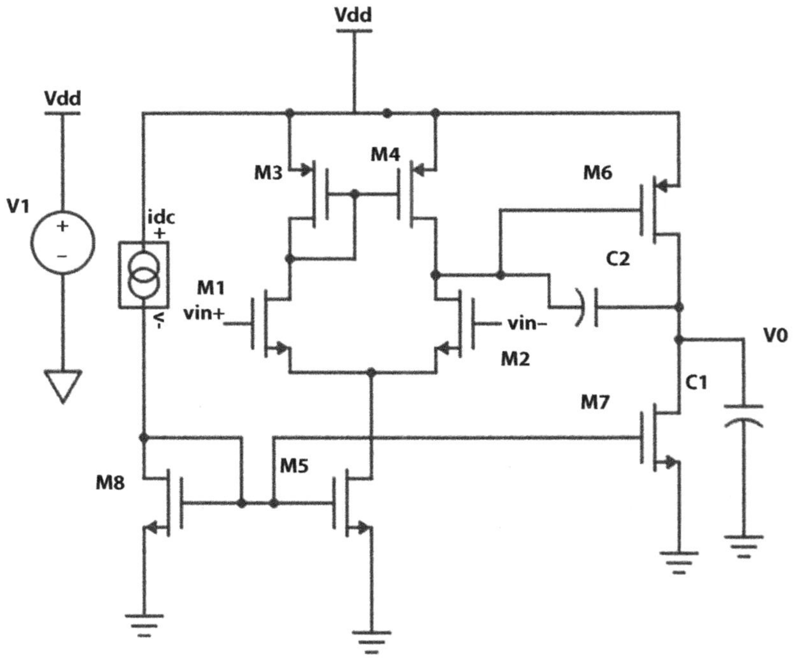


**Figure 16.1** Two-stage low noise biomedical amplifier.

inverting and noninverting, respectively, which serve as their gates. The signal that is received from the inputs M1 and M2 will be enhanced in accordance with the first stage's gain, which is the differential stage. There is a relationship between the M2's transconductance and the total resistance measured at the M2's drain, which determines gain of the stage. The current mirror (CM) at load offers three different benefits. In the first place, the resistance of the CM load is rather more. Furthermore, CM topology allows for the conversion of differential input to single-ended output, that may be carried out. Furthermore, it contributes to the CMRR enhancement. The M3 and M4 devices used as CM load.

In reality, the objective is to enhance the amplifier's output voltage swing in this design phase but also to increase amplifier's extra gain. The second stage, which has a common-source (CS) design and is made up of transistors M6 and M7, is the one that obtains the output from drain node of M2. Both M6 and M7 are responsible for amplifying the output of the first stage. Again, in a manner that is analogous to the previous stage, an active transistor M7 acts as a load component for an M6 transistor. This stage gain may be determined by multiplying the M6's Gm (transconductance) by the output resistances of M6 and M7 MOSFETs [11–13, 15]. This calculation helps to determine the gain of this stage.

The formulae for saturation drain current are provided as follows:

$$I_{Dn} = \frac{\mu_n C_{ox}\left(\frac{W}{L}\right)(V_{gs} - V_t)^2}{2} \tag{16.1}$$

$$I_{Pn} = \frac{\mu_p C_{ox}\left(\frac{W}{L}\right)(V_{gs} - V_t)^2}{2} \tag{16.2}$$

Transconductance is given by:

$$\frac{\partial I_D}{\partial V_{gs}} = g_m \tag{16.3}$$

From Equations (16.1) and (16.3), we can obtain

$$\left(\frac{W}{L}\right)_{(1,2)} = \frac{g_m^{\,2}}{2 I_D \mu_n C_{ox}} \tag{16.4}$$

Using Equation (16.2), we obtain the $\left(\frac{W}{L}\right)$ values of transistors M3 and M4, we obtained $\left(\frac{W}{L}\right)$ values of transistors M6 and M7 using:

$$\left(\frac{W}{L}\right)_{(6)} = \frac{g_{m6}}{g_{m4}}\left(\frac{W}{L}\right)_{(4)} \tag{16.5}$$

$$\frac{I_7}{I_5} = \frac{\left(\frac{W}{L}\right)_{(7)}}{\left(\frac{W}{L}\right)_{(5)}} \tag{16.6}$$

Here,

$I_7$ and $I_5$ are the currents passing through transistors M7 and M5 respectively

$g_{m6}$ and $g_{m4}$ are the transconductances of the transistors M6 and M4 respectively

For calculating CMRR, we use

$$\text{CMRR} = \text{gain } (A_D \text{ in dB}) - \text{common mode gain(dB)} \quad (16.7)$$

## 16.4 Design Specifications

Table 16.1 summarizes the key design specifications adopted for the low-power EEG amplifying unit.

**Table 16.1** Design specifications.

| Specification (units) | Values |
|---|---|
| VDD (V) | 1.8 |
| -3dB (kHz) | ≈ 10 |
| GBW (MHz) | 3500 |
| Phase Margin (0) | 75 |
| Gain Margin (dB) | 80 |
| CMRR (dB) | 120 |
| Delay (ns) | 30 |
| Slew Rate (V/μs) | 35 |
| Offset, O/P (mV) | 0.05 |
| Settling Time (ns) | 35 |
| SNR (dB) | 40 |
| Power (mW) | 0.1 |
| ICMR (V) | 0.38-1.4 |
| Output Swing (V) | 0.27-0.73 |
| C, Load (pF) | 1-2 |

## 16.5 Simulation and Results

Various aspect ratios (W/L) considered for the simulation of the device, are presented in Table 16.2. As shown in Table 16.3, variations in the supply voltage significantly influence key performance parameters such as noise,

**Table 16.2** Aspect ratios.

| Device | $\frac{W}{L}$ |
|---|---|
| M1 | 12 |
| M2 | 12 |
| M3 | 6 |
| M4 | 6 |
| M5 | 15 |
| M6 | 15 |
| M7 | 40 |
| M8 | 50 |

**Table 16.3** Monte Carlo simulation of transient response with respect to Vdd.

| Vdd(V) | Noise ($\mu V/\sqrt{Hz}$) | Gain (dB) | Power ($\mu W$) | common mode gain (dB) | CMRR (dB) |
|---|---|---|---|---|---|
| 0.4 | 0.1 | -15.596 | 1.8106 | -9.9257 | -5.670 |
| 0.5142 | 851.5 | -1.1622 | 2.3917 | 3.0878 | -4.25 |
| 0.6287 | 51.925 | -3.2699 | 2.9474 | -7.1831 | 3.91 |
| 0.7428 | 324.7 | 10.1561 | 3.5008 | -21.2005 | 31.35 |
| 0.8573 | 1050.6 | 39.2025 | 4.0568 | -14.5453 | 53.74 |
| 0.9714 | 791.5 | 79.8443 | 4.5724 | 0.7564 | 79.08 |
| 1.0857 | 1264.16 | 81.8881 | 5.1046 | -8.6192 | 90.501 |
| 1.2 | 1172.5 | 81.5668 | 5.6502 | -10.032 | 91.598 |
| 1.3142 | 1051.7 | 81.0994 | 6.2026 | -10.5479 | 91.64 |
| 1.4285 | 945.0 | 80.6394 | 6.7615 | -10.9313 | 91.56 |
| 1.5428 | 856.2 | 80.2148 | 7.3270 | -11.285 | 91.403 |
| 1.6571 | 783.13 | 79.8304 | 7.8994 | -11.6483 | 91.478 |
| 1.7714 | 725.35 | 79.4978 | 8.4796 | -12.0396 | 91.52 |
| 1.8857 | 674.24 | 79.176 | 9.0663 | -12.4832 | 91.65 |
| 2 | 638.007 | 78.9246 | 9.6637 | -12.9433 | 91.86 |

**Table 16.4** Monte Carlo simulation of transient response with respect to Idc.

| Idc (μA) | Noise ($\mu V/\sqrt{Hz}$) | Gain (dB) | Power (μW) | common mode gain (dB) | CMRR (dB) |
|---|---|---|---|---|---|
| 1 | 1084 | 81.2139 | 4.704 | -0.356 | 81.569 |
| 2 | 861.24 | 8.853 | 9.319 | 1.6171 | 7.21 |
| 3 | 63704 | 81.552 | 14.089 | 2.4633 | 79.16 |
| 4 | 470.91 | 80.970 | 18.7946 | 2.9325 | 78.04 |
| 5 | 351.07 | 80.2567 | 23.5049 | 3.22177 | 77.031 |
| 6 | 264.33 | 79.4689 | 28.2194 | 3.4043 | 76.064 |
| 7 | 202.22 | 78.629 | 32.937 | 3.5161 | 75.113 |
| 8 | 170.7 | 77.7457 | 37.6587 | 3.5569 | 74.009 |
| 9 | 151.9 | 76.8239 | 42.3826 | 3.5456 | 73.29 |
| 10 | 117.24 | 75.865 | 47.1089 | 3.4782 | 72.387 |

**Table 16.5** Monte Carlo simulation of transient response with respect to temperature.

| Temperature (C) | Noise ($\mu V/\sqrt{Hz}$) | Gain (dB) | Power (μW) | common mode gain (dB) | CMRR (dB) |
|---|---|---|---|---|---|
| -77 | 630.129 | 81.979 | 5.17559 | -9.318 | 91.3 |
| -67.375 | 704.46 | 82.183 | 5.1622 | -9.567 | 91.75 |
| -57.75 | 781.018 | 82.343 | 5.1511 | -9.785 | 92.13 |
| -48.125 | 855.84 | 82.449 | 5.1411 | -9.971 | 92.42 |
| -38.5 | 931.87 | 82.524 | 5.1330 | -10.121 | 92.65 |
| -28.875 | 1004.03 | 82.552 | 5.1258 | -10.239 | 92.79 |
| -19.25 | 1071.57 | 82.540 | 5.1197 | -10.312 | 92.85 |
| -9.625 | 1135.47 | 82.499 | 5.1151 | -10.335 | 92.83 |
| 0 | 1187.62 | 82.405 | 5.1110 | -10.322 | 92.73 |

*(Continued)*

**Table 16.5** Monte Carlo simulation of transient response with respect to temperature. (*Continued*)

| Temperature (C) | Noise $(\mu V/\sqrt{Hz})$ | Gain (dB) | Power (μW) | common mode gain (dB) | CMRR (dB) |
|---|---|---|---|---|---|
| 9.625 | 1228.69 | 82.267 | 5.1079 | -10.266 | 92.53 |
| 19.25 | 1255.44 | 81.080 | 5.1058 | -10.166 | 91.25 |
| 28.875 | 1264.24 | 81.835 | 5.1047 | -10.022 | 91.86 |
| 38.5 | 1251.16 | 81.519 | 5.1046 | -9.8361 | 91.36 |
| 48.125 | 1213.03 | 81.120 | 5.1055 | -9.6112 | 90.73 |
| 57.75 | 1146.99 | 80.618 | 5.1075 | -9.3511 | 89.97 |
| 67.375 | 1049.08 | 79.976 | 5.1103 | -9.0586 | 89.03 |
| 77 | 920.713 | 79.759 | 5.1143 | -8.7424 | 88.5 |

**Table 16.6** Comparison of designed amplifier with existing literatures.

| **Parameters** | [1] | [2] | [3] | [4] | [5] | **This work** |
|---|---|---|---|---|---|---|
| Technology (nm) | 90 | 90 | 90 | 90 | 90 | **90** |
| Vdd (V) | 1.5 | 0.5 | 1.8 | 2.5 | 1.5 | **1.085714** |
| No. of stages | 2 | 2 | 2 | 2 | 2 | **2** |
| CMRR | 103 | 102.6 | - | 43 | 104 | **90.1** |
| Slewrate $(V/\mu S)$ | 34.7 | - | 5 | 10 | 34.7 | **28** |
| ICMR+ | 1.4 | - | 1.6 | - | - | **0.8** |
| ICMR- | 0.38 | - | 0.8 | - | - | **1.6** |
| Noise $(\mu V/\sqrt{Hz})$ | - | 4.366 | - | - | - | **1.2m** |
| Power $(\mu W)$ | 210 | 28.42 | 300 | 221 | 219 | **5.1** |
| Phase Margin | 66.3 | 56 | - | 60 | 66.6 | **48.6** |
| Gain (dB) | - | 45.3 | - | - | 78.2 | **81.88** |
| GBW (MHz) | 3.3 | - | 5 | 5 | 3.3 | **2.51** |
| CL (pF) | 1 | 20 | 10 | 10 | 1 | **2** |

gain, power consumption, common-mode gain, and CMRR. To evaluate robustness against fluctuations in current, Monte Carlo simulations were conducted for multiple Vdd points, and the resulting noise, gain, and CMRR trends are summarized in Table 16.4. Table 16.5 details how changes in temperature impact the amplifier's transient behavior, highlighting notable variations in noise performance, gain, and common-mode rejection. Table 16.6 provides a comparison of the proposed amplifier with representative designs reported in existing literature.

## 16.6 Output Waveforms

Figure 16.2 illustrates the Monte Carlo simulations conducted for multiple Vdd points with Idc=1uA, and the resulting noise, gain, and CMRR trends. Figure 16.3 details the Monte Carlo simulations conducted for multiple Idc points with Vdd=1.085714V, and the resulting noise, gain, and CMRR trends. Figure 16.4 illustrates the gain analysis of the proposed amplifier. Figure 16.5 presents the power analysis, demonstrating how the designed amplifier maintains low power operation across the specified voltage and load ranges. The detailed noise characteristics of the designed amplifier, including its suitability for EEG signal acquisition, are depicted in Figure 16.6. Figure 16.7 presents the Monte Carlo–based transient response analysis, highlighting how temperature changes influence the amplifier's behaviour under Vdd = 1.085714 V and Idc = 1 μA.

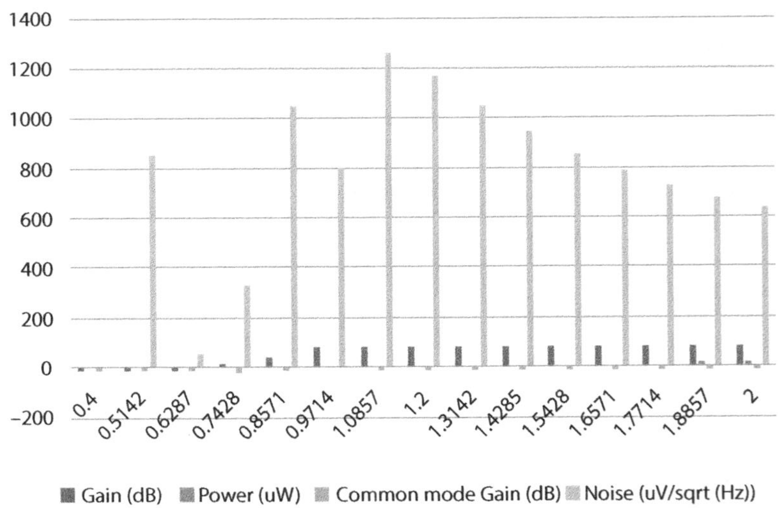


**Figure 16.2** Monte Carlo simulation of transient response with respect to Vdd (@ Idc=1$\mu$A).

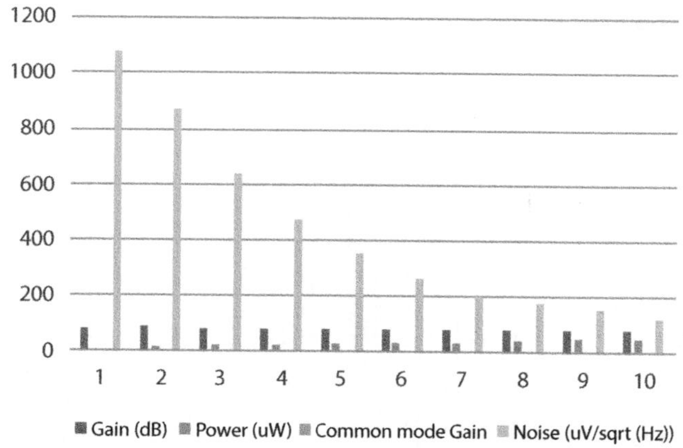


**Figure 16.3** Monte Carlo simulation of transient response with respect to Idc(@ Vdd=1.085714V).

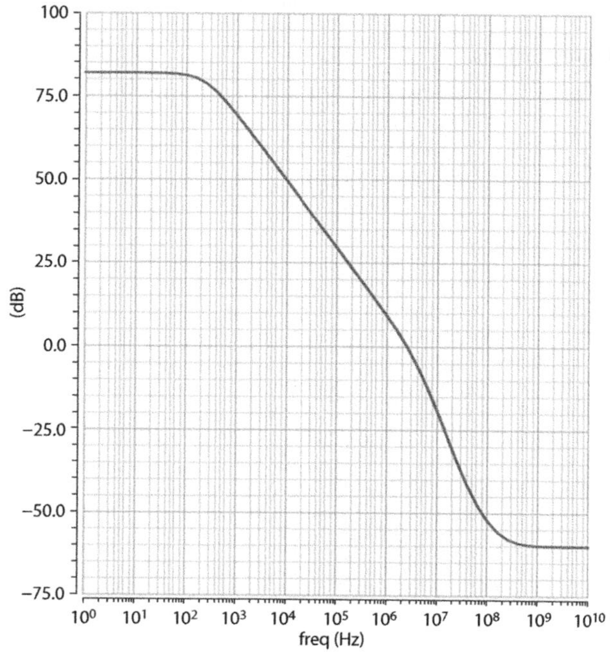


**Figure 16.4** Gain analysis.

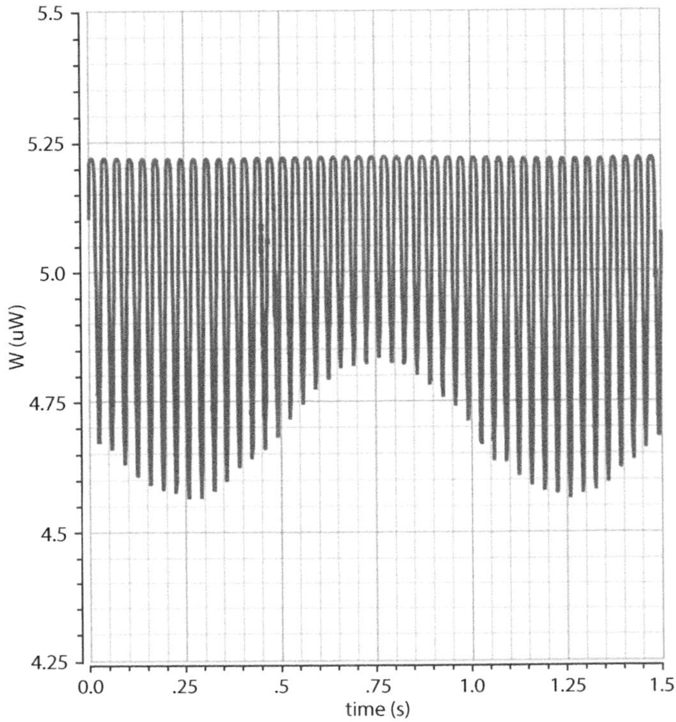


**Figure 16.5** Power analysis.

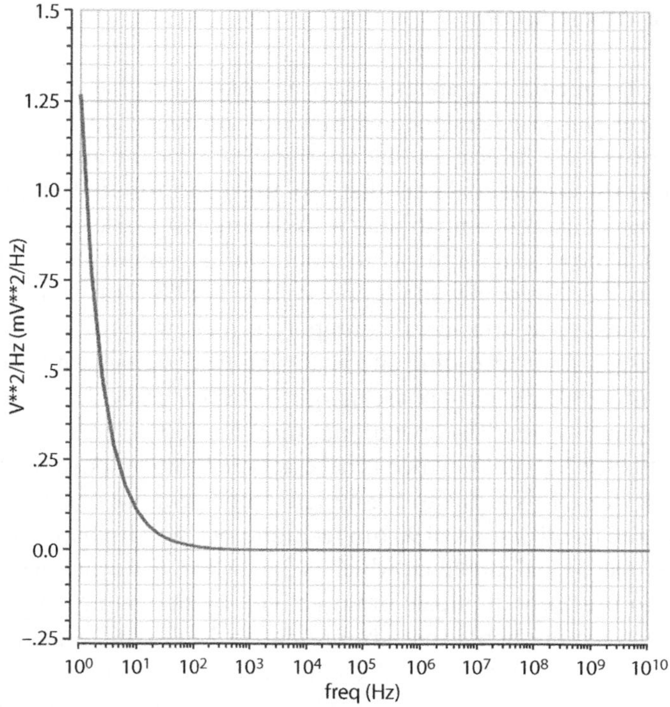


**Figure 16.6** Noise analysis.

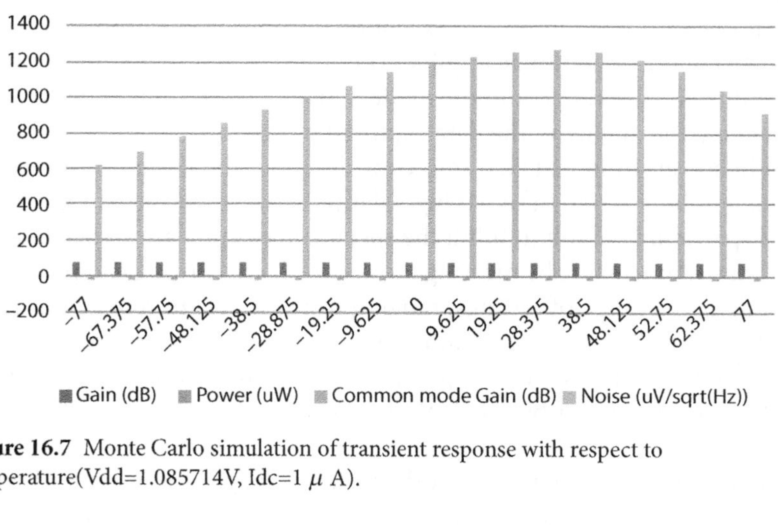


**Figure 16.7** Monte Carlo simulation of transient response with respect to temperature(Vdd=1.085714V, Idc=1 $\mu$ A).

## 16.7 Conclusion

With the aforementioned specs, it can be concluded that the operational amplifier (op-amp) was designed with careful consideration given to a number of performance indicators that are significant for the application it will be used for. The operational amplifier (op-amp) performs within predefined voltage ranges, ensures compatibility with common signal levels encountered in EEG signal processing, and has an input common-mode range (ICMR) of 0.8 V to 1.6 V. Additionally, the supply voltage is 1.085714 V. This amplifier exhibits its ability to considerably amplify signals without losing stability, which is essential for recording and analyzing EEG signals with accuracy. It has a gain of 81.881 dB and a phase margin of 48.6 degrees, demonstrating its proficiency in this area. The operational amplifier (op-amp) has a low power consumption of 5.104 μW, which is necessary for battery-powered or portable EEG devices. This is due to the design's intelligent use of semiconductor resources and its utilization of 90 nm technology. The gain bandwidth product of the operational amplifier, which is 2.51 MHz, is further evidence that it is able to precisely duplicate brain activity by managing a wide variety of frequency components that are present in EEG data. Signal integrity, minimum signal distortion, and compatibility with external devices are all guaranteed when the 1 pF load capacitance as well as other factors are taken into consideration. There are two stages in the design process that create a compromise between the level of complexity and the level of performance by matching the requirements of performance with the practical considerations. In addition, the exceptional common-mode rejection ratio (CMRR) of the operational amplifier (op-amp) of 90.1 dB demonstrates its capability to suppress common-mode signals that are not sought, hence enhancing the quality of the data captured by the EEG.

When everything is taken into consideration, the design of the operational amplifier strikes a satisfying balance between performance metrics. It was particularly built to fulfill the specific needs that are associated with EEG signal processing by meeting those criteria. Recording and processing EEG data in a range of settings, ranging from clinical to research, is made possible by its low power consumption, optimum performance, and compatibility with EEG system requirements. This makes it an effective and trustworthy design for biomedical EEG application.

**AI disclosure statement:** The author acknowledges the use of AI-based tools for minor language editing or grammatical correction. However, all intellectual contributions, analyses, and conclusions presented in this chapter are entirely the author's own.

## References


1. Banik, S., Mahmud, T., Rasel, M.M., Hasanuzzaman, M., A high-performance low-power two-stage OPAMP realized in 90nm CMOS process for biomedical application. *2020 IEEE Region 10 Symposium (TENSYMP)*, 2020, doi: 10.1109/tensymp50017.2020.9230957.
2. Wan Jusoh, W.M.E.A. and Ruslan, S.H., Design and analysis of current mirror OTA in 45 nm and 90 nm CMOS technology for bio-medical application, in: *Bulletin of Electrical Engineering and Informatics*, vol. 9, pp. 221–228, February 2020, ISSN: 2302-9285, DOI: 10.11591/eei.v9i1.1860.
3. Kavyashree, C.L., Hemambika, M., Dharani, K., Naik and Sunil MP, A.V., Design and implementation of two stage CMOS operational amplifier using 90nm technology. *IEEE International Conference on Inventive Systems and Control (ICISC)*, Coimbatore, pp. 1–4, 2017.
4. Satyanarayana, M., Nageshwar Rao, D., Aravindkumar, K., Vamshikrishna, K., Vinitha, K., Two Stage Operational Amplifier Using 90nm Technology.
5. Saidulu, R., Purna Chandar, B., Ravi Kumar, K., Design and Simulation of a High Gain Two-Stage Op-Amp Realised in 90nm CMOS Process.
6. Gupta, P. and Tripathi, S.L., Low power design of bulk driven operational transconductance amplifier, in: *2nd Int. Conf. 2017 Devices for Integrated Circuit (DevIC)*, pp. 241–246, 2017.
7. Mythry, S.V. and Moni, D.J., High flat gain low noise neural signal recording amplifier. *International Conference on Signal Processing and Communication (ICSPC'17)*, pp. 12–17, 2017.
8. Rajput, S.S., Singh, A., Chandel, A.K., Chandel, R., Design of Low-Power High-Gain Operational Amplifier for Bio-Medical Applications. *2016 IEEE Computer Society Annual Symposium on VLSI (ISVLSI)*, Pittsburgh, PA, pp. 355–360, 2016.
9. Bashir, M., Patri, S.R., KrishnaPrasad, K.S.R., 0.5 V, high gain two-stage operational amplifier with enhanced transconductance. *Int. J. Electron. Lett.*, 6, 1, 80–89, 2018.
10. Wang, Y., Zhao, X., Zhang, Q., Lv, X., Adjustably transconductance enhanced bulk-driven OTA with the CMOS technologies scaling. *Electron. Lett.*, 54, 5, 276–278, 2018.
11. Veldandi, H. and Rafi Aahmed, S., Design procedure for multifinger MOSFET two-stage OTA with shallow trench isolation effect. *IET Circuits Devices Syst.*, 129, 5, 513–522, 2018.
12. Shruit and Suman, Two stage CMOS operational amplifier: analysis and design. *Mody Univ. Int. J. Comput. Enginnering Res.*, 40–44, 2019.
13. Aqeel, S., Deva, P.K., Babu, C.M., Ganesh, R., Design of High Gain Two stage Op-Amp using 90nm Technology, in: *International Journal of Engineering Research & Technology (IJERT)*, ISSN: 2278-018, 3, May–2014.

14. Kakoty, P., Design of a high frequency low voltage CMOS operational amplifier. *Int. J. VLSI Design Commun. Syst. (VLSICS)*, 2, 1, Mar. 2011.
15. Banik, S. and Mahmud, T., ] A High-Performance Low–Power Two-Stage Operational Amplifier realized in 90nm CMOS process for Bio-Medical Applications. *2020 IEEE Region 10 Symposium (TENSYMP)*, Dhaka, Bangladesh, 5–7 June 2020.

17

# Heterojunction Tunnel FET with Graphene Nanoribbon

**Ritam Dutta[1]* and J. Ajayan[2]**

[1]*Dept. of CSE (IoT), Poornima Institute of Engineering & Technology, Jaipur, Rajasthan, India*
[2]*Dept. of ECE, SR University, Warangal, Telangana, India*

## *Abstract*

The continuous scaling of transistors into the nanoscale regime has led to the exploration of novel device architectures beyond conventional MOSFETs, with tunnel field-effect transistors (TFETs) emerging as promising candidates due to their steep subthreshold swing and low-power operation. In this work, a progressive modeling approach has been adopted to enhance TFET performance. Initially, a conventional dual-gate TFET is modified by introducing heterojunction material at the source and channel interface, resulting in a heterojunction TFET (HJ-TFET). To further improve carrier transport across the tunneling region, the device is engineered with heterogate dielectrics, forming a heterogate dielectric heterojunction TFET (HGD-HJ-TFET). Building on this structure, an ultra-thin graphene nanoribbon layer of thickness 0.5 nm is deployed over the intrinsic silicon channel of length 20 nm, yielding the proposed graphene-based heterojunction heterogate dielectric TFET (GR-HGD-HJ-TFET). The incorporation of graphene enhances carrier mobility and tunneling efficiency, thereby reducing leakage current and improving ON-state current. Device parameters such as gate oxide thickness, body thickness, and heterogate dielectric lengths are optimized through analytical modeling and simulation. The results confirm that the proposed GR-HGD-HJ-TFET exhibits superior electrostatic control and performance compared to conventional and previously reported TFET structures, establishing its potential for future low-power nanoelectronic applications.

**Corresponding author*: ritamdutta1986@gmail.com

P. Suveetha Dhanaselvam, Naima Guenifi, Shiromani Balmukund Rahi and J. Ajayan (eds.) Field-Effect Transistor Devices: Post-CMOS Theory and Applications, (333–346) © 2026 Scrivener Publishing LLC

***Keywords*:** Heterojunction TFET, graphene nanoribbon, GR-HGD-HJ-TFET, carrier tunneling, subthreshold swing, nanoscale devices, low-power electronics, device modeling

## 17.1 Introduction

In this chapter, at first a conventional dual gate tunnel field effect transistor (TFET) has been modified with heterojunction material at source and channel regions. Secondly the developed heterojunction TFET (HJ-TFET) has been further modified with heterogate dielectrics to investigate the mobility of carriers across the tunneling region. Finally, the heterojunction heterogate dielectric TFET (HGD-HJ-TFET) model has been aimed by deploying an ultra-thin carbon material i.e., graphene in the form of ribbon over the intrinsic silicon channel to investigate the influence of it over the proposed device structure i.e., graphene-based heterojunction heterogate dielectric tunnel FET (GR-HGD-HJ-TFET).

In Figure 17.1, the proposed device model is shown. The channel length ($L_{ch}$) is considered as 20 nm, over which a 0.5 nm ultra-thin graphene layer is deployed as nanoribbon. The body thickness ($t_{si}$) and the gate dielectric/ oxide thickness ($t_{ox}$) are considered as 10 nm and 2 nm respectively.

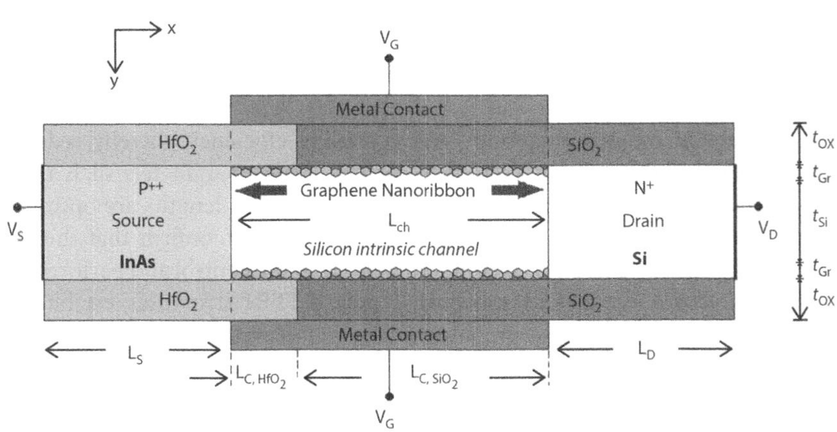


**Figure 17.1** 2-D Schematic diagram of proposed GR-HGD-HJ-TFET device structure.

To investigate the effect of hetero gate dielectric length, has been varied suitably and finally considered as $L_{C, HfO2}$= 4 nm and $L_{C, SiO2}$=16 nm.

The necessary device and electrical parameters are tabulated in Table 17.1 and Table 17.2. The dataset is considered for this research work taken from recent established literature survey [1–16].

**Table 17.1** Different parameters of proposed GR-HGD-HJ-TFET device model.

| **Device parameters** | **Values (in nm)** |
|---|---|
| Channel Length ($L_{ch}$) | 20 |
| Length of Source ($L_S$) | 30 |
| Length of Drain ($L_D$) | 30 |
| Gate Dielectric ($HfO_2$) Length ($L_{C,HfO2}$) | 2 to 18 (variable) |
| Gate Dielectric ($SiO_2$) Length ($L_{C,SiO2}$) | 2 to 18 (variable) |
| Body Thickness ($t_{Si}$) | 5 to 10 |
| Gate Dielectric/Oxide layer Thickness ($t_{ox}$) | 1 to 2 |
| Graphene nanoribbon Thickness ($t_{GR}$) | 0.5 to 1.5 (variable) |

**Table 17.2** The typical electrical parameters used for proposed GR-HGD-HJ-TFET device model.

| **Electrical parameters** | **Values** |
|---|---|
| Doping at source ($N_S$) | $1 \times 10^{20}$ cm$^{-3}$ |
| Doping at drain ($N_D$) | $5 \times 10^{17}$ cm$^{-3}$ |
| Doping at channel ($N_{ch}$) | $1 \times 10^{16}$ cm$^{-3}$ |
| Metal Work function ($\phi_m$) | 4.5 eV |
| Gate Voltage ($V_G$) | -1V to 1.5 V |
| Supply Voltage ($V_{DD}$) | 0.5 V |

In today's cutting edge semiconductor technology, TFET model can provide better device performances viz. better subthreshold swing (SS << 60mV/decade), high ON current ($I_{ON}$), low OFF current ($I_{OFF}$) and minimum ambipolarity ($I_{amb}$) compared to orthodox CMOS transistors [17]. In addition to this, graphene has been incorporated as ribbon structure to ensure better tunneling across tunneling junction. Here for n-type TFET model, the source – channel interface acts as tunneling junction. Better control over ambipolar conduction can result better TFET device for low power application [18–24].

Moreover, since graphene has its excellent carrier mobility with smaller effective mass at room temperature, results better ON-state current while implementing in TFETs. Additionally, having high thermal conductivity with high tensile strength can be effective for low power applications. Here in our work the graphene has been used as ribbon type replacing the sheet type. Because, graphene as a sheet type, provides no energy gap between the valence and conduction bands, whereas, the ribbon type graphene can easily tune their transport gaps (Figure 17.2) and improve the switching ratio i.e. ratio between $I_{ON}$ and $I_{OFF}$ current. This technique is called as quantum confinement effect (QCE) that can be effective in graphene as nanoribbon type material.

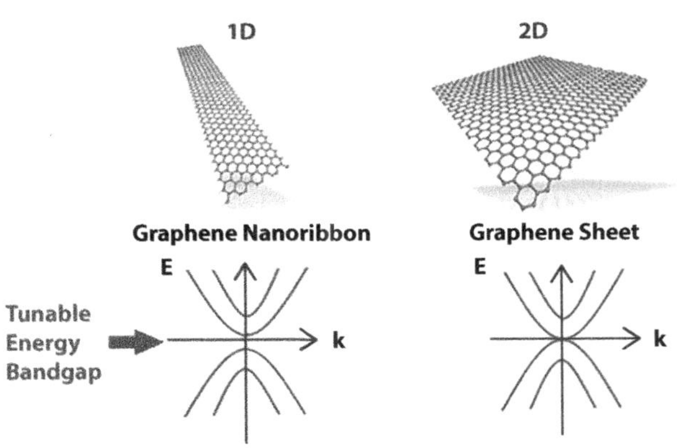


**Figure 17.2** Graphene as nanoribbon type used in our proposed GR-HGD-HJ-TFET device.

## 17.2 Simulation Strategy and Calibration

Silvaco TCAD is a numerical device simulator that is successful in validating the data by comparing several pre-stored data in their library. Moreover, this simulator has special talent to predict the electrical behaviour of any given device structure, and thereby producing a real insight to its internal physical mechanism. The simulation works are carried out using Silvaco TCAD in this paper [25].

Firstly, accurate mesh analysis is to be made by mentioning the x and y locations of every individual node points by identifying their locations as "x.mesh loc" and "y.mesh.loc". Here we have considered nonlocal BTBT modelling, so "qtx.mesh" with location details need to be defined. Then to satisfy boundary conditions, specific regions need to be identified along x and y axis for two-dimensional structure. Similarly, for three-dimensional structure, the regions must be selected along x, y and z axis. After this step, the electrodes must be defined for gate(s), source and drain regions of the structure. The tunneling parameter must be selected and materials used for this BTBT need to be defined. Now for n-type and p-type structure, respective doping concentrations must be decided and to be defined categorically. Since our model is n-type, therefore P type source, N type drain, and less doped intrinsic channel must be defined suitably. Next step is defining the contacts. In this block, the gate metal work function must be defined. The physical models viz. band gap narrowing (BGN), Shockley-Read-Hall (SRH), Local/Non-local tunneling model, Quantum model need to be defined and incorporated. The Fermi–Dirac statistics, SRH model has been used for utilizing recombination effect. The BGN model is used to develop C-V responses at highly doped source/drain regions. The nonlocal Band to Band Tunneling (BTBT) model based on Wentzel–Kramer–Brillouin (WKB) approximation has been inculcated to achieve the accurate simulation. Here the supply voltage has been kept at 0.5V to investigate the model's tunability for low power applications. Finally, need to apply biasing at respective gate, source and drain nodes for simulation of various device characteristics. Tony plot is used for plotting these characteristics to obtain graphical representations. For minute calibration of the device simulator, the experimental data from relevant literature has been extracted and compared with simulated data shows perfect agreement of both, shown in Figure 17.3.

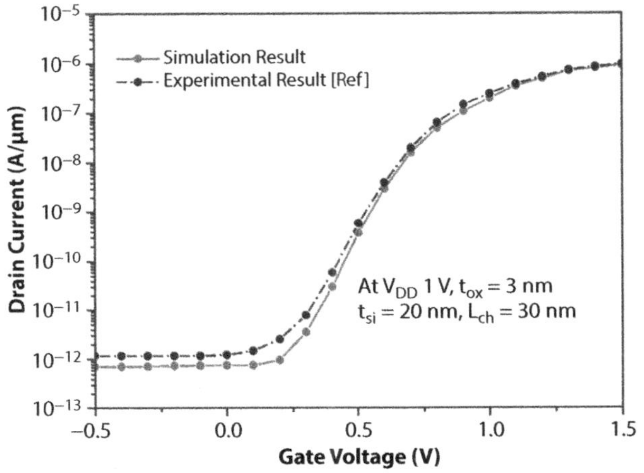


**Figure 17.3** Comparison between simulation and experimental results.

## 17.3 Results and Discussion

First of all, the analysis of transfer characteristics of different tunnel FET device models can illume the influence of graphene material as nanoribbon type deployed over proposed GR-HGD-HJ-TFET device model, shown in Figure 17.4 (a). The proposed GR-HGD-HJ-TFET device has been compared with conventional TFET, HJ-TFET and HGD-HJ-TFET models and results steeper subthreshold slope i.e., SS = 22.38 mV/decade compared to all other combinations.

Secondly, the proposed GR-HGD-HJ-TFET device model has also ensured best drive current of $1.38 \times 10^{-5}$ A/μm with controlled ambipolar conduction of $1.41 \times 10^{-16}$ A/μm, eventually results better $I_{ON}/I_{amb}$ ratio. Moreover, the OFF-state current for our proposed device has also been recorded lowest as $2.74 \times 10^{-17}$ A/μm compared to other models endorses better $I_{ON}/I_{OFF}$ switching ratio. The comparison results are shown in Figure 17.4(b).

Thirdly, the turn on voltage ($V_{ON}$) for the proposed model has been recorded much less compared to all other TFET device models.

To investigate the band 2 band tunneling rate both at source – channel as well as channel – drain junctions of our proposed n-type TFET device, the energy band diagram (EBD) analysis plays pivotal role. Figures 17.5(a) and 17.5(b) depicts the EBD, where the improved tunneling window has been comprehensively identified, due to incorporation of graphene nanoribbon layer over silicon channel.

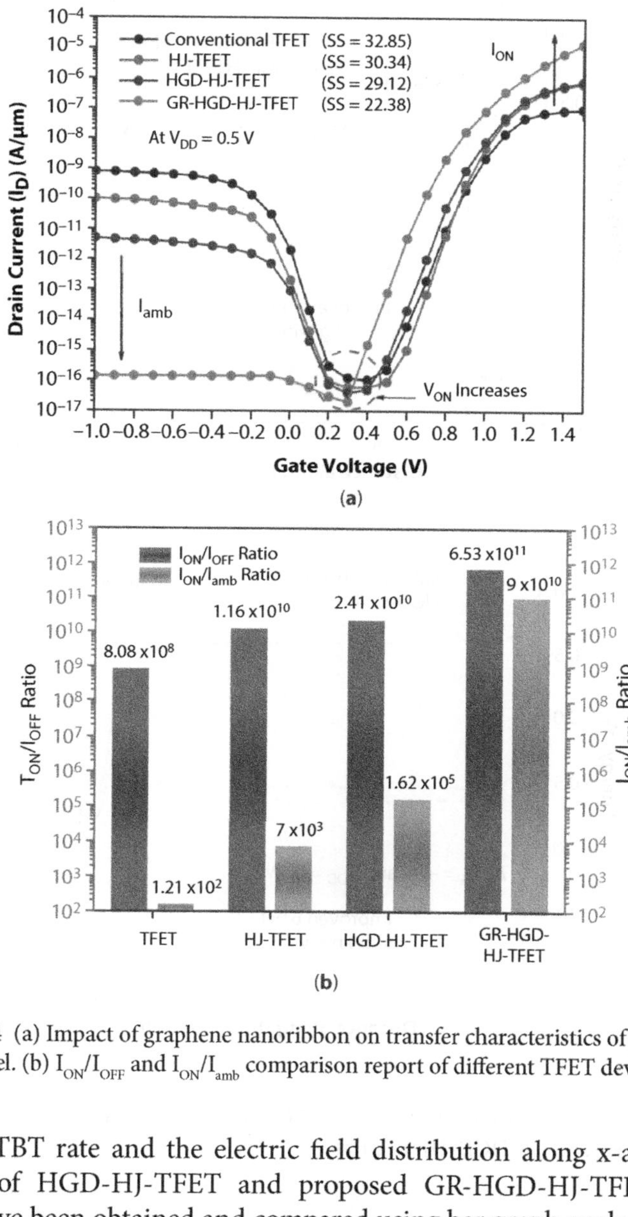


**Figure 17.4** (a) Impact of graphene nanoribbon on transfer characteristics of proposed TFET model. (b) $I_{ON}/I_{OFF}$ and $I_{ON}/I_{amb}$ comparison report of different TFET device models.

The BTBT rate and the electric field distribution along x-axis of the channel of HGD-HJ-TFET and proposed GR-HGD-HJ-TFET device model have been obtained and compared using bar graph analysis, shown in Figures 17.6(a) and 17.6(b) for ambipolar and ON-state conduction respectively.

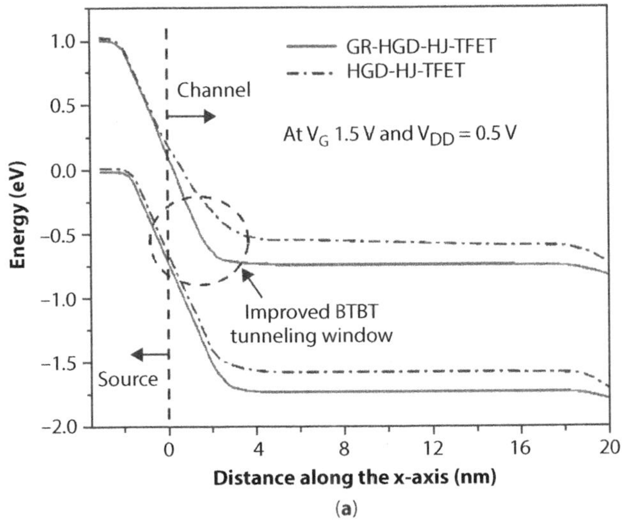


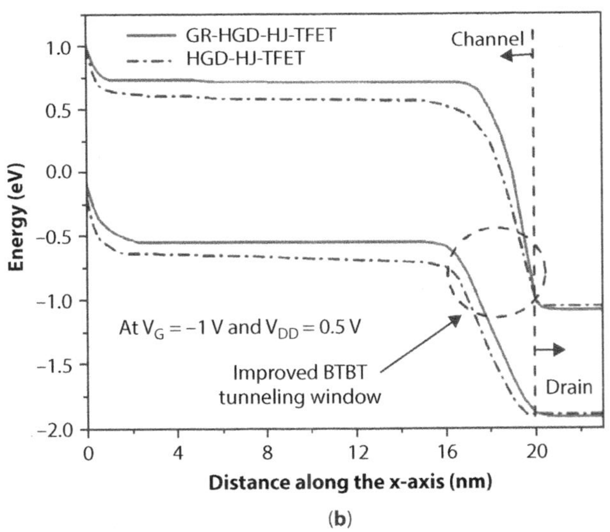


**Figure 17.5** (a) EBD of HGD-HJ-TFET and GR-HGD-HJ-TFET at $V_G$ = 1.5V. (b) EBD of HGD-HJ-TFET and GR-HGD-HJ-TFET at $V_G$ = - 1V.

In this section, the influence of heterogate dielectric over homogate dielectric layered as oxide material on our proposed GR-HGD-HJ-TFET device, has been analysed. From Figure 17.7 (a) it can be clearly identified that the $HfO_2$ – $SiO_2$ heterogate dielectric results much controlled ambipolar conduction in comparison with other homogate dielectric materials

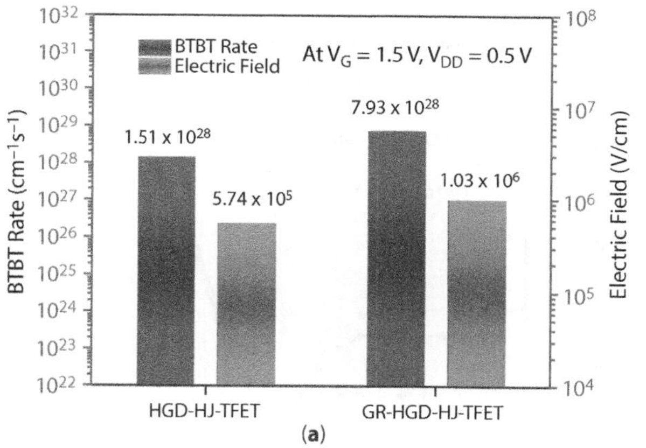


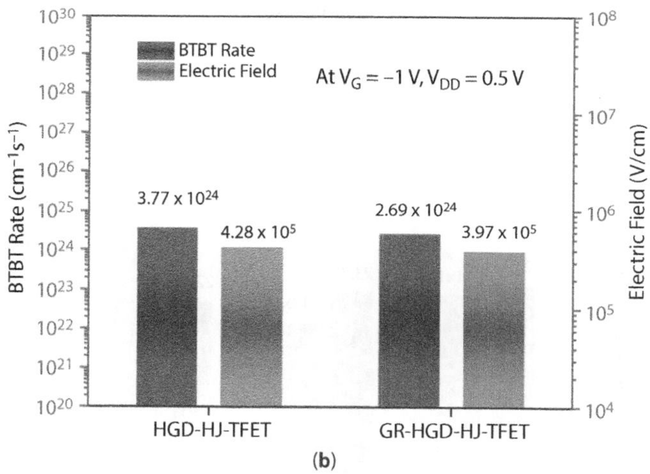


**Figure 17.6** (a) BTBT rate and electric field comparison report of different TFET device models at $V_G$= 1.5V. (b) BTBT rate and electric field comparison report of different TFET device models at $V_G$= - 1V.

viz. $SiO_2$. $HfO_2$. However, the ON-state current remains almost same for each of three cases. In bar graph comparison shown in Figure 17.7 (b) it is also observed that our proposed device produces much lowered SS as 22.38 mV/decade and considerable improvement in $I_{ON}/I_{amb}$ ratio as 3.74 × $10^{11}$.

Moreover, the applied $HfO_2$ – $SiO_2$ heterogate dielectric produces varied $I_{ON}/I_{amb}$ current ratio for different heterogate dielectric lengths, with unaltered SS. The typical heterogate dielectric lengths of $HfO_2$ = 4 nm and $SiO_2$ = 16 nm laid over 20 nm silicon channel eventually results better device performance.

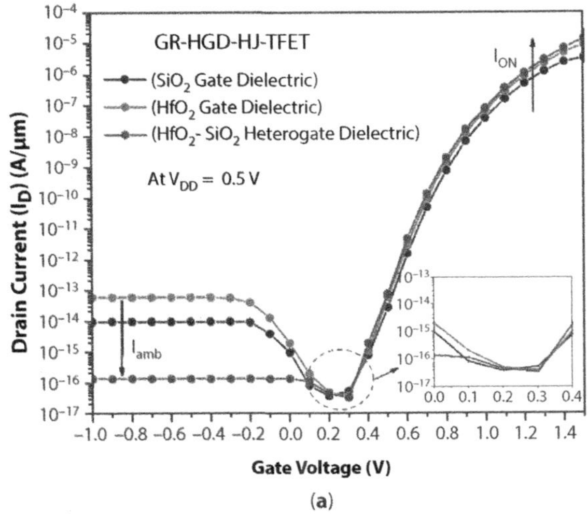


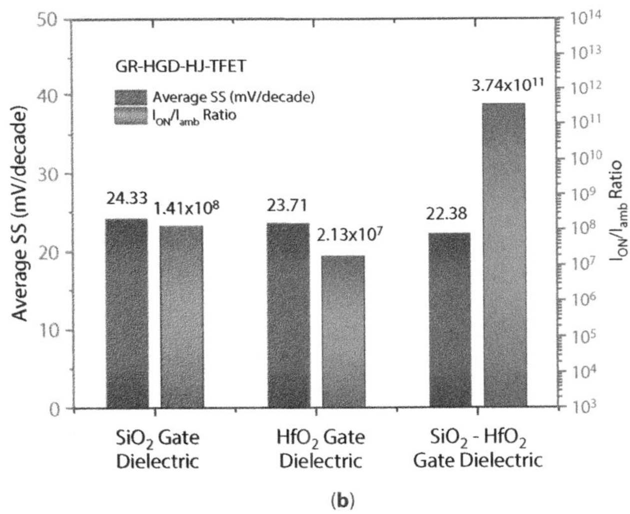


**Figure 17.7** (a) Transfer characteristics of proposed GR-HGD-HJ-TFET on different gate dielectrics. (b) $SS_{Avg}$ and $I_{ON}/I_{amb}$ comparison report of GR-HGD-HJ-TFET on different gate dielectric applied. (c) Transfer characteristics of proposed GR-HGD-HJ-TFET on heterogate dielectric length variation. (*Continued*)

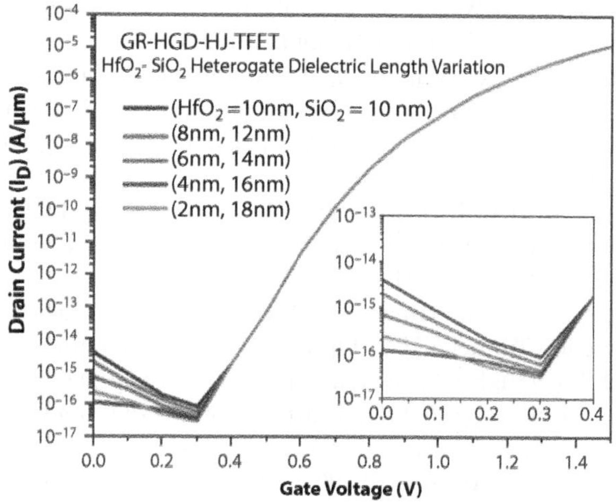


**Figure 17.7 (Continued)** (a) Transfer characteristics of proposed GR-HGD-HJ-TFET on different gate dielectrics. (b) $SS_{Avg}$ and $I_{ON}/I_{amb}$ comparison report of GR-HGD-HJ-TFET on different gate dielectric applied. (c) Transfer characteristics of proposed GR-HGD-HJ-TFET on heterogate dielectric length variation.

**AI disclosure statement:** The author acknowledges the use of AI-based tools for minor language editing or grammatical correction. However, all intellectual contributions, analyses, and conclusions presented in this chapter are entirely the author's own.

# References


1. Abdi, D.B. and Jagadesh Kumar, M., Suppressing Ambipolar Conduction Using Dual Material Gate in Tunnel-FETs Having Heavily Doped Drain. *Int. J. Electron. Commun. Eng.*, 10, 5, 2016, doi:10.5281/zenodo.1124043.
2. Dutta, R., Konar, S.C., Paitya, N., Influence of Gate and Channel Engineering on Multigate Tunnel FETs: A Review, in: *Computational Advancement in Communication Circuits and Systems. Lecture Notes in Electrical Engineering*, vol. 575, K. Maharatna, M. Kanjilal, S. Konar, S. Nandi, K. Das (Eds.), Springer, Singapore, 2020, https://doi.org/10.1007/978-981-13-8687-9_31.
3. Ahmed, M., Hammam, M., Schmidt, M.E., Muruganathan, M., Suzuki, S., Mizuta, H., Sub-10 nm graphene nano-ribbon tunnel field-effect transistor. *Carbon*, 126, 588–593, 2018.
4. Anju, T.S., Nigam, K., Pandey, S., Sharma, D., Kondekar, P., Investigation of gate material engineering in junctionless TFET to overcome the trade-off

between ambipolarity and RF/linearity metrics. *Superlattices Microstruct.*, 109, 307–315, 2017, doi:10.1016/j.spmi.2017.03.059.

5. Dutta, R., Paitya, N., Majumdar, A., Ambipolar Reduction Methodology for SOI Tunnel FETs in Low Power Applications: A Performance Report. *International Journal of Recent Technology and Engineering (IJRTE)*, vol. 8, January 2020, ISSN: 2277-3878.
6. Avci, U.E., Morris, D.H., Young, I.A., Tunnel Field-Effect Transistors: Prospects and Challenges. *J. Electron Devices Soc.*, 3, 3, 88–95, 2015, doi: 10.1109/JEDS.2015.2390591.
7. Dutta, R., Guha, A., Rahaman, M., Paitya, N., DC Performance Analysis of Heterojunction Tunnel FET by Optimizing Various High-κ Materials: $HfO_2$/$ZrO_2$ with Low/High-κ Spacer, in: *Intelligent Techniques and Applications in Science and Technology. ICIMSAT 2019. Learning and Analytics in Intelligent Systems*, vol. 12, S. Dawn, V. Balas, A. Esposito, S. Gope (Eds.), Springer, Cham, 2020, https://doi.org/10.1007/978-3-030-42363-6_109.
8. Dutta, R. and Paitya, N., Novel InAs/Si Heterojunction Dual-Gate Triple Metal P-i-N Tunneling Graphene Nanoribbon Field Effect Transistor (DG-TM-TGNFET) For High-Frequency Applications, in: *Generation, Detection and Processing of Terahertz Signals. Lecture Notes in Electrical Engineering*, vol. 794, A. Acharyya, A. Biswas, P. Das (Eds.), Springer, Singapore, 2022, https://doi.org/10.1007/978-981-16-4947-9_17.
9. Biswas, C. and Lee, Y.H., Graphene versus carbon nanotubes in electronic devices. *Adv. Funct. Mater.*, 21, 3806–3826, 2011.
10. Dutta, R., Nonconventional Heterostructure Tunnel FET for Sensitive Tera Hertz Detection, in: *Terahertz Wireless Communication Components and System Technologies*, M. El Ghzaoui, S. Das, T.R. Lenka, A. Biswas (Eds.), Springer, Singapore, 2022, https://doi.org/10.1007/978-981-16-9182-9_11.
11. Dutta, R. and Paitya, N., Analysis of InN-Based Surrounded Gate Tunnel Field-Effect Transistor for Terahertz Applications, in: *Emerging Trends in Terahertz Solid-State Physics and Devices*, A. Biswas, A. Banerjee, A. Acharyya, H. Inokawa, J. Roy (Eds.), Springer, Singapore, 2020, https://doi.org/10.1007/978-981-15-3235-1_6.
12. Cai, J., Han, X., Wang, X., Meng, X., Atomic Layer Deposition of Two Dimensional Layered Materials: Processes, Growth Mechanisms, and Characteristics. *Matter*, 2, 3, 587–630, 2020, doi: 10.1016/j.matt.2019.12.026.
13. Chandan, V., Nigam, K., Sharma, D., Approach on electrically doped TFET for suppression of ambipolar and improving RF performance. *IET Circuits Devices Syst.*, 13, 6, 787–792, 2019, doi: 10.1049/iet-cds.2018.5394.
14. Dutta, R., Paitya, N., Rahaman, M., Guha, A., Kumari, P., Impact of Different III-V material used in Source of Dual Gate Tri-Metal n-type Tunnel FET for Improved RF Performance. *2021 Second International Conference on Electronics and Sustainable Communication Systems (ICESC)*, Coimbatore, India, pp. 333–338, 2021, doi: 10.1109/ICESC51422.2021.9532794.

15. Dutta, R. and Paitya, N., Influence of Triple Material on Performance Study of Double Gate PiN Tunneling Graphene Nanoribbon FET for Low Power Logic Applications. *J. Nano Electron. Phys.*, 13, 3, 03020, 2021, doi: https://doi.org/10.21272/jnep.13(3).03020.
16. Choi, W.Y. and Lee, H.K., Demonstration of hetero-gate-dielectric tunneling field effect transistors (HG TFETs). *Nano Convergence*, 3, 13, 2016, doi:10.1186/s40580016-0073-y.
17. Aswathy, M., Biju, N.M., Komaragiri, R., Comparison of a 30nm Tunnel Field Effect Transistor and CMOS Inverter Characteristics, in: *2013 Third International Conference on Advances in Computing and Communications (ICACC)*, Cochin, pp. 149–152, 2013, doi: 10.1109/ICACC.2013.36.
18. Vijh, M., Gupta, R.S., Pandey, S., Graphene Based Tunnel Field Effect Transistor for RF Applications. *2019 PhotonIcs & Electromagnetics Research Symposium - Spring (PIERS-Spring)*, Rome, Italy, pp. 256–259, 2019, doi: 10.1109/PIERS-Spring46901.2019.9017872.
19. Dhanaselvam, P., Karthikeyan, B., Kavitha, K., Kavitha, P., Basic Concepts of Heterojunction Tunnel Field-Effect Transistors. *Field Effect Transistors*, 221-234, 2025.
20. Shamloo, H. and Yazdanpanah Goharrizi, A., Computational study of bilayer armchair graphene nanoribbon tunneling field-effect transistors for digital circuit design. *Diamond Relat. Mater.*, 151, 111759, 2025, ISSN 0925-9635, Doi:10.1016/j.diamond.2024.111759.
21. Shamloo, H. and Yazdanpanah Goharrizi, A., Performance study of tunneling field effect transistors based on the graphene and phosphorene nanoribbons. *Micro Nanostruct.*, 169, 207336, 2022, ISSN 2773-0123, https://doi.org/10.1016/j.micrna.2022.207336.
22. Dhanaselvam, P.S., Karthikeyan, B., Anand, P., Future Prospective Beyond CMOS Technology Design. *Field Effect Transistors*, 73–80, 2025.
23. Dutta, R., Paitya, N., Subash, T.D., Electric Field and Surface Potential Analytical Modeling of Novel Double Gate Triple Material PiN Tunneling Graphene Nano Ribbon FET (DG-TM-PiN-TGNFET). *Silicon*, 13, 719–727, 2021, https://doi.org/10.1007/s12633-020-00474-6.
24. Dutta, R. and Paitya, N., Effect of Pocket Intrinsic Doping on Double and Triple Gate Tunnel Field Effect Transistors, in: *Proceedings of the 2nd International Conference on Communication, Devices and Computing. ICCDC 2019. Lecture Notes in Electrical Engineering*, Springer, Singapore, vol. 602, 2020, https://doi.org/10.1007/978-981-15-0829-5_25.
25. Kampen, C., Burenkov, A., Lorenz, J., Challenges in TCAD simulations of tunneling field effect transistors. *IEEE Eur. Solid-State Device Res. Conf. Proc. (ESSDERC)*, 139–142, 2011.

# 18

# Hybrid Field Effect Transistor

**K. Kavitha[1], P. Suveetha Dhanaselvam[1], B. Karthikeyan[1*], S. Nagarajan[2], G. Pradeep Kumar[1] and P. Karthikeyan[1]**

[1]*Department of Electronics and Communication Engineering, Velammal College of Engineering and Technology, Madurai, Tamilnadu, India*
[2]*Department of Electronics and Communication Engineering, Easwari Engineering College, Chennai, Tamilnadu, India*

***Abstract***

The relentless pursuit of miniaturization and energy efficiency in semiconductor devices has pushed conventional MOSFETs to their physical and operational limits, signaling the end of Moore's Law. To address these challenges, researchers have turned to innovative Field-Effect Transistor (FET) architectures that leverage novel materials and operating principles. Among these, Ferroelectric FETs (FeFETs), Tunnel FETs (TFETs), and Junctionless FETs (JL-FETs) have emerged as promising candidates, each offering unique advantages such as non-volatility, steep subthreshold slopes, and simplified fabrication. However, no single FET technology can simultaneously address all the requirements of modern electronics, including low power consumption, high performance, and scalability. This has led to the development of hybrid FET architectures, which combine the strengths of multiple FET technologies to create devices that surpass the capabilities of their individual counterparts. This chapter provides a comprehensive exploration of hybrid FETs, beginning with an overview of the limitations of conventional MOSFETs and the motivations for adopting advanced FET architectures. The foundational principles of FeFETs, TFETs, and JL-FETs are discussed in detail, highlighting their unique mechanisms such as polarization-induced threshold control in FeFETs, band-to-band tunneling in TFETs, and electric field modulation in JL-FETs. By bridging the gap between conventional and emerging transistor technologies, hybrid FETs represent a transformative approach to semiconductor design, offering a pathway to next-generation electronics that are faster, smaller, and more energy-efficient.

***Keywords*:** MOSFET, FeFET, TFET, JL-FET, Hybrid FET, semiconductor technology

**Corresponding author*: om.bkarthikeyan@gmail.com

P. Suveetha Dhanaselvam, Naima Guenifi, Shiromani Balmukund Rahi and J. Ajayan (eds.) Field-Effect Transistor Devices: Post-CMOS Theory and Applications, (347–380) © 2026 Scrivener Publishing LLC

## 18.1 Introduction

The relentless pursuit of miniaturization in semiconductor technology has driven remarkable advances in microelectronics over the past few decades. As traditional MOSFET architectures approach their physical and performance limits, the semiconductor industry faces critical challenges in sustaining power efficiency, switching speed, and device reliability. The demand for high-performance, low-power electronics in emerging applications such as edge computing, artificial intelligence, and energy-efficient IoT devices has intensified the search for alternative transistor designs that can extend or even replace conventional CMOS technologies. Among the various alternatives being explored, several next-generation field-effect transistor (FET) architectures have shown promise by addressing specific bottlenecks in power scaling, electrostatic control, and fabrication complexity. For instance, ferroelectric FETs exploit the spontaneous polarization of ferroelectric materials to modulate channel conduction, enabling non-volatile behavior and nearly zero standby power. Tunnel FETs, on the other hand, leverage quantum tunneling for carrier injection, allowing operation below the thermal limit of subthreshold slope and significantly reducing leakage current. Junctionless FETs simplify doping profiles by using a uniformly doped channel, reducing fabrication complexity and short-channel effects.

Each of these device types presents unique advantages, but they also come with inherent trade-offs. Ferroelectric materials may suffer from endurance degradation, tunnel FETs often exhibit low ON currents, and junctionless devices may face scaling limitations in controlling electrostatics effectively. Recognizing that no single architecture may fully satisfy the multifaceted requirements of future semiconductor applications, this chapter introduces the concept of a hybrid FET design that unifies the key strengths of these advanced transistor types. By combining the memory-retention capability of ferroelectric FETs, the ultra-low-power potential of tunnel FETs, and the fabrication simplicity of junctionless structures, a new class of hybrid FETs can be envisioned. This integrated approach not only offers a path to optimize energy efficiency and switching behavior simultaneously, but it also opens up new possibilities in circuit-level design, especially for logic-in-memory systems, neuromorphic platforms, and edge-AI accelerators. Moreover, the feasibility of such hybrid structures is increasingly supported by advances in material engineering, nanoscale device modeling, and 3D integration techniques. Through an in-depth exploration of these hybrid architectures, this chapter aims to

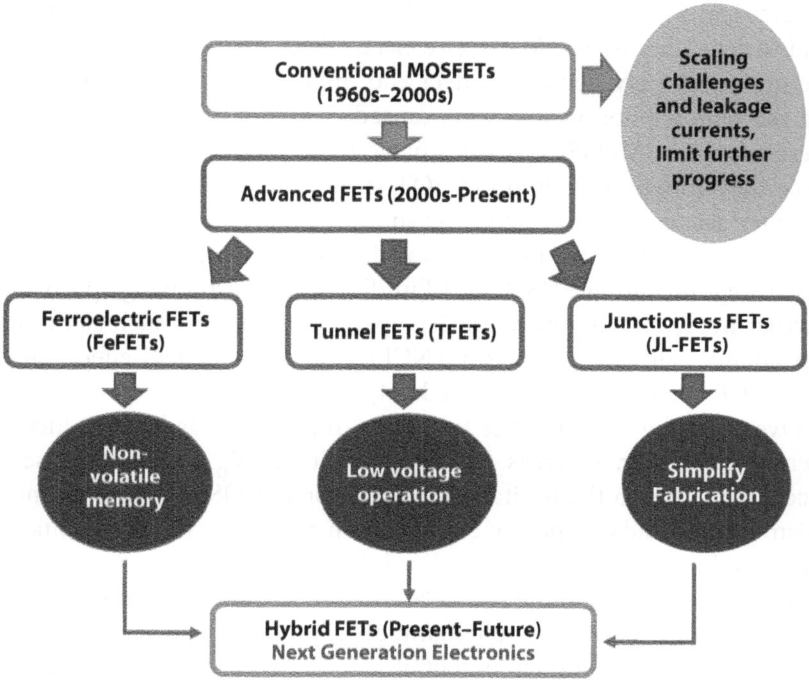


**Figure 18.1** Evolution of FET architectures.

present a forward-looking perspective on the evolution of transistor design in the post-CMOS landscape. Figure 18.1 depicts the evolution of FET architectures.

Over the past few decades, the dimension of complementary MOSFET (CMOS) has been aggressively reduced with the guidance of Moore's law. The scaling enables much denser circuit and more functional chip in a given area and decreases the manufacturing cost. However, simple scaling has encountered more and more challenges due to fundamental device physics limitations. Due to bad electrostatic gate control, short channel effects (SCEs) lead to threshold voltage ($V_{TH}$) roll-off and drain-induced barrier lowering (DIBL). In addition, the velocity saturation and ballistic transport further limit the improvement in device performances as the gate length is shrunk down below 100 nm or even smaller. Note that the total capacitance can be also scaled, and thus the speed of circuit is improved even without enhancement in the drive current. Gate tunneling leakage also becomes an issue in power consumption when the gate oxide thickness is down below 3 nm. The high- k dielectric was introduced to suppress the gate leakage by keeping the effective oxide thickness (EOT) but it could lead to reliability issues. Metal gate electrode is also adopted to eliminate the undesired poly-depletion effect and improve the mobility by screening the remote phonon scattering caused in high-k dielectrics. Even with these advances, MOSFET is still unable to be further aggressively scaled since the limitation of thinning EOT is inevitable. Therefore, to provide stronger gate control to resolve the short channel effects, a new approach is required to allow future reduction in channel length. The multi-gate transistors, including FDSOI and FinFET, are the promising approaches. Furthermore, applying novel device operation concepts, such as negative capacitance field-effect transistor (NCFET) and tunnel field-effect transistor (TFET), has been also proposed [34–43].

Over past few decades, the transistor dimension has been reduced as mentioned earlier, whereas the supply voltage ($V_{DD}$) cannot be scaled accordingly due to the limitation of traditional MOSFET turn-on mechanism which leads to power consumption problem. The gate voltage is capacitively coupled to the channel and changes the surface potential of the channel. The carriers in the source allowed to surmount the potential barrier statistically follow the Boltzmann distribution, which is exponentially proportional to the surface potential [24, 25, 43–61]. The drain current in the subthreshold region thus is an exponential function of the surface potential ($\Phi_S$) follows Equation 18.1.

$$\varphi_S \propto exp\left(\frac{q\varphi_S}{k_B T}\right) \tag{18.1}$$

Even the gate voltage is able to perfectly couple to the surface potential; one order magnitude of drain current only can be obtained by 60 mV or more at room temperature. In other words, the subthreshold slope (SS) cannot exceed 60 mV/decade. This is called the Boltzmann Tyranny. In real devices, the gate voltage cannot be perfectly coupled to the channel due to the additional interfacial capacitance and drain-coupled capacitance (SCEs), which is also true for the state-of-the-art multi-gate transistor [24, 25, 43–56]. TFET and NCFET are promising candidates for future technologies, such as low power Internet of Things (IoT), since they are fully compatible with the modern semiconductor processes.

## 18.2 Ferroelectric FETs (FeFETs)

Ferroelectric Field-Effect Transistors (FeFETs) represent a transformative approach in transistor engineering, primarily by integrating a ferroelectric material within the gate stack of a conventional MOSFET structure. This material has a unique property: it can "remember" the direction of an electric field applied to it, even after the field is removed. The core innovation lies in the utilization of remanent polarization, the ability of the material to retain an electric dipole moment even after the external electric field is removed, to modulate the channel conductance [1]. This property not only enables standard transistor switching but also introduces non-volatile memory capabilities, making FeFETs highly attractive for future logic-in-memory and low-power embedded systems. At a fundamental level, the FeFET operates similarly to a traditional MOSFET, with source, drain, and channel regions formed in a semiconductor substrate as shown in Figure 18.2 [2]. The critical distinction lies in the gate dielectric, where the conventional silicon dioxide or high-k dielectric is replaced or supplemented with a ferroelectric material commonly doped hafnium oxide ($Hf_XZr_{1-X}O_2$, abbreviated as HZO) in scaled devices due to its compatibility with CMOS processes.

When a voltage is applied to the gate of the FeFET, the ferroelectric layer inside it switches its polarization, much like flipping a switch. This change in polarization affects how easily electricity can flow through the transistor. If the polarization is in one direction, the transistor allows current to flow (it's "on"). If it's in the opposite direction, the transistor blocks the current

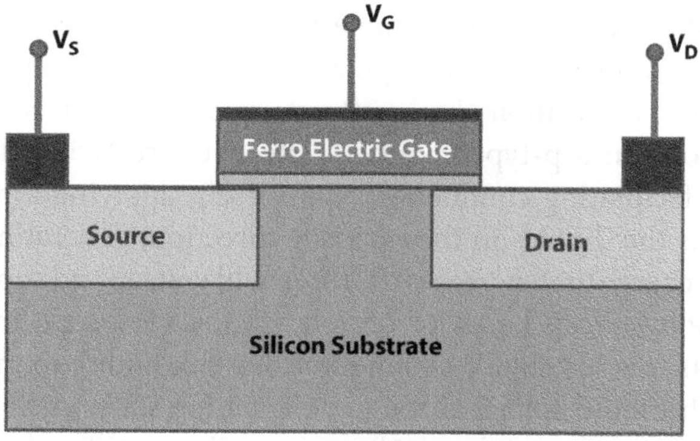


**Figure 18.2** Ferro electric field effect transistor (FeFET).

(it's "off"). The ferroelectric layer exhibits a characteristic double-well energy landscape, with two stable polarization states corresponding to the "up" and "down" orientation of dipoles. These states result in distinct threshold voltages for the device: one associated with accumulation or enhancement mode (device ON), and the other with depletion or inversion mode (device OFF). The double-well energy landscape describes how the material's electric polarization can switch between two stable states. Ferroelectric materials have tiny regions called domains, where the electric dipoles (like tiny arrows representing the direction of charge separation) can align in one direction or the opposite direction. These two alignment directions correspond to the two "valleys or wells" in the energy landscape. The hill between the valleys represents the energy barrier that must be overcome to switch the polarization from one direction to the other. When an external electric field is applied to the ferroelectric material, it tilts the energy landscape, making one valley lower in energy than the other. This encourages the dipoles to align in the direction of the lower-energy valley. Once the field is removed, the material stays in that state because the energy barrier prevents it from spontaneously switching back. This is why ferroelectric materials have a "memory" effect and they retain their polarization state without needing continuous power. Unlike in conventional MOSFETs, the position of the threshold voltage in an FeFET is not fixed but depends on the polarization state of the ferroelectric material, which can be altered *via* the applied gate voltage and retained even after the voltage is removed [3, 4].

The dynamic control of threshold voltage ($V_{th}$) in an FeFET arises directly from the ability of the ferroelectric layer to influence the charge environment at the channel interface. When a positive gate voltage exceeding the coercive field of the ferroelectric material is applied, the dipoles align in a direction that induces a positive polarization charge at the gate-channel interface. This polarization charge attracts electrons in an n-type channel (or repels holes in a p-type channel), effectively reducing the threshold voltage and facilitating conduction. Conversely, applying a negative gate voltage aligns the dipoles in the opposite direction, generating a negative polarization charge that increases the threshold voltage and suppresses conduction. The hysteresis behavior of polarization versus electric field (P-E curve) results in a bistable $V_{th}$ configuration. This ability to modulate and lock-in the threshold voltage through polarization states enables FeFETs to store binary logic values ('0' and '1') intrinsically, providing both switching and memory functions within the same device. Importantly, this threshold voltage modulation is non-volatile: once the polarization is set, it remains stable over time and does not require continuous power to maintain. This contrasts with dynamic RAM or standard CMOS logic, where charge storage is volatile and must be refreshed or constantly powered [5].

FeFETs bring substantial advantages in memory applications due to their inherent non-volatility, fast switching speed, and low power operation. In memory configurations, the two polarization states can represent logical '0' and '1', enabling FeFETs to function as single-transistor memory elements (1T). This stands in contrast to traditional non-volatile memory technologies such as Flash, which rely on charge trapping and require complex structures and high write voltages. One of the key benefits of FeFET-based memory lies in its ability to perform write and read operations rapidly (typically in nanoseconds) and with significantly lower energy per bit compared to Flash. Furthermore, due to the CMOS-compatible nature of modern ferroelectric materials like HZO, FeFETs can be integrated monolithically onto logic platforms, allowing the implementation of in-memory computing architectures. This integration drastically reduces the latency and energy associated with data movement between separate logic and memory blocks, addressing one of the critical bottlenecks in modern computing known as the von Neumann bottleneck [6, 7].

Another promising application area is in neuromorphic and machine learning hardware. FeFETs can be configured as synaptic devices in artificial neural networks, where the polarization state encodes synaptic weights. Due to their analog-like behavior and multi-level polarization potential, FeFETs can store more than two states, enabling high-density, low-power learning systems that mimic the brain's processing style. Despite their promise, FeFETs also face challenges. These include endurance degradation due to repeated polarization switching (fatigue), retention loss from depolarization fields, and variability in threshold voltage due to fabrication inconsistencies. Nonetheless, ongoing advancements in material science, such as dopant engineering and interface control, continue to improve the reliability and scalability of FeFETs [8, 9]. FeFETs represent a powerful convergence of logic and memory functions, driven by the physical properties of ferroelectric polarization. Their ability to retain state without power, modulate threshold voltage dynamically, and operate at low voltages positions them as a cornerstone technology for future computing systems, particularly in domains where energy efficiency and memory-logic co-integration are paramount.

Figure 18.3(a) shows the metal-ferroelectric-insulator-semiconductor (MFIS) FeFET structure, which dopes $HfO_2$ as ferroelectric (Fe) layer in the gate stack of the transistor. The equivalent circuit is also shown in Figure 18.3(b), where the ferroelectric capacitance ($C_{FE}$) couples with the capacitance of the underlying MOSFET (CMOS). When the ratio of $C_{FE}$/CMOS is sufficiently low, the polarization state of the ferroelectric layer

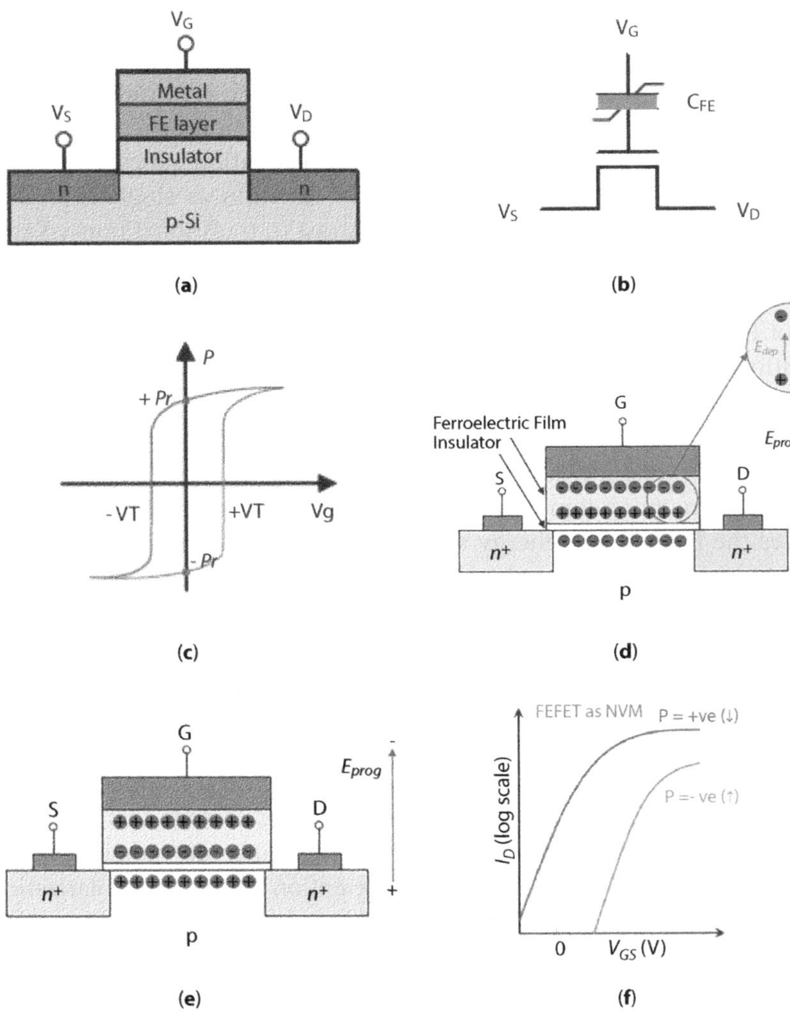


**Figure 18.3** Structure and properties of FeFETs: (a) FeFET device structure; (b) the equivalent circuit is composed of a nonlinear ferroelectric capacitor in series with a MOSFET; (c) hysteresis loop and corresponding voltage operating range.

can be retained, resulting in a hysteresis characteristic curve, as shown in Figure 18.3(c), which generates a storage window for the device.

The working principle of the FeFET is that the ferroelectric layer can exert a field effect on the channel through polarized charges, thereby controlling the threshold voltage of the transistor. For N-type FeFET, when a positive gate-source voltage Vgs > +VT is applied, the ferroelectric domain in the ferroelectric layer will flip to the positive polarization state, attracting electrons from the drain and source to the channel, resulting in an "ON" state corresponding to a low threshold voltage. Similarly, applying a negative gate-source voltage Vgs < −VT will flip the ferroelectric domain to the negative polarization state, increase the holes provided by the substrate, increase the threshold voltage, and the FeFET is in the "OFF" state. As shown in Figure 18.3c, this writing process has hysteresis characteristics, so when the gate-source voltage is not enough to flip the polarization state of the ferroelectric domain, the remnant polarization in the ferroelectric layer will maintain the device switching state, which enables NV functionalities. Due to the inherent gain of the underlying MOSFET, the $I_{ON}/I_{OFF}$ ratio of FeFETs corresponding to the two switching states can be up to $10^6$. In addition, the three-terminal structure of FeFETs separates the state writing (by applying sufficient positive/negative Vgs) and state reading paths (*via* drain-source current).

The LK equation 18.2 describes the relationship between the applied electric field and the internal polarized charges within the gate. This equation is nonlinear with ferroelectric material parameters, $\alpha$, $\beta$, and $\gamma$ are and the static coefficients and $\rho$ is a kinetic coefficient associated with the ferroelectric material.

$$E = \alpha P + \beta P^3 + \gamma P^5 + \rho \frac{dP}{dt} \tag{18.2}$$

A ferroelectric material (FE) exhibits an S-shaped polarization (P) versus electric field (E) characteristic as shown Figure 18.3. Since the polarization is related to charge and the field to voltage, the *Q-V* curve is also S-shaped with a negative capacitance region known as Landau switch. The concept of the FeFET consists in integrating the ferroelectric material (Fe) in the gate stack. In OFF- state the ferroelectric material is negatively polarized with negative charges on top of the semiconductor interface. This causes VT is high and $I_{OFF}$ is reduced. It increases $V_G$ switches the ferroelectric polarization, bringing positive charges on top of the electron channel at low VT state. The buildup of the inversion charge with $V_G$ is accelerated by the concomitant decrease in threshold voltage. More recently, sub-$kT/q$ swing was demonstrated in long MOSFETs with HfZrO-based gate dielectrics, operated in deep subthreshold and saturation regions ($V_G$ << VT and $|V_{DD}| > 0.5$ V). The sharp switch disappeared in the ohmic region and there was no evidence for negative capacitance effect.

FeFET is basically a regular transistor which has a ferroelectric layer in its gate stack, as shown in (Figure 18.3(a)). As a result of its simple structure, it provides excellent CMOS compatibility and scalability. Besides its advantages FeFET has also drawbacks. One of the issues of FeFET is gradual loss polarization on its ferroelectric layer due its physical features. Also programming the device so many times or programming it to its highest possible polarization level reduces the endurance. The only difference between a FeFET and a conventional MOSFET is the additional ferroelectric layer placed between the metal and the insulator layers at the gate stack of the FeFET, as shown in Figure 18.3(a)). Data is stored by programming the ferroelectric layer to different polarization states. During write operations, an electrical field higher than a certain threshold (i.e., coercive force), is applied. The direction of the electrical field determines the polarization of the ferroelectric layer, known as, the final position of the dipoles. The physical operation of FeFET is shown in Figure 18.3(d) and Figure 18.3(e). Figures 18.3(d) and 18.3(e) show the biasing conditions when programming logic `1' and `0', respectively. The remnant polarization is used to denote the remaining polarization after the programming voltages are removed from all the terminals of the FeFET. The two opposite remnant polarization states creates two deferent VT states corresponding to logic `1' and `0', as shown in Figure 18.3(c). Due to the VT difference, the FeFET requires specific gate voltages based on its logic state to turn on, and the difference between those gate voltages is called memory window. The polarization of the ferroelectric layer determines the conductivity of the FeFET, which represents the data stored on the FeFET. Since the body of the FeFET is p-type, the more negative ions are attracted to the FeFET's channel, the higher conductivity will be demonstrated when the same gate voltage is applied. Although there exist the maximum levels that the two polarization states of the ferroelectric layer can reach, the ferroelectric layer can be exibly programmed to any intermediate polarization level between these two maximum levels through adjusting the amplitude and/or the pulse width of the programming voltage. Figure 18.3(e), shows the I-V characteristics of FeFET.

Figures 18.3(d) and 18.3(e) show the biasing conditions when programming logic `1' and `0', respectively and 18.3(e) transfer characteristic FeFET and picturization of non-volatile memory (NVM) in device.

## 18.3 Tunnel Field-Effect Transistors

Tunnel Field-Effect Transistors (TFETs) have emerged as a promising alternative to conventional MOSFETs for ultra-low-power applications, primarily due to their fundamentally different carrier injection mechanism. While MOSFETs rely on thermionic emission over a potential barrier, TFETs operate based on band-to-band tunneling (BTBT), a quantum mechanical phenomenon that enables subthreshold operation beyond the classical thermal limit of 60 mV/dec at room temperature. This unique transport mechanism offers a pathway toward achieving steep subthreshold slopes, reduced leakage currents, and energy-efficient switching, particularly critical as device dimensions approach the nanometer regime and supply voltages continue to scale [10–12].

TFETs are composed of four fundamental components source, drain, gate, and channel. Each of these components plays a crucial role in the operation of the TFET, working together to enable the tunneling mechanism that distinguishes TFETs from traditional transistors like MOSFETs. The working principle of tunnel field-effect transistors (TFETs) is fundamentally different from that of conventional MOSFETs. TFETs follow the quantum mechanical phenomenon of band-to-band tunneling (BTBT) to control the flow of current, enabling them to operate efficiently at lower voltages. When a voltage is applied to the gate (Vgs), it modulates the energy bands in the source and the channel. If the gate voltage is high enough, it aligns the energy bands of the source and channel such that the conduction

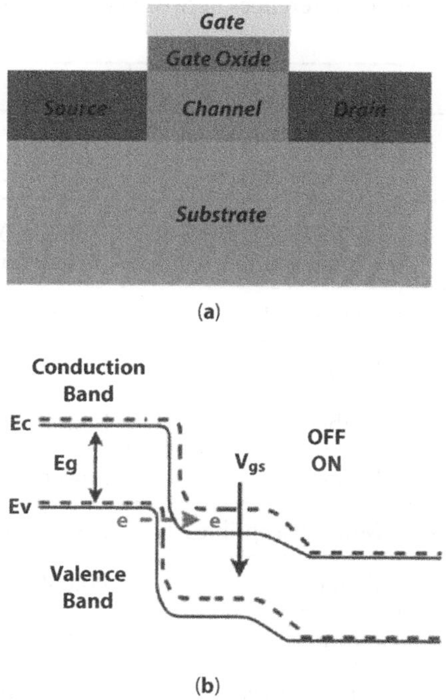


**Figure 18.4** (a) Device schematic of tunnel FET, (b) tunnel FET operation.

band of the source overlaps with the valence band of the channel shown in Figure 18.4(b). This alignment creates a narrow barrier through which electrons can tunnel as shown in Figure (18.4(b)). TFETs can achieve a subthreshold swing below the 60 mV/decade limit of conventional MOSFETs. This results in faster switching times and improved performance at low operating voltages. The unique characteristics of TFETs are particularly advantageous in high-speed computing applications and for maintaining performance as devices continue to shrink in size. Table 18.1 shows the summary and of differences between conventional MOSFETs vs. TFETs [28, 56, 62–67].

The core structure of a TFET as shown in Figure 18.4 resembles that of a reverse-biased p-i-n diode with a gate terminal placed over the intrinsic (or lightly doped) channel region to modulate the tunneling junction. The source region is heavily doped with opposite polarity to the drain—typically $p^+$ for n-type TFETs and $n^+$ for p-type devices. Unlike a MOSFET, where the gate voltage modulates a thermionic barrier, in a TFET the gate controls the alignment of energy bands between the source and the channel to enable or suppress quantum tunneling. When a gate voltage is applied, it lowers the

**Table 18.1** Summary and of differences between conventional MOSFETs vs. TFETs [28, 56, 62–67].

| Feature | MOSFETs | TFETs |
|---|---|---|
| Working principle | Thermionic emission over a barrier | Quantum tunneling through a barrier |
| Structure | Single semiconductor material | Potentially different materials (heterojunction) |
| Subthreshold swing | ≥ 60 mV/decade | < 60 mV/decade |
| On-state current | Higher | Lower |
| Power consumption | Higher | Lower |
| Leakage current | Higher | Lower |
| Short-channel effects | More susceptible | Less susceptible |
| Switching speed | Faster | Potentially slower |
| Scalability | Limited | Better suited for scaling |
| Applications | Broad range, high-performance | Low-power, emerging technologies |

energy barrier in the intrinsic channel such that the valence band of the $p^+$ source aligns with the conduction band of the channel. This alignment creates a narrow tunneling window, allowing electrons to directly tunnel from the valence band of the source into the conduction band of the channel. The current resulting from this process is band-to-band tunneling current, which is highly sensitive to the electric field and energy overlap, and largely independent of thermal energy. As a result, TFETs can achieve a subthreshold swing lower than the MOSFET thermal limit, enabling switching at much lower supply voltages ($V_{DD}$), often below 0.5 V [13].

One of the most distinguishing features of TFETs is their ability to attain subthreshold slopes (SS) below 60 mV/dec, which is the theoretical minimum for MOSFETs (as shown in Figure 18.5) at room temperature due to the Boltzmann distribution of carrier energy. In TFETs, since the current

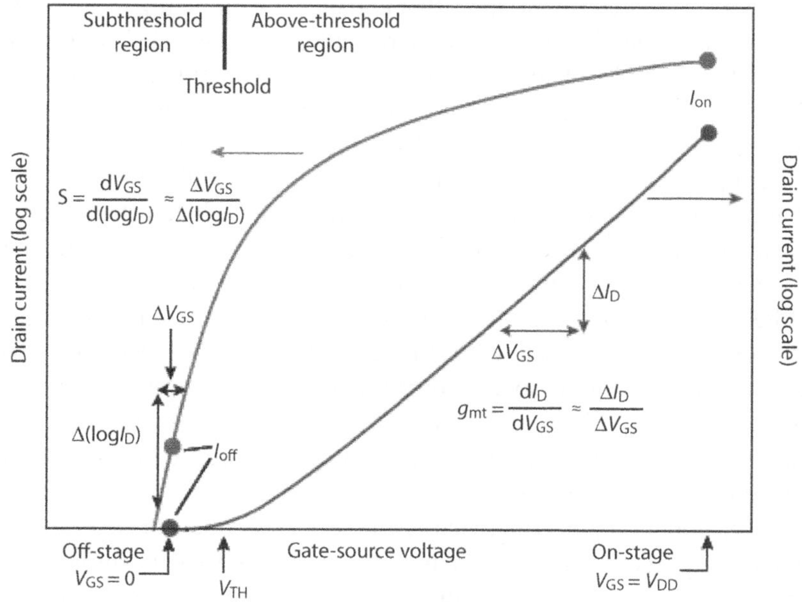


**Figure 18.5** FET transfer characteristics showing $I_D$ (on a logarithmic scale on the left and a linear scale on the right) versus the gate–source voltage, $V_{GS}$. The transistor is considered to be switched on when $V_{GS}$ is equal to the maximum voltage supplied to the device, $V_{DD}$. The higher the slope in the subthreshold region ($V_{GS} < V_{Th}$), the better the transistor switch-on characteristics become. In conventional MOSFET is limited subthreshold slopes (SS) and the thermionic emission that makes a minimum SS of 60 mV/dec to metal–oxide–semiconductor (MOS) at room temperature the complementary metal oxide semiconductor (CMOS) technologies do not support much design space in the sub 0.5 V operating regime.

is driven by tunneling rather than thermal excitation, the SS can, in principle, be reduced to values as low as 20–30 mV/dec, depending on the material system and device engineering. This steep switching behavior is crucial for reducing dynamic power consumption, as it allows the device to turn ON and OFF effectively even with a significantly lowered supply voltage. Consequently, TFETs can operate in the subthreshold region while still delivering acceptable ON current levels, making them highly attractive for energy-constrained applications such as wearable electronics, implantable medical devices, and edge-AI sensors. However, achieving a balance between low OFF-state leakage and adequate ON-state current ($I_{ON}$) remains a challenge. Since tunneling is a relatively weak process compared to thermionic emission, TFETs often exhibit lower I_ON than MOSFETs, which limits their use in high-performance logic. To address this, researchers have explored material systems with narrower bandgaps and higher tunneling probabilities, such as III–V semiconductors (e.g., InAs, GaSb), 2D materials (e.g., $MoS_2$), and heterostructures that enhance the tunneling window [14].

Material engineering plays a pivotal role in optimizing TFET performance. For effective band-to-band tunneling, the source-channel junction must exhibit abrupt band alignment and minimal tunneling distance. This necessitates high-quality interfaces and bandgap engineering. Heterojunction TFETs, for example, use source and channel materials with different bandgaps to enhance tunneling probability. GaSb/InAs and Ge-SiGe heterostructures have shown particular promise due to their staggered (Type-III) band alignments that facilitate efficient tunneling. Apart from material choices, novel device architectures are being explored to improve electrostatic control and scaling behavior. Vertical TFETs allows the tunneling direction to be perpendicular to the substrate, enabling compact layouts and better control over channel length. Double-gate or gate-all-around (GAA) TFETs enhance gate control and reduce leakage. In Dielectric-engineered TFETs, the gate dielectric thickness is non-uniformly modulated to improve the electric field in the tunneling region. Moreover, doping engineering, such as source pocket doping or charge plasma-based doping-free TFETs, has been investigated to fine-tune the tunneling barrier and optimize performance without relying on abrupt dopant profiles.

Due to their ultra-low power consumption and steep switching capabilities, TFETs are ideal candidates for energy-sensitive applications where battery life and thermal management are critical. TFETs are also gaining interest for security-sensitive hardware, as their low leakage and unique switching characteristics make them less susceptible to side-channel attacks that exploit power signatures. Despite their promise, several challenges remain before widespread deployment is feasible. These include variability in threshold voltage due to sensitivity to interface traps, process-induced defects, and limited ON current in silicon-based implementations. Continued research into material synthesis, junction formation techniques, and scalable fabrication methods is essential to bring TFETs closer to commercial reality. Tunnel FETs offer a compelling path forward in the post-CMOS era, particularly for low-power applications where energy efficiency is prioritized over raw speed. Their quantum-driven operation breaks the fundamental limits imposed on MOSFETs, positioning them as a key technology in the next wave of semiconductor innovation [15].

Power dissipation has emerged as a critical concern in contemporary times. The faster scaling of the complementary metal-oxide-semiconductor (CMOS) sector has resulted in greater power loss, mostly as a result of the reduced feature size regulation. Since dynamic power dissipation is proportional to the square of the supply voltage ($P_{dyn} \propto V^2_{DD}$) lowering the supply voltage is one practical way to mitigate these problems. Lowering the supply voltage, however, also lowers the ON current, which decreases the transistor speed $I_{on} \propto (V_{DD} - V_T)^\alpha$. To maintain sufficient ON current and operating speed, the threshold voltage $V_T$ must be reduced. Unfortunately, lowering the threshold voltage increases the OFF current ($I_{OFF} \propto 10^{-V_T/SS}$), which raises the static power dissipation of the transistor ($P_{OFF} \propto I_{OFF}$). Effectiveness and power dissipation are always traded off, according to this contradictory criterion. The inherent 60 mV/decade limit of the subthreshold swing makes it difficult to reduce the operating voltage even while transistor dimensions continue to decrease as CMOS technology advances. The negative capacitance (NC) characteristic of ferroelectric materials facilitates a decrease in the subthreshold swing.

The most difficult challenge in modern semiconductor industry for integrated circuits (ICs) and systems are power consumption thermal management. Although with the advancement of material science and engineering technology an individual transistor produces a very small amount of heat, however integrated billions of components in a single chip, ICs and system produces huge amount of heat, is always a big challenge and becoming bottleneck issues for scientists and engineers. The supply voltage ($V_{DD}$) reduction is most effect tool and technique to resolve this issue, due to square dependency of power factor ($P_{dyn} \propto V_{DD}^2$). But practically, rigorous scaling of supply voltage ($V_{DD}$) technique reached to fundamental limits of 60mV/decade, so called so-called "Boltzmann Tyranny". To resolve the power issue and thermal management of large-scale application in integrated circuits (ICs) by reducing the power supply ($V_{DD}$). Because total power consumption, $P_{total} = P_{static} + P_{dynamic}$ and dynamicpower consumption followed the relation, $P_{dynamic} = \alpha \times C_{total} \times V_{DD}^2 \times f$. The used variable in dynamic power consumption having usual meaning. The low power supply ($V_{DD}$) scaling also helps to reduces the switching energy, followed by $\propto V_{DD}^2\left(\in + {}^{I_{OFF}}/_{I_{ON}}\right)$, here $\in$ is active time ratio. The standard $\in$ value is chosen $1/800$ for logic circuit and $1/10000$ for memory.

Subthreshold Swing S, written in Equation 18.3 is used to understand how sharply the inversion region is formed with the control of the gate voltage Vg, and it is inversely proportional to the subthreshold slope. This is strictly related to the depletion capacitance $C_d$ and $C_{ox}$

$$S \equiv \left(\frac{d[log(I_D)]}{dV_{GS}}\right)^{-1} = (ln10)\left(\frac{kT}{q}\right)\left(\frac{dV_{GS}}{d\varphi_S}\right) = (ln10)\left(\frac{kT}{q}\right)\left(\frac{C_{ox}+C_d}{C_{ox}}\right) \tag{18.3}$$

In Equation 18.3 $\psi_s$ is the *surface potential.* Also in this case, we got a problem with a capacitive voltage divider $C_{ox}$. To get a steeper transcharacteristic we have to minimize the subthreshold slope. For subthreshold slope minimize, first increasing $C_{ox}$ by minimizing the oxide thickness, leading to a certain increasing of gate-leakage, setting a lower bound for the $t_{ox}$. In second approach, decreasing $C_d$ by reducing the doping concentration. Conventional planar MOSFETs scaling stopped to be convenient when the physical constraints did not allow further improvements without worsening too much subthreshold slope, so stand-by power. In conventional MOSFET device technology, this physical device barrier is treated as fundamental thermionic limitation and poses barrier for aggressive $V_{DD}$ scaling down in VLSI circuits. The numeric value of this fundamental of conventional CMOS technology is ~ 60 mV/decade at room temperature. The numeric value of this room temperature is taken as 300K. The minimum possible, $V_{DD}$ is limited due to subthreshold swing(S) of $V_{DD} \geq S \times \log_{10}(I_{ON}/I_{OFF})$. In this conceptual $V_{DD}$ scaling limitation, $I_{ON}$ and $I_{OFF}$ having usual meaning, denoted current in on and off-state. To break the "Boltzmann Tyranny" limitation of conventional CMOS devices.

The S of classical MOS devices is modeled as:

$$\begin{aligned} S &= \left(\frac{d(log_{10}I_D)}{dV_{GS}}\right)^{-1} = \frac{dV_{GS}}{d\psi_S} \times \frac{d\psi_S}{d(log_{10}I_{DS})} = \left(1+\frac{C_d}{C_{ox}}\right) \times \frac{d\psi_S}{d(log_{10}I_{DS})} \\ &= m \times n \end{aligned} \tag{18.4}$$

As written in Equation 18.4, subthreshold swing, (S=Body Factor × Transport Factor), denoted as "m" and "n" of classical MOSFET. The term "m" defined as $\left(1+{}^{C_d}/_{C_{ox}}\right) \gg 1$, due to "positive" series connected gate oxide capacitance "$C_{OX}$". This shows that, the body factor for classical CMOS technology is greater than 1. In Equation 18.4, the term "n" is basically using for determining the band bending due to applied gate volatge, $V_{GS}$. Figure 18.6 shows modeling of conventional subthreshold swing(S), written in Equation 18.3 and Equation 18.4. This scientific cartoon clearly explore future scope ultra low power research and development with ferroelectric material having negative capacitance.

In recent years, negative capacitance (NC) has been widely studied due to its potential application in both organic and inorganic semiconductor devices. It is well known that the ongoing scaling of field effect transistors (FETs) urges people to find effective ways to reduce the power dissipation in the process of switching events. A key factor limiting the low voltage operation is the subthreshold swing S, which means that the voltage is needed for changing an order of magnitude of drain current. Recently, it

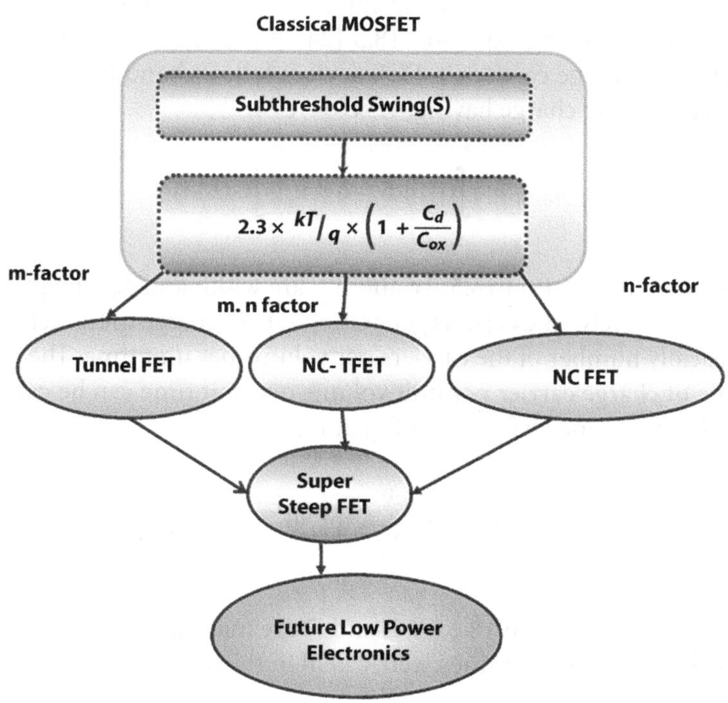


**Figure 18.6** Modeling of conventional subthreshold swing(S), written in Equation 18.3 and Equation 18.4.

has been suggested that the minimum voltage for traditional standard FETs at room temperature could be overcome by replacing the ordinary gate oxide by a ferroelectric oxide providing an effective negative capacitance (NC). The idea of overcoming the Boltzmann limit(S=60 mV/decade) by negative capacitance (NC) lies in the fact that in a series combination of a negative and a positive capacitor, the total capacitance becomes larger than its constituent positive capacitor. Interestingly and particularly, this is just the opposite of two positive capacitors, where the total capacitance is always reduced. For a ferroelectric gate FET, the negative gate capacitance can make the total capacitance, looking into the gate, larger than the semiconductor capacitance. Thus, to induce the same amount of charge in the channel, the ferroelectric gate FET would require a smaller gate voltage than what required for a traditional FET. That is to say, the gate voltage could be reduced below the Boltzmann limit(S<60 mV/decade).

### 18.3.1 Kane's Model

TFET follows band to band tunneling (BTBT) transport of charge carriers. The band-to-band (*B2B*) tunneling current ($I_{BTBT}$) in tunneling *FET* can be obtained the following Equation 18.5 [24–30]. In Equation 18.5, G(E) is generation of charge carriers in FET devices follows band to band tunneling, q is electronic charge having, $1.6x10^{-19}$ C

$$I_{BTBT} = q\int G(E)dv = qWL_{gate}\int G(E)dx \tag{18.5}$$

As shown in Equation 18.5, *W* and $L_{gate}$ are width and gate length TFET device respectively. The *G (E)* is generation rate of charge in tunneling FET. It is basically number of the carriers per volume per unit time. The generation rate of charge carrier per unit volume per unit time can be calculated by well known Kane's has written Equation 18.6

$$G(E) = A\frac{E^D}{\sqrt{E_g}}\exp\left(-B\,{E_g^{3/2}}/{E}\right) \tag{18.6}$$

As shown in Equation 18.6, A, B and D are material dependent parameters of Kane's model. E is the electric field in the device.

### 18.3.2 Tunneling Current Model

Another approach for determining the tunneling current in Tunnel FET is Wentzel–Kramers–Brillouin (WKB) approximation method. The WKB approach has modeled as (Eq. (18.7)).

$$I_{ON} \propto T_{WKB} \approx exp\left(-\frac{4\lambda E_g^{3/2}\sqrt{2m^*}}{3q\hbar\left(E_g + \Delta\varnothing\right)}\right) \tag{18.7}$$

In Eq. 18.7, m* is the effective mass and $E_g$, is the band gap. λ is the tunneling distance between the source and the channel, ΔΦ is the band energy difference between the conduction band of source and valence band of the channel, ħ, q are reduced the Plank's constant and electronic charge respectively, while $t_{ox}$ and $t_{si}$ are physical oxide thickness and semiconductor thickness respectively. The differents heterostructures have been proposed based on III/V or Si/Ge nanowires obtained by top-down, and bottom-up approaches in order to reduce the sub-threshold slope (SS) and increase the $I_{on}$ current. Moreover, group IV semiconductor is a preferred material for making TFET because they are easily integrable on a Si platform. In (Equation 18.7), can with measure with the help of device parameters. In Figure 18.7, band to band tunneling () can be measured by model, written in Equation 18.8. The tunneling distance in equilibrium

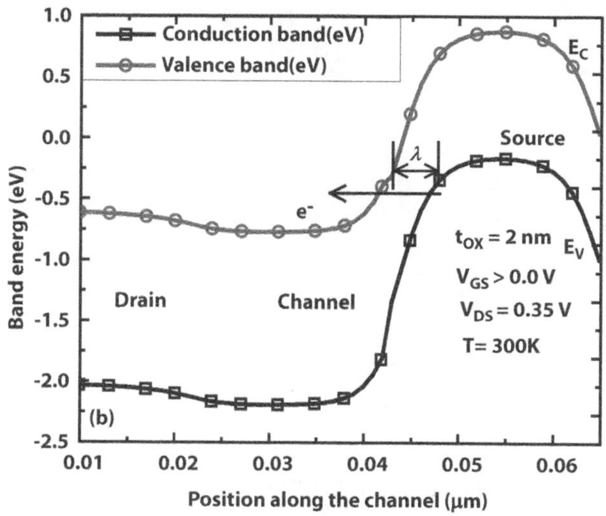


**Figure 18.7** Investigation of tunneling distance λ in on-state device.

at $V_{GS}$=0.0 V is purely physical parameter depends on $t_{ox}$ and $t_{si}$ are physical oxide thickness and semiconductor thickness and dielectric constant ($\varepsilon_{Si}$ $\varepsilon_{ox}$) respectively. But in non-equilibrium at $V_{GS}$ > 0.0 V for n-type device can be modulate with help of applied, gate to source voltage ($V_{GS}$). Basically BTBT is a quantum transport phenomena governed by WKB approximation. For various dielectric ($\varepsilon_{ox}$), band to band tunneling rate has been determine with the help of simulation and modeling approach for n-type TFET is shown in Figure 18.8. This simulation and modeling based analysis strongly support the fundamental device physics indicates higher dielectric induced larger value of electric field inside device causes more band bending, results larger tunneling of charge carriers inside device.

$$\lambda = \sqrt{\left(\frac{\epsilon_{Si}}{\epsilon_{ox}}\right) \times t_{ox} t_{Si}} \tag{18.8}$$

In the OFF state, the potential barrier between the source and the channel is so wide that no tunneling occurs. Only a very small leakage current exists. In the ON state, when the gate voltage exceeds the threshold voltage, the potential barrier between the channel and the source becomes narrow

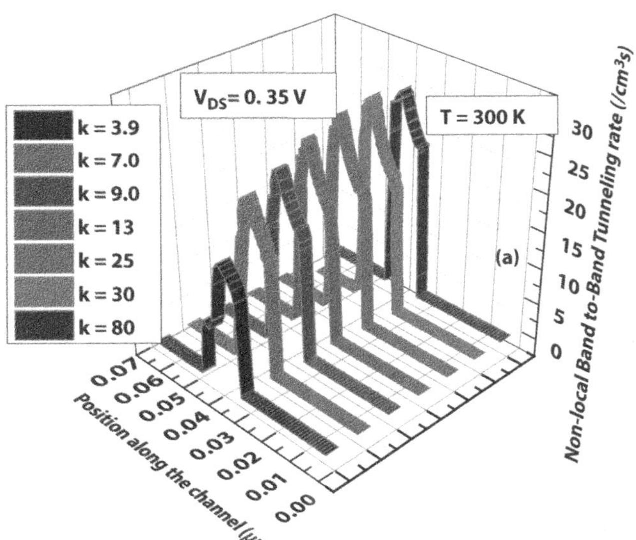


**Figure 18.8** Band to band tunneling in TFET for dielectrics constants for $t_{ox}$ = 2.0 nm and $t_{si}$ = 5.0 nm.

enough to allow a significant tunneling current. This allows realization of high ON-state current ($I_{ON}$) to OFF-state current ($I_{OFF}$) ratio at a low power supply ($V_{DD}$). However, $I_{ON}$ depends on the band alignment at source/channel interface and the energy band gap ($E_G$) in the source material, and Si-based TFETs still has much lower $I_{ON}$ than state-of-the-art Si MOSFETs. To improve $I_{ON}$ in Si-based TFETs, heterostructures with a type-II staggered band alignment at the interface between source and channel, e.g., SiGe/Si and Ge/Si, have been investigated, Ge source TFET with good performance has been realized on Si(100) [24–32]. Since the TFET has a different source carrier injection mechanism than does the MOSFET, it can achieve sub-60-mV/dec SS, showed the feasibility of the TFET for low-power applications [24–33].

TFET was proposed, many researchers have been done on applying it as an ultra-low power logic device. It is p-i-n configuration; many new devices can be constructed by combining TFET with MOSFET, to solve different problems to realize a subthreshold swing (*SS*) steeper than that achievable by MOSFETs. One example is the application of TFET in embedded DRAM. Usually, the high-speed System-on-Chip (SoC) needs integrated embedded 1T-1C (1 transistor - 1 capacitor) DRAM. Embedded DRAM (eDRAM) with TFET can be formed by using the floating junction gate memory cell configuration.

As device dimensions decrease, conventional MOS devices face challenges such as short-channel effects (SCEs), which degrade performance. TFETs, however, are less affected by these SCEs issues due to their tunneling mechanism. This makes TFETs more scalable, supporting the continued miniaturization of electronic components. This FET device can achieve subthreshold swing lower than 60 mV/decade due to the tunneling mechanism, leading to more efficient switching at low voltages. Generally this FET device having lower on current than MOSFETs due to the limitations in tunneling efficiency. TFETs are helpful in wearable devices, such as fitness trackers, smart watches, and medical monitors, benefit significantly from the low power consumption and compact size. These FET device support the development of lightweight, efficient, and long-lasting wearable electronics that can operate for extended periods on a single charge. Smartphones and tablets are continually evolving to include more features while maintaining or improving battery life. TFETs can play a critical role in achieving this balance by providing efficient power management and performance improvements, enabling the integration of more advanced functionalities without compromising battery life. Continuous health monitoring devices, such as heart rate monitors and glucose sensors, require low-power operation to ensure long-term usage without frequent battery replacement. TFETs' low power characteristics are ideal for these applications, making medical monitoring more reliable and user-friendly [28, 56, 62–67].

## 18.4 Junctionless FETs

Junctionless Field-Effect Transistors (JL-FETs) represent a significant paradigm shift in transistor design, eliminating the need for traditional p-n junctions by employing a uniformly doped channel. In contrast to conventional MOSFETs, which require precisely formed source and drain junctions with opposite doping polarity to the channel, JL-FETs use the same doping type across the source, drain, and channel regions typically heavily n-type or p-type. This architectural simplification addresses several scaling-related challenges and opens up pathways for reduced fabrication complexity, enhanced electrostatic control, and improved device variability at nanometer nodes [16, 17].

The cornerstone of JL-FET operation lies in the uniform doping profile across the entire active region. This uniformity eliminates the abrupt doping transitions that are characteristic of junction-based transistors, thereby mitigating many of the problems associated with short-channel effects (SCEs), junction leakage, and dopant diffusion. In conventional scaled MOSFETs, maintaining abrupt source/drain junctions is increasingly difficult due to dopant diffusion during high-temperature processing steps, especially as channel lengths approach sub-10 nm dimensions. JL-FETs bypass this issue entirely by relying on electrostatic depletion of a uniformly doped channel, rather than modulation of junction barriers, to control current flow. Because the channel is heavily doped (often in the range of $10^{18}$ - $10^{19}$ $cm^{-3}$), it would naturally conduct current. However, in JL-FETs, when a gate voltage is applied, the gate electric field depletes the carriers in the channel, effectively turning the device OFF. This behavior is opposite to enhancement-mode MOSFETs, where the channel forms only upon gate biasing. Thus, JL-FETs are intrinsically depletion-mode devices, and their ability to switch relies on the gate's capacity to deplete the majority carriers within the thin channel body. The efficiency of this depletion and thereby the switching behaviour is critically dependent on the channel thickness and the gate work function [18]. Figure 18.9 illustrates the schematic diagram of a Junctionless Field-Effect Transistor (JL-FET).

One of the most compelling advantages of JL-FETs is the simplification of the fabrication process. Traditional CMOS transistors require multiple,

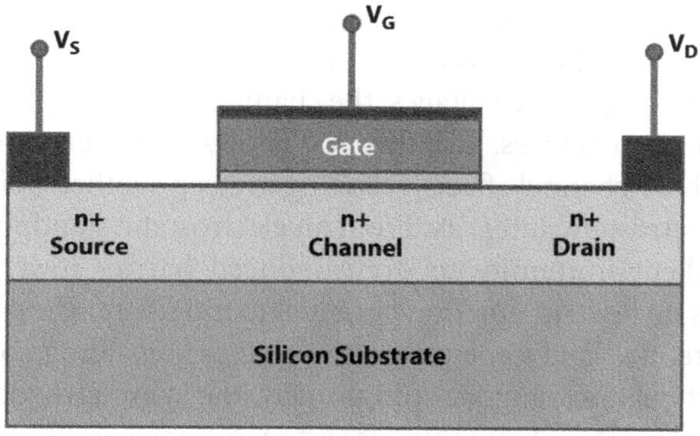


**Figure 18.9** Junctionless FET (JL-FET).

tightly controlled steps to form source/drain junctions, including ion implantation, annealing for dopant activation, and junction engineering (such as halo implants or shallow extensions) to suppress SCEs. These steps become increasingly complex and error-prone as device dimensions shrink and the thermal budget must be tightly managed. JL-FETs remove the need for these junction formation processes, as the doping is introduced uniformly across the entire active region typically during the early stages of substrate preparation or *via in-situ* doping during channel deposition. This not only reduces the overall number of fabrication steps but also minimizes thermal cycling, which in turn helps to preserve the structural integrity of ultra-thin channels and gate dielectrics. Additionally, the absence of junctions mitigates problems such as dopant clustering, random dopant fluctuation (RDF), and diffusion-induced variability, contributing to more predictable electrical behaviour and improved yield at scale. Moreover, JL-FETs are well-suited for emerging 3D integration technologies, such as gate-all-around (GAA) nanowire FETs and nanosheet FETs. These architectures benefit from the symmetric electrostatics and doping uniformity of JL-FETs, which simplify the challenge of forming consistent junctions in vertical or stacked configurations [19, 20].

The switching mechanism in JL-FETs is fundamentally electrostatic in nature. Because the channel is uniformly and heavily doped, current conduction is naturally present in the absence of a gate field. To turn the device OFF, a gate voltage is applied such that it creates a strong electric field across the gate dielectric. This field depletes the free carriers from the channel region, effectively reducing its conductivity. When the gate voltage is sufficiently negative (for n-type JL-FETs), the depletion region extends fully across the thin channel, pushing the device into the OFF state.

The ability to fully deplete the channel is a function of both the doping concentration and the physical thickness of the channel. To ensure full depletion at practical gate voltages, the channel must be thin—often below 10 nm for modern devices. This thinness allows for better gate control and suppresses short-channel effects, resulting in steep subthreshold slopes and improved electrostatic integrity. From an electrostatic standpoint, JL-FETs exhibit excellent immunity to drain-induced barrier lowering (DIBL), since there is no built-in junction potential to modulate. The gate fully governs the potential landscape across the channel, enabling robust control even at short channel lengths. Additionally, the symmetric doping profile allows for ambipolar behavior under certain biasing conditions, which can be leveraged in novel circuit applications, including reconfigurable logic.

In terms of device scaling, JL-FETs present a compelling alternative to conventional junction-based devices, particularly when integrated with advanced channel materials such as silicon nanowires, germanium, or 2D materials like $MoS_2$. These materials offer high carrier mobility and excellent electrostatic properties, which, when combined with the junctionless concept, can result in devices that are both high-performing and easy to fabricate. JL-FETs redefine transistor design by eliminating the need for abrupt source/drain junctions and instead relying on uniform doping and gate-induced depletion for switching. This approach simplifies fabrication, reduces process variability, and enhances device scalability. By utilizing electric field modulation across a heavily doped but ultra-thin channel, JL-FETs achieve competitive switching performance while avoiding many of the limitations inherent to traditional MOSFETs. As semiconductor technology continues to push toward atomic-scale dimensions, JL-FETs stand out as a viable and potentially disruptive technology for both digital and analog applications in the post-CMOS era.

## 18.5 Concept of Hybrid FET Architecture

A hybrid FET is a field-effect transistor that combines two or more different materials or technologies to leverage their complementary advantages, such as a GaN/SiC HyFET that uses high-mobility GaN for the channel and a SiC JFET structure for voltage blocking. Other examples include hybrid CMOS-Ferroelectric FETs for image sensors with tunable dynamic range, and heterostructure FETs for sensors like graphene/ReS2 sensors for arsenic detection. The term can also refer to hybrid integrated circuits, hybrid circuit designs, or even hybrid overdrive pedals in audio equipment. Some popular examples of hybrid FETs GaN/SiC. This is a power device that combines Gallium Nitride (GaN) for its high-mobility channel with Silicon Carbide (SiC) for high-voltage blocking, creating a device with high efficiency and voltage blocking capability. In hybrid CMOSr-Ferroelectric FET, CMOS combine with Ferroelectric FET (FeFET) to achieve a tunable dynamic range, which is an improvement over traditional CMOS sensors. It is basically image sensor based FET, overshot the limitations of convention CMOS image sensor. In heterostructure FET, in channel different materials form the channel, such as a Graphene/ReS2 FET used for detecting specific analytes like arsenic in water monitoring systems. In hybrid integrated FET, semiconductor expert combines different types of FETs, like nanosheet (NS) GAA-FETs and SiGe/Si FinFET at manufacturing level to overcome the conventional manufacturing of limitations.

As semiconductor devices approach the physical and functional limitations of conventional scaling, the push toward hybrid architectures that amalgamate multiple device principles has gained significant momentum. The conceptual convergence of FeFETs, TFETs and JL-FETs represents a forward-looking strategy to harness the distinct strengths of each device class, while compensating for their individual limitations. The result is a theoretically compelling hybrid FET architecture that seeks to deliver ultra-low-power operation, simplified fabrication, and embedded non-volatility in a single, multifunctional transistor design. At the heart of this hybrid concept is the recognition that each of the three FET types solves a specific bottleneck in advanced device scaling. TFETs are known for their sub-60 mV/decade subthreshold swing enabled by quantum tunneling, which makes them ideal for ultra-low-voltage logic. JL-FETs, in contrast, simplify fabrication by eliminating junctions and rely solely on gate electrostatics to modulate current flow, which is advantageous for 3D stacking and uniform doping control [21]. FeFETs introduce the third dimension of functionality: non-volatile memory capabilities, made possible by remanent polarization in a ferroelectric gate dielectric that modulates the threshold voltage even after power is removed. The convergence of these principles offers a blueprint for a new transistor class capable of both logic and memory, operating at ultra-low power, with reduced process complexity.

The plot in Figure 18.10 provides a comparative visualization of the energy band diagrams for three advanced FET architectures: FeFETs, TFETs and JL-FETs. Each curve in the plot represents the energy band profile of a specific FET type, highlighting its unique operational characteristics. The FeFET is represented by a red sinusoidal curve, which illustrates the polarization-induced shift in energy bands due to the ferroelectric layer in the gate stack. This periodic variation in energy levels demonstrates how the polarization state of the ferroelectric material can modulate the threshold

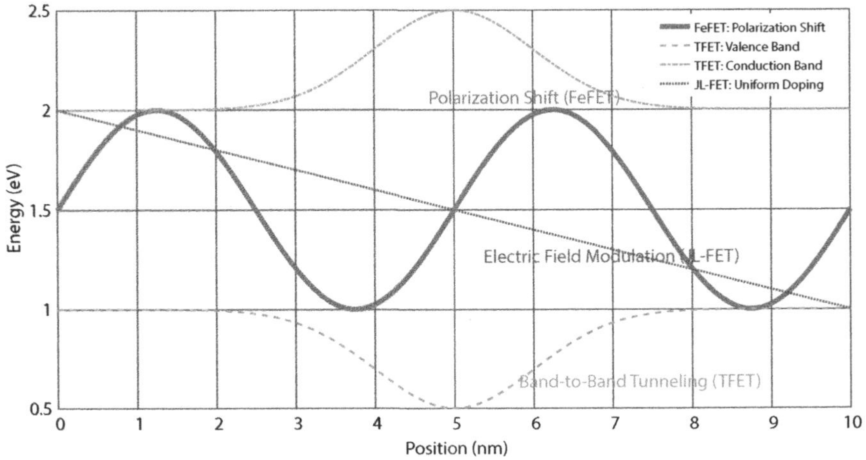


**Figure 18.10** Comparative energy band diagrams for FeFET, TFET and JL-FET.

voltage, enabling non-volatile memory capabilities. The TFET is depicted by two green curves: a dashed line for the valence band and a dash-dot line for the conduction band. These curves show the BTBT mechanism, where electrons tunnel from the valence band to the conduction band at the interface between the p-type and n-type regions. The overlap between the valence and conduction bands at a specific point (e.g., x=5) highlights the tunneling junction, which is critical for achieving low-voltage operation and a steep subthreshold slope. Finally, the JL-FET is represented by a blue linear curve, showing the energy band profile in a uniformly doped channel. The linear slope indicates the electric field modulation that controls the current flow in the absence of junctions, simplifying fabrication and reducing leakage currents. Together, these energy band diagrams provide a clear and intuitive understanding of the fundamental principles behind each FET type, emphasizing their distinct advantages and operational mechanisms. The FeFET's polarization-induced shift, the TFET's band-to-band tunneling, and the JL-FET's electric field modulation collectively showcase the innovations driving the development of next-generation transistors. This visualization serves as a valuable tool for comparing and contrasting these technologies, aiding in the design and analysis of hybrid FET architectures that combine their strengths for advanced electronic applications.

In a hybrid FET design [22, 23], the gate stack is envisioned as a ferroelectric–high-k bilayer, drawing from the FeFET paradigm as shown in Figure 18.11. The ferroelectric layer (e.g., doped hafnium oxide) introduces

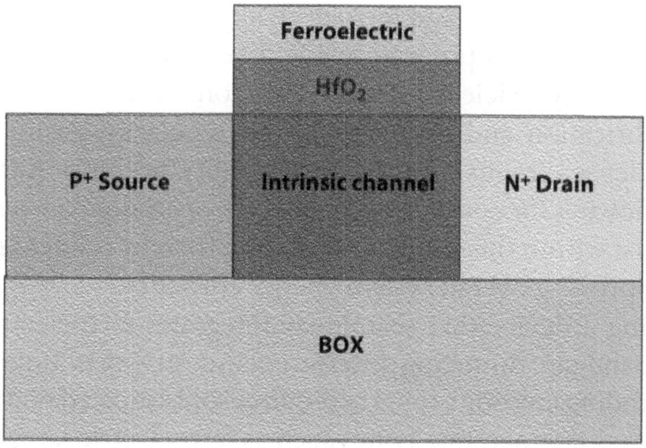


**Figure 18.11** Ferroelectric tunnel FET (Fe-TFET).

programmable polarization states, enabling bistable control of the device's threshold voltage. This polarization can be used to shift the alignment conditions required for band-to-band tunneling in the TFET portion of the device. In this configuration, the ferroelectric polarization dynamically modulates the tunneling window between the source and the channel, acting as a non-volatile selector for TFET activation. Such a mechanism provides an elegant way to combine logic operations with memory retention, without requiring a separate memory element. From the TFET perspective, the hybrid device retains a tunnel junction interface at the source-channel region, possibly leveraging a heterojunction or a steep-bandgap transition to maximize tunneling efficiency. The inclusion of ferroelectric gating can overcome one of the main limitations of TFETs, the need for precise electrostatic control to achieve acceptable ON-current levels. The polarization-induced threshold voltage shift can lower the gate voltage required to bring the valence and conduction bands into alignment, thereby enhancing tunneling probability at reduced bias.

The junctionless principle can be integrated through uniform doping across the source, drain, and channel. This eliminates the need for abrupt doping profiles at the tunnel junction interface, allowing the hybrid structure to be fabricated with fewer thermal steps and enhanced scalability. A uniformly doped channel also benefits from reduced variability, a critical factor for devices at sub-7 nm nodes. Moreover, the absence of junctions allows for better electrostatic symmetry and compatibility with 3D architectures such as gate-all-around nanowires or stacked nanosheets, where traditional junction engineering is increasingly impractical. This tripartite synergy yields several notable advantages. First, the combination of tunneling-based transport and ferroelectric gating enables both steep subthreshold operation and programmable threshold control features highly desirable for energy-efficient computing. Second, the JL-FET like structure simplifies fabrication and promotes layout regularity, easing integration into highly dense circuits. Third, non-volatility becomes intrinsic to the transistor, allowing for computing-in-memory paradigms where logic and storage coexist within the same physical unit, thereby reducing data movement and improving overall system energy efficiency.

Engineering a device that seamlessly integrates ferroelectric materials with steep-bandgap tunneling interfaces and supports uniform doping without degrading any one property requires sophisticated material synthesis and interface engineering. Issues such as ferroelectric fatigue, reliability under high-field tunneling conditions, and temperature-dependent behavior must be addressed. However, with ongoing advances in atomic-layer deposition, doped $HfO_2$ ferroelectrics, and heterostructure epitaxy, the hybrid architecture is transitioning from theoretical curiosity to practical feasibility. In essence, this emerging hybrid FET concept represents the culmination of multiple parallel innovations in transistor design. It reflects a broader shift from one-size-fits-all transistors toward application-specific device engineering, where logic speed, energy efficiency, and functional density are co-optimized at the device level. By coalescing the steep switching of TFETs, the manufacturing ease of JL-FETs, and the non-volatility of FeFETs, this architectural fusion could serve as a foundational building block for post-CMOS computing platforms, including neuromorphic systems, edge-AI accelerators, and secure hardware primitives.

Tunnel field-effect transistors (TFETs) rely on quantum-mechanical tunnelling and, unlike conventional metal–oxide–semiconductor field-effect transistors (MOSFETs), require less than 60 mV of gate voltage swing to induce one order of magnitude variation in the drain current at ambient temperature. The III–V heterostructure TFETs are promising for low-power applications, but are outperformed by MOSFETs in terms of speed and energy efficiency when high performance is required at higher drive voltages. Hybrid technologies—combining both TFETs and MOSFETs—could enable low-power and high-performance applications, but require the co-integration of different materials in a scalable complementary metal–oxide–semiconductor (CMOS) platform [15–40].

The hybrid phase transition FET (Hyper-FET) is a very recent inclusion in the family of steep (< 60 mV/decade) switching transistors. This novel transistor comprises of a correlated material or in general term, threshold switch augmented to the source terminal of a regular transistor (MOSFET or FinFET). The current driven abrupt phase transition of the threshold switch leads to abrupt (not instantaneous) switching in the transistor itself and can overcome the Boltzmann limit [10–40].

In summary, the hybrid field-effect transistor (Hybrid FET) integrates disparate materials or device types together (e.g., combining CMOS with Tunnel FETs, Vanadium dioxide for Phase-FETs, or Graphene) to overcome conventional transistor limitations like the Boltzmann limit (60 mV/decade) or to improve low-power, high-frequency, and steep-switching performance. These devices often use novel materials in the channel, source, or contacts to achieve superior $I_{ON}/I_{OFF}$, $g_m$, and higher energy efficiency.

## 18.6 Conclusion

The development of hybrid FET architectures marks a significant milestone in the evolution of semiconductor technology, offering innovative solutions to the challenges posed by the end of Moore's Law and the limitations of conventional MOSFETs. By combining the unique strengths of FeFETs, TFETs and JL-FETs, hybrid FETs provide a versatile platform for achieving low-power operation, high performance, and enhanced scalability. Together, these technologies address the critical needs of modern electronics, from energy-efficient logic and memory to neuromorphic computing. The exploration of hybrid FETs has revealed their potential to overcome the fundamental limitations of traditional transistors, such as high power consumption, SCEs and fabrication complexity. By integrating multiple operating principles into a single device, hybrid FETs offer a pathway to next-generation electronics that are faster, smaller, and more energy-efficient. However, significant challenges remain, including material compatibility, interface engineering, and the precise control of device architectures. Addressing these challenges will require continued innovation in materials science, nanofabrication techniques, and device modeling. As we continue to explore and refine these devices, they hold the promise of revolutionizing the semiconductor industry and enabling the next wave of technological breakthroughs.

**AI disclosure statement:** The author acknowledges the use of AI-based tools for minor language editing or grammatical correction. However, all intellectual contributions, analyses, and conclusions presented in this chapter are entirely the author's own.

## References


1. Dhanaselvam, P.S., Vasuki, S., Karthikeyan, B., Sriram Kumar, D., Nanoscale High-K Tri-Material Surrounding-Gate MOSFET—An Insight Analysis. *Field Effect Transistors,* 433-442, 2025.
2. Dhanaselvam, P.S., Karthikeyan, B., Nagarajan, S., Padmanaban, B., Nanowire Transistors. *Field Effect Transistors,* 81-92, 2025.
3. Salahuddin, S. and Datta, S., Use of negative capacitance to provide voltage amplification for low power nanoscale devices. *Nano Lett.*, 8, 2, 405–410, 2008.
4. Kong, Q., Liu, L., Zheng, Z., Sun, C., Zhou, Z., Jiao, L., Kumar, A., *et al.*, Back-end-of-line-compatible fin-gate ZnO ferroelectric field-effect transistors. *IEEE Trans. Electron Devices*, 70, 4, 2059–2066, 2023.
5. Böscke, T.S., Müller, J., Bräuhaus, D., Schröder, U., Böttger, U. J. A. P. L., Ferroelectricity in hafnium oxide thin films. *Appl. Phys. Lett.*, 99, 10, 2011.
6. Schroeder, U., Hwang, C.S., Funakubo, H. (Eds.), *Ferroelectricity in doped hafnium oxide: materials, properties and devices,* Woodhead Publishing, 2025.
7. Ali, T., Polakowski, P., Riedel, S., Büttner, T., Kämpfe, T., Rudolph, M., Pätzold, B., *et al.*, Silicon doped hafnium oxide (HSO) and hafnium zirconium oxide (HZO) based FeFET: A material relation to device physics. *Appl. Phys. Lett.*, 112, 22, 2018.
8. Jerry, M., Chen, P.-Y., Zhang, J., Sharma, P., Ni, K., Yu, S., Datta, S., Ferroelectric FET analog synapse for acceleration of deep neural network training, in: *2017 IEEE international electron devices meeting (IEDM)*, IEEE, pp. 6–2, 2017.
9. Khan, A., II, Chatterjee, K., Pablo Duarte, J., Lu, Z., Sachid, A., Khandelwal, S., Ramesh, R., Hu, C., Salahuddin, S., Negative capacitance in short-channel FinFETs externally connected to an epitaxial ferroelectric capacitor. *IEEE Electron Device Lett.*, 37, 1, 111–114, 2015.
10. Ionescu, A.M. and Riel, H., Tunnel field-effect transistors as energy-efficient electronic switches. *Nature*, 479, 7373, 329–337, 2011.
11. Dhanaselvam, P., Karthikeyan, B., Vanitha, P., Anand, P., Scope with TFET-based circuit and system design, in: *Device Circuit Co-Design Issues in FETs*, pp. 181–18, Taylor and Francis, 1st Edition, CRC Press, 2023.
12. Dhanaselvam, P.S., Sriram Kumar, D., Karthikeyan, B., Vanitha, P., Basic Science and Development of Subthreshold Swing Technology, in: *Tunneling Field Effect Transistors*, pp. 17–27, CRC Press, 2023.
13. Dhanaselvam, P.S., Karthikeyan, B., Anand, P., Future Prospective Beyond CMOS Technology Design. *Field Effect Transistors,* 73-80, 2025.
14. Dhanaselvam, P., Karthikeyan, B., Kavitha, K., Kavitha, P., Basic Concepts of Heterojunction Tunnel Field-Effect Transistors. *Field Effect Transistors,* 221–234, 2025.

15. Choi, W.Y., Park, B.-G., Duk Lee, J., King Liu, T.-J., Tunneling field-effect transistors (TFETs) with subthreshold swing (SS) less than 60 mV/dec. *IEEE Electron Device Lett.*, 28, 8, 743–745, 2007.
16. Dhanaselvam, P.S., Balamurugan, N.B., Nithya Ananthi, A., A surface potential model of junctionless surrounding gate (JLSG) MOSFET, in: *2013 International Conference on Green Computing, Communication and Conservation of Energy (ICGCE)*, IEEE, pp. 237–239, 2013.
17. Colinge, J.-P., Lee, C.-W., Afzalian, A., Dehdashti Akhavan, N., Yan, R., Ferain, I., Razavi, P., *et al.*, Nanowire transistors without junctions. *Nat. Nanotechnol.*, 5, 3, 225–229, 2010.
18. Rios, R., Cappellani, A., Armstrong, M., Budrevich, A., Gomez, H., Pai, R., Rahhal-Orabi, N., Kuhn, K., Comparison of junctionless and conventional trigate transistors with $ L_ {g} $ down to 26 nm. *IEEE Electron Device Lett.*, 32, 9, 1170–1172, 2011.
19. Manikandan, S., Suveetha Dhanaselvam, P., Karthigai Pandian, M., A 2D unified subthreshold drain current investigation for junctionless cylindrical surrounding gate (JCSG) silicon nanowire transistor. *Silicon*, 14, 10, 5311–5318, 2022.
20. Lee, C.-W., Ferain, I., Afzalian, A., Yan, R., Dehdashti Akhavan, N., Razavi, P., Colinge, J.-P., Performance estimation of junctionless multigate transistors. *Solid State Electron.*, 54, 2, 97–103, 2010.
21. Kumar, L.V., Kumar, G.P., Karthikeyan, B., Scaling of Field-Effect Transistors. *Field Effect Transistors*, 63–72, 2025.
22. Saha, R., Simulation study on ferroelectric layer thickness dependence RF/Analog and linearity parameters in ferroelectric tunnel junction TFET. *Microelectron. J.*, 113, 105081, 2021.
23. Dhanaselvam, P.S., Rao, K.S., Rahi, S.B., Yadav, D.S. (Eds.), *Field Effect Transistors*, John Wiley & Sons, 2025.
24. Bind, U.C., Rahi, S.B., Kumar, K., Identification of Negative Capacitance in Ferroelectric in FET Devices. *FeFET Devices, Trends, Technol. App.*, 155–184, 2025.
25. Guenifi, N., Rahi, S.B., Benmahdi, F., Chaabane, H., Optimization for Device Figure of Merit of Ferroelectric Tunnel FET using Genetic Algorithm. *ECS J. Solid State Sci. Technol.*, 12, 2, 023001, 2023.
26. Rahi, S.B., Tayal, S., Upadhyay, A.K., A review on emerging negative capacitance field effect transistor for low power electronics. *Microelectron. J.*, 116, 10542, 2021.
27. Suzuki, R., Toprasertpong, K., Nakane, R., Nako, E., Takenaka, M., Takagi, S., Reservoir computing utilizing a complementary combination of n-and p-channel FeFETs. *IEEE Electron. Device Lett.*, 2024.
28. Song, Y.S., Thoutam, L.R., Tayal, S., Rahi, S.B., Arun Samuel, T.S., *Handbook of Emerging Materials for Semiconductor Industry*, Springer, 2024.
29. Gnani, E., Gnudi, A., Reggiani, S., Baccarani, G., Theory of the junctionless nanowire FET. *IEEE Transact. Electron Dev.*, 58, 9, 2903–2910, 2011.

30. Rahi, S.B., Asthana, P., Gupta, S., Heterogate junctionless tunnel field-effect transistor: Future of low-power devices. *J. Comput. Electron.*, 16, 1, 30–38, 2017.
31. Rahi, S.B. and Ghosh, B., High-k double gate junctionless tunnel FET with a tunable bandgap. *RSC Adv.*, 5, 67, 54544–54550, 2015.
32. Nowbahari, A., Roy, A., Marchetti, L., Junctionless transistors: State-of-the-art. *Electronics*, 9, 7, 1174, 2020.
33. Raj, B., Rahi, S.B., Yadav, N., *FeFET Devices, Trends, Technology and Applications*, John Wiley & Sons, 2025.
34. Hutchby, J.A., Bourianoff, G.I., Zhirnov, V.V., Brewer, J.E., Extending the road beyond CMOS. *IEEE Circuits Devices Mag.*, 18, 2, 28–41, 2002.
35. Iwai, H., Future of CMOS technology, in: *2004 Semiconductor Manufacturing Technology Workshop Proceedings (IEEE Cat. No. 04EX846)*, IEEE, pp. 5–17, 2004.
36. Chowdhury, N., Azad, S.F., Khosru, Q.D., Negative capacitance tunnel field effect transistor: A novel device with low subthreshold swing and high on current. *ECS Trans.*, *58*, 16, 1, 2014.
37. You, W. and Su, P., Design Space Exploration Considering Back-Gate Biasing Effects for 2D Negative-Capacitance Field-Effect Transistors. *IEEE Trans. Electron Devices*, 64, 8, 3476–3481, Aug. 2017.
38. Jimenez, D., Miranda, E., Godoy, A., Analytic Model for the Surface Potential and Drain Current in Negative Capacitance Field-Effect Transistors. *IEEE Trans. Electron Devices*, 57, 10, 2405–2409, Oct. 2010.
39. Jiang, C., Liang, R., Xu, J., Investigation of Negative Capacitance Gate-all-Around Tunnel FETs Combining Numerical Simulation and Analytical Modeling. *IEEE Trans. Nanotechnol.*, 16, 1, 58–67, Jan. 2017.
40. Bidenko, P., Lee, S., Han, J., Song, J.D., Kim, S., Simulation Study on the Design of Sub- kT/q Non-Hysteretic Negative Capacitance FET Using Capacitance Matching. *IEEE J. Electron Devices Soc.*, 6, 910–921, 2018.
41. Jo, J., Woo, Y.C., Park, J.-D., Shim, J.W., Yu, H.-Y., Shin, Negative capacitance in organic/ferroelectric capacitor to implement steep switching MOS devices. *Nano Lett.*, 15, 7, 4553–4556, Jun. 2015.
42. Liao, Y.-H., *et al.*, Electric Field-Induced Permittivity Enhancement in Negative-Capacitance FET. *IEEE Trans. Electron Devices*, 68, 3, 1346–1351, March 2021.
43. Ota, H., Migita, S., Hattori, J., Fukuda, K., Torikumi, A., Structural advantages of silicon-on-insulator FETs over FinFETs in steep subthreshold-swing operation in ferroelectric-gate FETs. *Jpn. J. Appl. Phys.*, 56, 4, Mar. 2017.
44. Dasgupta, S., *et al.*, Sub-kT/q Switching in Strong Inversion in PbZr0.52Ti0.48O3 Gated Negative Capacitance FETs. *IEEE J. Explor. Solid-State Comput. Devices Circuits*, 1, 43–48, Dec. 2015.
45. Ota, H., Migita, S., Hattori, J., Fukuda, K., Toriumi, A., Material and device engineering in fully depleted silicon-on-insulator transistors to realize a

steep subthreshold swing using negative capacitance. *Jpn. J. Appl. Phys.*, 55, Apr. 2016.

46. Li, Y., Yao, K., Samudra, G.S., Effect of Ferroelectric Damping on Dynamic Characteristics of Negative Capacitance Ferroelectric MOSFET. *IEEE Trans. Electron Devices*, 63, 9, 3636–3641, Sept. 2016.
47. Li, Y., Kang, Y., Gong, X., Evaluation of Negative Capacitance Ferroelectric MOSFET for Analog Circuit Applications. *IEEE Trans. Electron Devices*, 64, 10, 4317–4321, Oct. 2017.
48. Yeung, C.W., Khan, A.I., Sarkar, A., Salahuddin, S., Hu, C., Low power negative capacitance FETs for future quantum-well body technology. *2013 International Symposium on VLSI Technology, Systems and Application (VLSI-TSA)*, pp. 1–2, 2013.
49. Rahi, S.B., Upadhyay, A.K., Lal, H., Karumuri, S.R., Negative-Capacitance Field-Effect Transistor for Optimization of Power Factor for Modern Applications, in: *Field Effect Transistors*, pp. 417–431, 2025.
50. Guenifi, N., Rahi, S.B., Chabane, H., Dibi, K., Ferroelectric Tunnel FET: Next Generation of Classical Low Power CMOS Technology, in: *FeFET Devices, Trends, Technology and Applications*, pp. 131–153, 2025.
51. Upadhyay, A.K., Reniwal, B.S., Rahi, S.B., Beohar, A., Negative Capacitance Tunnel Field-Effect Transistor: Impact and Future Scope, in: *Handbook of Emerging Materials for Semiconductor Industry*, pp. 555–564, Springer Nature Singapore, Singapore, 2024.
52. Upadhyay, A.K., Rahi, S.B., Tayal, S., Song, Y.S., Recent progress on negative capacitance tunnel FET for low-power applications: Device perspective. *Microelectron. J.*, 129, 1055835, 2022.
53. Tayal, S., Smaani, B., Rahi, S.B., Upadhyay, A.K., Bhattacharya, S., Ajayan, J., Jena, B., Myeong, I., Park, B.-G., Song, Y.S., Incorporating bottom-up approach into device/circuit co-design for SRAM-based cache memory applications. *IEEE Trans. Electron Devices*, 69, 11, 6127–6132, 2022.
54. Smaani, B., Rahi, S.B., Labiod, S., Analytical compact model of nanowire junctionless gate-all-around MOSFET implemented in Verilog-A for circuit simulation. *Silicon*, 14, 16, 10967–10976, 2022.
55. Sathishkumar, M., Arun Samuel, T.S., Ramkumar, K., Vivek Anand, I., Rahi, S.B., Performance evaluation of gate engineered InAs–Si heterojunction surrounding gate TFET. *Superlattices Microstruct.*, 162, 1070995, 2022.
56. Smaani, B., Paras, N., Rahi, S.B., Song, Y.S., Yadav, R., Tayal, S., Impact of the self-heating effect on nanosheet field effect transistor performance. *ECS J. Solid State Sci. Technol.*, 12, 2, 021005, 2023.
57. Naima, G. and Rahi, S.B., Design and optimization of heterostructure double gate tunneling field effect transistor for ultra low power circuit and system, in: *Electrical and Electronic Devices, Circuits, and Materials: Technological Challenges and Solutions*, pp. 19–36, 2021.

58. Paras, N., Rahi, S.B., Upadhyay, A.K., Bharti, M., Song, Y.S., Design and analysis of novel La: HfO2 gate stacked ferroelectric tunnel FET for non-volatile memory applications. *Mem.-Mater. Devices Circuits Syst.*, 7, 1001015, 2024.
59. Pandey, C.K., Das, D., Nanda, U., Dash, D., Chaudhury, S., Song, Y.S., Rahi, S.B., DC Analysis and Analog/HF Performances of GAA-TFET with Dielectric Pocket, in: *Tunneling Field Effect Transistors*, pp. 149–168, CRC Press, 2023.
60. Pathakamuri, A.K., Pandey, C.K., Das, D., Nanda, U., Rahi, S.B., A Ferroelectric Negative-Capacitance TFET with Extended Back Gate for Improvement in DC and Analog/HF Parameters, in: *Field Effect Transistors*, pp. 205–220, 2025.
61. Rahi, S.B., Song, Y.S., Kandpal, J., Advanced Operational Amplifier with Emerging CMOS Technology, in: *Circuit Design for Modern Applications*, pp. 120–136, CRC Press, 2025.
62. Wang, P.-Y. and Tsui, B.-Y., Investigation Into Gate-to-Source Capacitance Induced by Highly Efficient Band-to-Band Tunneling in p-Channel Ge Epitaxial Tunnel Layer Tunnel FET. *IEEE Trans. Electron Devices*, 63, 4, 1788–1790, April 2016, doi: 10.1109/TED.2016.2535373.
63. Tayal, S., Upadhyay, A.K., Rahi, S.B., Song, Y.S. (Eds.), *Advanced Ultra Low-power Semiconductor Devices: Design and Applications*, John Wiley & Sons, 2023.
64. Zeng, Y., *et al.*, Quantum Well InAs/AlSb/GaSb Vertical Tunnel FET With HSQ Mechanical Support. *IEEE Trans. Nanotechnol.*, 14, 3, 580–584, May 2015, doi: 10.1109/TNANO.2015.2419232.
65. Samuel, T.A., Song, Y.S., Tayal, S., Vimala, P., Rahi, S.B. (Eds.), *Tunneling Field Effect Transistors: Design, Modeling and Applications*, CRC Press, 2023.
66. Goel, E., Pandey, A., Rahi, S.B., Samuel, A. (Eds.), *Handbook of Advanced Semiconductor Field Effect Transistors*, John Wiley & Sons, 2025.
67. Tayal, S., Upadhyay, A.K., Kumar, D., Rahi, S.B. (Eds.), *Emerging low-power semiconductor devices: Applications for future technology nodes*, CRC Press, 2022.

# Index

# Also of Interest

## Check out these other related titles from Scrivener Publishing

*Handbook of Advanced Semiconductor Field Effect Transistors,* Edited by Ekta Goel, Archana Pandey, Shiromani Balmukund Rahi, and Arun Samuel, ISBN: 9781394412570. Advance your understanding of semiconductor technology with this indispensable handbook, offering an in-depth look at the modeling, simulation, and fabrication of advanced nanoscale semiconductor field-effect transistors (FETs).

*Advanced Ultra Low-Power Semiconductor Devices: Design and Applications* Edited by Shubham Tayal, Abhishek Kumar Upadhyay, Shiromani Balmukund Rahi, and Young Suh Song, ISBN: 9781394166411. Written and edited by a team of experts in the field, this important new volume broadly covers the design and applications of metal oxide semiconductor field effect transistors.

*Field Effect Transistors,* Edited by P. Suveetha Dhanaselvam, K. Srinivasa Rao, Shiromani Balmukund Rahi, and Dharmendra Singh Yadav, ISBN: 9781394248476. *Field Effect Transistors* is an essential read for anyone interested in the future of electronics, as it provides a comprehensive yet accessible exploration of innovative semiconductor devices and their applications, making it a perfect resource for both beginners and seasoned professionals in the field.

*Integrated Devices for Artificial Intelligence and VLSI: VLSI Design, Simulation and Applications,* Edited by Balwinder Raj, Suman Lata Tripathi, Tarun Chaudhary, K. Srinivasa Rao, and Mandeep Singh, ISBN: 9781394204359. With its in-depth exploration of the close connection between microelectronics, AI, and VLSI technology, this book offers valuable insights into the cutting-edge techniques and tools used in VLSI design automation, making it an essential resource for anyone seeking to stay ahead in the rapidly evolving field of VLSI design.

*Controller Design for Industrial Applications,* Edited by Arindam Mondal & Souvik Ganguli, ISBN: 9781394287079. *Controller Design for Industrial Applications* is essential for anyone looking to master the advanced techniques of intelligent controller design, enabling you to effectively tackle the complexities of modern industrial processes and optimize performance in an ever-evolving landscape.

*Special Electrical Machinery,* Edited by Dr. Jigneshkumar P. Desai, ISBN: 9781394193912. This book is a comprehensive guide to specialized motors, covering everything from switched reluctance motors to axial flux ironless PM motors, giving in-depth information on their operating principles, applications, and controls.

*NANODEVICES FOR INTEGRATED CIRCUIT DESIGN,* Edited by Suman Lata Tripathi, Abhishek Kumar, K. Srinivasa Rao, and Prasantha R. Mudimela, ISBN: 9781394185788. Written and edited by a team of experts in the field, this important new volume broadly covers the design of nano-devices and their integrated applications in digital and analog integrated circuits (IC) design.

*Electrical and Electronic Devices, Circuits, and Materials: Technical Challenges and Solutions,* edited by Suman Lata Tripathi, Parvej Ahmad Alvi, and Umashankar Subramaniam, ISBN: 9781119750369. Covering every aspect of the design and improvement needed for solid-state electronic devices and circuit and their reliability issues, this new volume also includes overall system design for all kinds of analog and digital applications and developments in power systems.

Printed and bound by CPI Group (UK) Ltd, Croydon, CR0 4YY
17/09/2026
14964264-0004